CROP IMPROVEMENT

Volume 1
Physiological Attributes

U.S. GUPTA

Department of Plant Science
Faculty of Agriculture
A.B.U., Zaria, Nigeria

Westview Press
BOULDER • SAN FRANCISCO • OXFORD

Published in 1992 in the United States by
  Westview Press
  5500 Central Avenue
  Boulder, Colorado 80301

Library of Congress Cataloging-in-Publication Data

Gupta, U.S., 1940-
    Crop improvement / U.S. Gupta
        P. cm.
    Includes bibliographical references and indexes.
    Contents: v. 1. Physiological attributes
    ISBN 0-8133-1380-5 : $36.50
    1. Crop improvement. 2. Crops–Physiology. 3. Crops–Effect of stress on. 4. Crops–Quality.
    I. Title
  SB106. I47G87 1992   91–26609
  831 — dc20   CIP

Originally published in India by Oxford & IBH Publishing Co. Pvt. Ltd., New Delhi. Printed in India.

PREFACE

For feeding the growing world population, increasing agricultural production is our primary goal. The land area cannot be increased much as the best productive land has already been brought under the plough. Less productive lands have to be cultivated, and crop productivity improved. Presently *Crop Improvement* continues to be the "breeders baby". But it must be borne in mind that yield is not under direct genetic control. It is the multitude of physiological and biochemical plant processes which interact to have an effect on yield. Before attempts for increasing productivity are made, crop physiological attributes leading to production have to be well understood and then the crop plants have to be improved for individual attributes. Thereafter, individual attributes from superior genotypes have to be tailored, and a grand synthetic high yielding variety has to be developed according to the computor model. Interaction of physiologists, genetists, modelists and breeders has become imperative for achieving better and faster results.

This superior high yielding variety developed under ideal cropping conditions is not expected to perform equally well under stress conditions which plants usually face in nature. The stress conditions are related with changes in latitude, altitude, soil type, environment — climatic, edaphic and biological, management, and resource availability. Thus this superior genotype has to be tested under individual stress conditions or in fact, high yielding genotypes have to be developed under each stress condition.

In this way, we can expect higher production under both agronomically normal or stress environments. Plants can be further improved for quality, that is, producing more protein, fat or energy from the same mass of produce, or improve their palatability or aroma, amino-acid profile, unsaturated fatty acids etc., and reduce the undesirable toxic principles as in mustard, cassava and lathyrus.

Keeping the above objectives in mind, a three-volume treatise has been planned covering the physiological attributes in volume 1, stress tolerance in volume 2, and quality characters in volume 3. Volume 1 embodies the aspects of general production physiology as basic background, in the Preamble. Then the aspects of photosynthetic efficiency, photoinsensitivity, determinate habit, dwarf stature, shortening growth duration, rate and duration of grain filling, harvest index, root type, nitrogen fixation in legumes and non-legumes, and yield stability have been discussed in different chapters. The information presented is recent, international, authoritative and is designed to cater to the needs of advanced students in Crop Physiology, Genetics, Plant Breeding, Agronomy, Soil Science, and progressive farmers, planners and crop managers. Both teachers and researchers can equally benefit from this treatise on 'Crop Improvement'. Any suggestion from our learned readers for further improvement of this volume will be highly appreciated.

The patience of and encouragement from each of my family members and friends during its preparation and production phases are gratefully acknowledged. Special thanks go to Mridula Gupta, for drawing all the figures in this volume.

U.S. GUPTA

CONTENTS

Contents

PRODUCTION PHYSIOLOGY

Identification of physiological components of yield and their genetic control will help in planning crosses to maximizing segregation of genotypes possessing the physiological complementation and balance required for high yield, thereby leading to more rapid and predictable yield improvement. A wide genetic variation exists in many of the yields determining physiological characters (*see* the various chapters in this book). The fundamentals of crop improvement are genetic variability, recombination, genotype selection and evaluation. Without genetic variability there is transgressive segregation resulting from recombination and genetic assortment, and therefore no opportunity for genotype improvement by differential selection.

One of the reasons of slow increase in grain productivity has been the indifferent attitude of plant physiologists who worked almost exclusively with single plants or even plant organs and organelles, materials well suited to the study of basic physiological processes, but not yield related results as regards crop productivity. The discipline of crop physiology emerged about three decades ago and the results are slowly becoming accessible to plant breeders. The reason of slow progress is the inaccessibility of the publications widely scattered among a vast number of scientific journals all over the world and thus known only to the specialists. They have not been reviewed and published in the form of a book for easy accessibility to breeders engaged in crop improvement and crop physiologists alike.

An increase in dry matter of any plant is a consequence of carbon assimilation (photosynthesis). The best known process of photosynthesis is the C_3 mechanism (Fig. 1) taking place through Calvin cycle in most crop plants, for example, most cereals (wheat, rice, barley, oats, etc.) and legumes. This process is not so efficient as all C_3 plants possess a mechanism called photorespiration (Fig. 2). Photorespiration takes place only in chlorophyllous organs in the presence of light together with photosynthesis, but in this process CO_2 is not fixed, rather it is evolved as in respiration. Photorespiration is at least three to five times faster than 'dark' respiration (Zelitch, 1976). Unlike respiration, NADPH is used, and ATP is consumed rather than synthesized. Oxidation of ribulose biphosphate (RuBP) to glycolate requires 8 ATP and 5 NADPH, a further 7 NADPH are consumed in conversion of 2 glycolate molecules to 1 glycerate and possibly 5 ATP needed for sucrose synthesis (Lawlor, 1979). These deficiencies are made up from photosynthetic or respiratory energy sources. In short, photorespiration is an unwanted and wasteful process in all C_3 plants occurring due to the presence of an isomer of RuBPCase, namely, RuBPOase. The two isomeric enzymes responsible for photosynthesis and photorespiration (see Fig. 2)

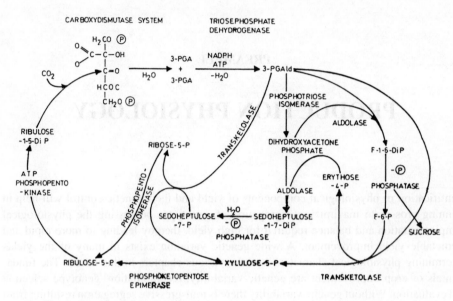

Fig. 1. Reaction of C_3 mechanism of photosynthesis (Calvin cycle / TCA cycle).

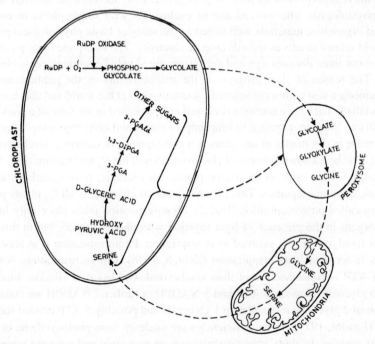

Fig. 2. Reactions of photorespiration. These are shown in respect of compartments within a cell.

respectively are so similar that no specific enzyme inhibitor could be found till today which could inhibit the activity of RuBPOase but not of the RuBPCase.

Another more efficient mechanism of photosynthesis occurs through organic acid metabolism (C_4 plants) in the bundle sheath cells of the plants like sorghum, pearl millet, maize and sugarcane. These plants have a negligible rate of photorespiration (about 10 per cent of C_3 species — Zelitch et al., 1977). Unlike the Calvin process where the first stable product is a 3-carbon sugar (C_3 plants), here the first product is a 4-carbon organic acid (malate or aspartate). A simplified scheme for C_4 pathway metabolism is given in Fig. 3 (Knai and Black, 1972). Carbon dioxide is initially fixed via PEPCase into the C_4 carboxyl of oxalacetate which then gives rise to malate (aspartate in some species) in mesophyll cells which is then transferred to the bundle sheath cells where CO_2 is released by C_4 decarboxy-lation, and refixed via Calvin cycle. The C_3 compound formed as the other product of this decarboxylation is then returned to mesophyll cells where it serves as a precursor for generating PEP.

C_4 photosynthesis is a little more energy-consuming mechanism than C_3; for one mol-ecule of CO_2 fixed, C_4 plants require 5 ATP and 2 NADPH molecules compared to 3 ATP and 2 NADPH molecules required by C_3 plants, have higher rates of net photosynthesis than C_3 plants, though considerable variations do occur among the groups and among different varieties of a crop (Table 1). But on the whole the following order of photosyn-thetic efficiency becomes apparent (Singh et al., 1974).

C_3 grasses < C_3 dicots < C_4 aspartate formers < C_4 malate formers

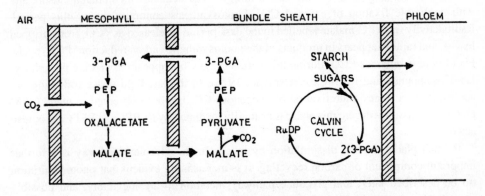

Fig. 3 : A simplified scheme for C_4 pathway of photosynthesis showing the basic reactions and their intercellular location (Adapted after Knai and Black, 1972; Hatch, 1976).

From the detailed data presented in Table 2 it becomes apparent that at constant LAI (4), total daily net CO_2 assimilation of C_4 crops is always higher than C_3 crops whether grown at low or high altitudes, in clear or overcast sky, at low or high CO_2 concentrations, and even under moisture stress conditions (van Keulen et al., 1980). Water requirement of C_4 plants, per kg of CO_2 assimilated, is about 25 per cent less. A further reduction in moisture requirement can be achieved by increasing the atmospheric CO_2 concentration.

Yet another mechanism of photosynthesis taking place in some plants like pine apple, *Agave, Aloe, Bryophyllum* and *Opuntia* is called crassulacean acid metabolism (CAM)

Table 1:   Typical rates of net photosynthesis in single leaves of various species at high illuminance and normal atmospheric CO_2.

C_4 crops	Net photosynthesis (mg CO_2 dm^{-2}h^{-1})	C_3 crops	Net photosynthesis (mg CO_2 dm^{-2}h^{-1})
Maize	46 – 63	Rice	34.5 – 62
Sorghum	55	Wheat	17 – 31
Sugar cane	42 – 49	Barley	17.7 – 21
		Sunflower	37 – 44
		Sugar-beet	24 – 48
		Soybean	29 – 43
		Beans	12 – 17
		Cotton	24 – 40
		Tobacco	16 – 21

Such plants generally inhabit in dry environments and so in order to save upon the transpirational loss of water, they keep their stomata closed during the day and open during the night. The dark fixation of CO_2 involves PEPCase and RuBPCase and is active in the dark. PEP carboxylation is the principal path of CO_2 fixation in the dark. Towards the end of the dark period, carboxylation reaction is inhibited by the end product, malate. Many CAM plants show an initial burst of CO_2 fixation in the light period that is independent of stomatal response (Fig. 4). RuBPCase may be active at this time. During the initial burst, exogenous and endogenous CO_2 may be fixed by either PEPCase or RuBPCase. In most CAM plants the phase of de-acidification in light is associated with stomatal closure and with negligible fixation of external CO_2. Photosynthesis continues during this period. Radioactivity from ^{14}C-malate labelled in the dark period is released as $^{14}CO_2$ from stripped leaves, and rapidly appears in products of the photosynthetic carbon reduction (PCR) cycle. Fixation of exogenous $^{14}CO_2$ during this period is slow but appears to involve phosphorylated compounds and RuBPCase (Osmond, 1976). In the late light phase following de-acidification, a steady-state fixation of exogenous CO_2 has been observed in many CAM plants. CO_2 fixation during this phase is primarily mediated by RuBPCase. PEPCase is also active.

In such plants, carbon dissimilation by means of the glycolate pathway is a further important component of carbon recycling. It is increasingly evident that photorespiration (to be described later) and glycolate pathway metabolism are inevitably, and possibly essentially, associated with carbon assimilation via RuBPCase and the PCR cycle. It is not surprising, then, that RuBPCase from *Kalanchöe daigremontiana* shows RuBPCase activity, that CAM plants contain glycolate pathway enzyme activities comparable with the levels of RuBPCase and that ultrastructural studies show well developed microbodies in green cells of CAM plants (Osmond, 1976).

Another type of important process taking place in all green and non-green cells and at all times (day and night) is called respiration. In respiration (so-called dark respiration), CO_2 fixed during photosynthesis (C_3, C_4, CAM) is liberated but the much needed energy to drive important reactions like that of photosynthesis, mineral uptake, cyclosis, chromosome migration during cell division, is generated. In photorespiration, however, CO_2 is

Table 2: Simulated values of total daily net CO_2 assimilation total daily transpiration and their ratios for a C_3-canopy (LAI=4) and C_4-canopy (LAI=4) and C_4-canopy under moisture stress, growing at different latitudes on completely clear and completely overcast days at two levels of external CO_2 concentration (Internal CO_2 conc. fixed) (After van Keulen et al., 1980)

Latitude	Type of day illumination	CO_2 conc. (ppm)	Total daily net assimilation (Kg CO_2/h/day)	Total daily transpiration (mm/day)	Transpiration/ assimilation (Kg H_2O/Kg CO_2)	Total daily net assimilation (Kg H_2O/Kg CO_2)	Total daily transpiration (mm/day)	Transpiration/ assimilation ratio (Kg CO_2/h/day)
				C_3			C_4	
10	Clear	330	659	4.2	64.2	874	3.6	41.7
	Overcast		298	1.9	62.7	321	1.3	39.6
30	Clear		753	4.8	64.3	1008	4.2	42.0
	Overcast		330	2.0	61.8	356	1.4	38.4
50	Clear		785	5.0	63.2	1039	4.3	41.1
	Overcast		329	2.1	63.2	351	1.4	39.6
10	Clear	430	663	2.6	39.8	874	2.7	31.3
	Overcast		298	1.2	40.3	321	1.0	30.5
30	Clear		759	3.0	39.8	1008	3.2	31.5
	Overcast		330	1.3	39.1	356	1.1	29.7
50	Clear		791	3.1	39.1	1039	3.2	30.9
	Overcast		329	1.3	40.7	351	1.1	30.7
10	Clear	330	Under moisture stress			810	3.3	40.6
30	Clear					929	3.8	40.6
50	Clear					988	4.0	40.2
10	Clear	430				873	2.7	31.4
30	Clear					1004	3.15	31.4
50	Clear					1032	3.2	30.9

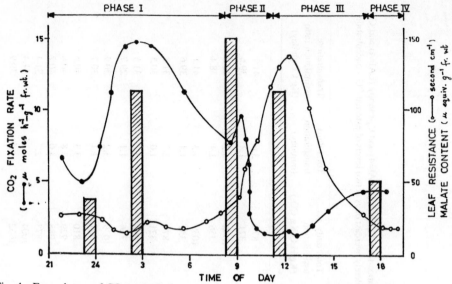

Fig. 4 : Four phases of CO_2 assimilation in *Kalenchoe diagremontiana*. CO_2 fixation rate $= \bullet\!\!-\!\!\bullet$ leaf resistance to water vapour exchange $= o\!\!-\!\!o$; and malic acid contant at various times is shown by vertical bars (Osmond, 1976).

evolved but no energy-rich compound synthesized. There are two mechanisms of aerobic respiration, namely, via TCA cycle and hexose monophosphate shunt (Figs. 5 and 6) operating. In the absence of oxygen, anaerobic respiration or fermentation takes place (Fig. 5).

A consolidated picture of the assimilatory and dissimilatory processes taking place in a green plant with regard to their cellular compartmentation is given in Fig. 7. Thus it becomes amply clear that growth or increase in dry matter is a consequence of balance between true or absolute photosynthesis (whatever mechanism), and respiration and pho-torespiration (apparent photosynthesis). Thus the growth rate at any given time can be predicted by the photosynthesis/respiration (P/R) ratio.

From the foregoing discussion, it is apparent that there is a need to eliminate/suppress the unwanted and wasteful process, photorespiration. A decade of efforts by several scientists to block the activity of RuBPOase without affecting the activity of RuBPCase unduly, has not proved successful. However, blocking the oxidation of glycolic acid with a suitable-α-hydroxysulphonate under conditions of rapid photorespiration in tobacco leaf tissue increased the net CO_2 uptake (Zelitch, 1976). Floating illuminated tobacco leaf discs on 20 mM potassium glycidate resulted in a 40 to 50 per cent inhibition of glycolic acid synthesis, a 40 per cent inhibition of photorespiration, and a 40 to 50 per cent increase in net photosynthetic CO_2 uptake. Treatment with glycidate also decreased the pool sizes of glycine and serine, but inhibited the oxidation of glycolic acid. Hence glycidate must function as an inhibitor of the glycolate pathway. Zelitch (1976) further stressed that photosynthesis of tobacco leaf discs can be increased by blocking glycolate synthesis with potassium glycidate (Fig. 8). In the presence of glycidate, aspartate and glutamate increased by two-fold to three-fold.

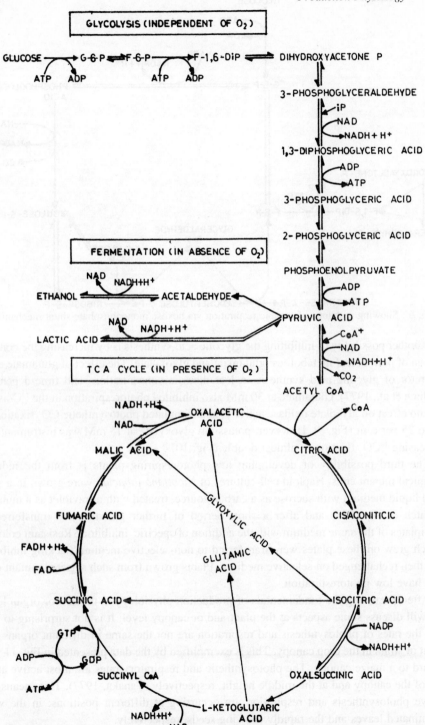

Fig. 5 : Showing reactions of glycolysis, and anaerobic (fermentation) and aerobic respiration via TCA cycle.

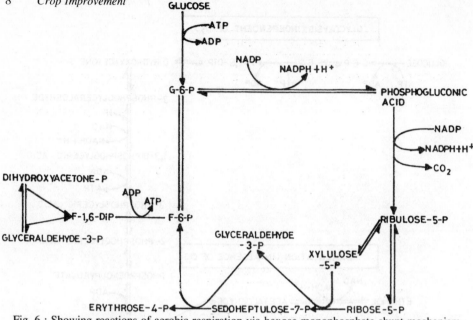

Fig. 6 : Showing reactions of aerobic respiration via hexose monophosphate shunt mechanism.

Another possibility of inhibiting the glycolic acid synthesis is by increasing the concentration of normal leaf metabolites: glutamate and glyoxylate. The effect of glutamate as an inhibitor of glycolic acid synthesis in leaf discs is concentration- and time-dependent (Zelitch et al., 1977). Glutamate at 30 mM also inhibited photorespiration in the ^{14}C-assay, had no effect on glycolate oxidase activity, and stimulated photosynthetic CO_2 fixation by 18 to 25 per cent (Fig. 9). However, potassium glyoxylate at 15 mM was instrumental in increasing $^{14}CO_2$ fixation to almost double (Fig. 10).

The third possibility of developing low photorespiring plants is from the induced chemical mutant cells. Haploid cell cultures of *Nicotiana tabacum* were grown in a standard liquid medium with sucrose as a carbon source, treated with ultraviolet as a mutagen (Zelitch et al., 1977) and after a short period of further growth were transferred to Petriplates of the same medium with the addition of specific inhibitors. Resistant colonies which grew on these plates were transferred to non-selective medium (lacking inhibitor) and then rechallenged on selective medium. Plants grown from such selected mutant cells will have low photorespiration.

After discussing the fundamentals of production physiology at the cell or organ level, we will discuss some aspects at the plant and/or canopy level: It is not surprising to note that the rates of photosynthesis and respiration are not the same in different organs of a plant or strata in the crop canopy. This is exemplified by the data presented in Fig. 11 with regard to a maize canopy. The photosynthetic and respiratory rates are most active at the top of the canopy and at the middle height, respectively (Tanaka, 1977). This means that active photosynthesis and respiration are occurring at different positions; in the well-illuminated leaves and the rapidly growing seeds, respectively.

Idle organs which make no contribution to production, such as excessively elongated

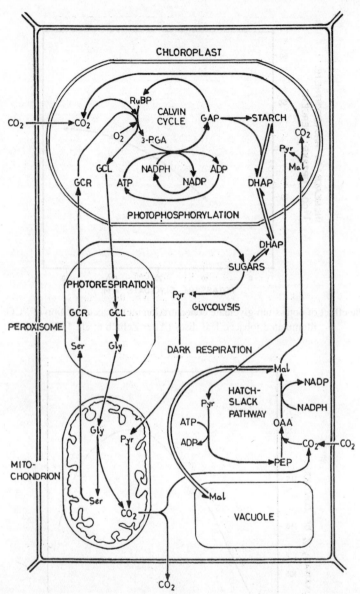

Fig. 7. Outline of the photosynthesis and respiration pathway in plants. These are shown in respect of compartments within a cell but for differences between cells and in time. DHAP = dihydroxyacetone phosphate; GAP = glyceral-dehyde phosphate; GCL = glycollate; GCR = glycerate; Gly = glycine; Mal = Malate; OAA = oxalacetate; PEP = phosphoenolpyruvate; 3-PGA = 3-phospho-glyceraldehyde; Pyr = pyruvate; RuBP = ribulose-1, 5-biphosphate.

internodes at the base of a culm or lower leaves which do not photosynthesize due to shade cast by the upper leaves, consume photosynthetic products in maintenance respiration. These organs may be called parasitic.

In plants growing rapidly by their own photosynthetic products, 35 to 40 per cent is

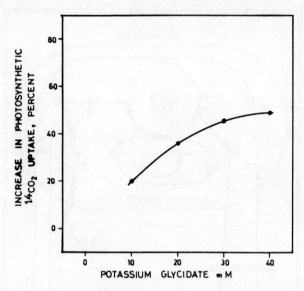

Figure 8. The effect of potassium glycidate concentration on the accumulation of $^{14}CO_2$ fixation by illuminated tobacco leaf discs (After Zelitch et al., 1977)

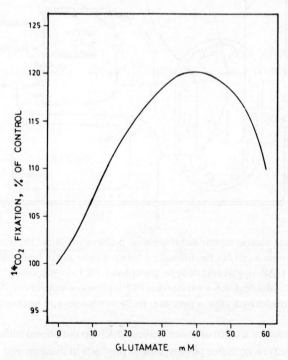

Fig. 9. The effect of L-amylase concentration on the stimulation of photosynthetic $^{14}CO_2$ fixation by tobacco leaf discs (After Zelitch et al., 1977)

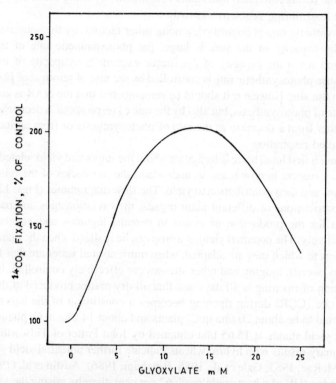

Figure 10. The effect of potassium glyoxylate (pH 4.5) on the stimulation of photosynthetic $^{14}CO_2$ fixation by tobacco leaf discs (After Zelitch et al., 1977)

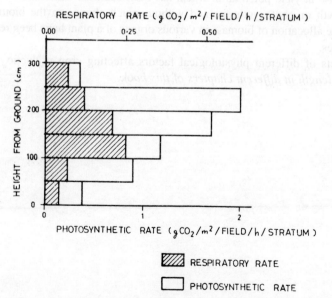

Fig. 11. Photosynthetic and respiratory rates of various strata of a maize canopy during grain filling (After Tanaka, 1977).

respired, and the rest is converted into stable constituents by using the energy generated in respiration, at least during vegetative growth.

The photosynthetic rate is controlled, among other factors, by the source-sink interaction. When the capacity of the sink is large, the photosynthetic rate of the source is accelerated, and when the capacity of the source exceeds the capacity of the sink, it is retarded, because photosynthetic rate is controlled by the rate of removal of photosynthetic products from the site. However, it should be remembered that the NAR is controlled not only by the rate of photosynthesis, but also by the rate of respiration. A decrease in the NAR may result either from a decrease in the rate of photosynthesis or from an increase in the rate of uncoupled respiration.

With this much first hand basic information about the important yield related physiological processes, a student is now ready to understand the intricacies of the relationships of different organs and their contribution to yield. The flow diagrammes (Figs. 12, 13 and 14) project the contributions of different plant organs, their relationships and partitioning of photosynthate for the production of grains in cereals, legumes and vegetative storage organs, respectively. The potential yield of a crop can be realized when the plants are grown in environments to which they are adapted, when nutrients and water are non-limiting, and pests, diseases, weeds, lodging and other stresses are effectively controlled. If we assume that the duration of ripening is 40 days and that all dry matter produced at the maximum crop growth rate (CGR) during ripening becomes a constituent of the harvesting organ, yield is expected to be about 20 t/ha in C_4 plants and about 14 t/ha in C_3 plants. The world record wheat yield stands at 15.65 t/ha obtained by John Potter of Tidworth, England in 1982. Preliminary results with hybrid wheats indicate further potential yield gains of 10 to 25 per cent (McRae, 1985; Gale et al., 1986; Bingham, 1986). Austin et al. (1980) estimate that harvest index (HI) of wheat could reach 62 per cent, thereby raising the yield potential by 25 per cent.

The increase in yield potential of wheat has been obtained without any increase in the rates of growth or photosynthesis, and without much change in the biomass. Clearly, changes in the allocation of biomass in various organs of a plant have been responsible for yield increases.

The aspects of different physiological factors affecting crop production will now be *discussed at length in different chapters of this bo*ok.

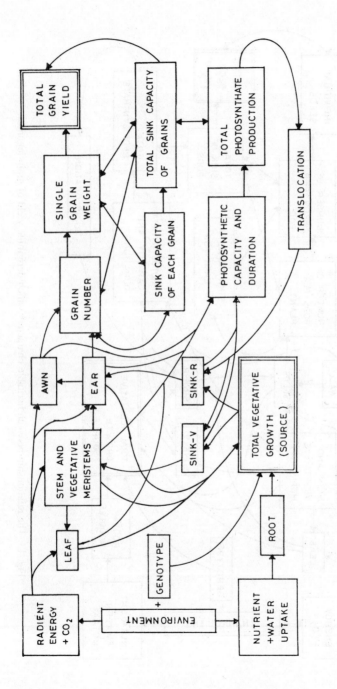

Fig. 12. The relationship between heritable characters which determine cereal grain yield.

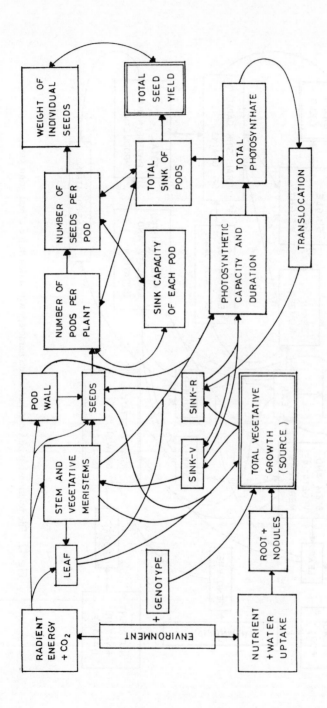

Fig. 13. The relationship between heritable characters which deremine yield of grain legumes.

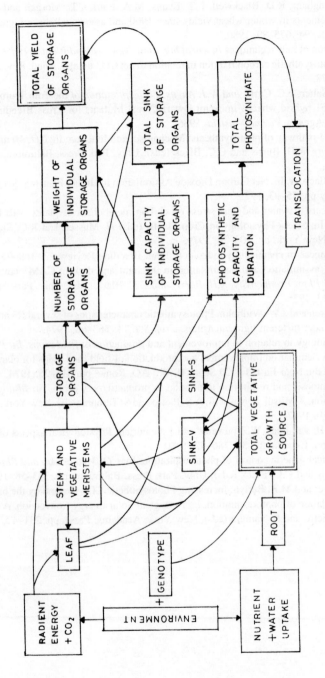

Fig. 14. The relationship between heritable characters which determine yield of storage organs (roots, tubers, corms, rhizomes etc.)

REFERENCES

Austin, R.B., J. Bingham, R.D. Blackwell, L.T. Evans, M.A. Ford, C.L. Morgan and M. Taylor. Genetic improvements in winter wheat yields since 1900 and associated physiological changes. J. agric. Sci. U.K. 94: 675–89, 1980.

Bingham, J. Adoption of new techniques *in wheat breeding. New Zealand Plant Breed. Conf.,* 1986.

Black, C.C., Jr. Photosynthetic carbon fixation in relation to net CO2 uptake. Annu. Rev. Pl. Physiol. 24: 253–86, 1973.

Gale, M.D., A.M. Salter; F.C. Curtis *and W.J. Augus. The exploitation of* the Tom Trumb dwarfing gene, Rht3, in F1 hybrid wheats. In: Semi Dwarf Cereal Mutants for Cross Breeding. IV. Int. Atomic Energy Age*ncy, Vienna, Austria,* 1986.

Hatch, M.D. The C4 pathway of photosynthesis: Mechanism and function. In: *CO₂ Metabolism and Plant Productivity.* R.H. Burris and C.C. Black (eds), Univ. Park Press, Baltimore, pp. 59–81, 1976.

Knai, R. and C.C. Black, Jr. In: Net Carbon Dioxide Assimilation in Higher Plants. S. Sect. Am. Soc. Plant Physiology pp. 75–93, 1972.

Lawlor, D. W. Effects of water and heat s*tress on* carbon metabolism of plants with C3 and C4 photosynthesis. In: Stress Physiology in Crop Plants. (eds) *H.* Mussell and R.C. Staples, John Wiley & Sons, New York. pp. 303–26, 1979.

Mc Rae, D.H. Advances in chemical hybridization. Plant Breeding Reviews. 3: 169–91, 1985.

Osmond, C.B. CO₂ assimilation and dissimilation in the light and dark in CAM plants. In: CO2 Metabolism *and Plant Productivity. R.H. Burris an*d C.C. Black (eds). Univ. Park Press, Baltimore. pp. 217–33, 1976.

Singh, M., W.L. Ogren and J.W. Widholm. Photosynthetic characteristics of several C3 and C4 plant species grown under different light intensities. *Crop Sci.* 14: 563–66, 1974.

Stoy, V. Crop physiology in relation to improvement and produc*tion of field crops. In: Proc. of the First* FAO/SIDA Seminar on Improvement and production of field food crops for plant scientists from Africa and the Near East. 1 – 20 Sept., 1973. FAO, Rome. pp. 315–25, 1974.

Tanaka, A Photosynthesis and respiration in relation to pro*ductivity of crops. In: Biological Solar* Energy Conversion. A Mitsui, S. Miyachi; A.S. Pietro and S. Tamura (eds), New York, Academic Press. pp. 213–29, 1977.

Van Keulen, H., H.H. van Laar, W. Louwerse and J. Goudrian. Physiological aspects of increased CO₂ concentration. Experiantia. 36: 786–92, 1980.

Zelitch, I. Biochemical genetic control of photorespiration. In: *CO₂ Metabolism and Plant Productivity.* R.H. Burris and C.C. Black (eds), Univ. Park Press, Baltimore. pp. 343–58, 1976.

Zelitch, I., D.J. Oliver and M.B. Berlyn. Increasing Photosynthetic CO2 fixation by the biochemical and genetic regulation of photorespiration. In: Biological Solar Energy Conversion. A. Mitsui, S. Miyachi, A.S. Pietro and S. Tamura (eds), New York, Academic Press. pp. 231–42, 1977.

CHAPTER 1

PHOTOSYNTHETIC EFFICIENCY

INTRODUCTION

Photosynthesis is the key to dry matter production and hence yield of economic organ(s). Thus increasing photosynthetic efficiency is the most important way of increasing productivity, but partitioning of photosynthate to the economic organ may not proportionately increase. This is the main reason why many photosynthesis research workers have obtained discouraging results. Improvement in photosynthetic rate can be brought about by identifying or producing genotypes with improved structure and activity of the photosynthetic apparatus while assimilate partitioning can be brought about by manipulating the growth functions of organs and the interactions between them by means of the genetically controlled system for endogenous regulation of physiological processes (Kumakov, 1982). Selection for improved photosynthetic activity and the identification of forms with high combining ability for photosynthetic characters have been indicated (Bykov and Zelenskii, 1982). Further, reduction in photorespiration by genetic engineering of ribulose 1,5 biphosphate carboxylase (RuBP Case) and increased interception of light by establishing ground cover early in the season, can improve photosynthetic efficiency (Moli, 1983; Walker and Sivak, 1986). Nasyrov (1978) had dealt at length with the genetics of chloroplast, and concedes that the chloroplast genome is not sufficient and that the photosynthetic enzymes / efficiency has to depend on nuclear genes. Still the breeding work carried out on photosynthetic improvement is minimal. The subject has been earlier reviewed by Liu and Liu (1984) and Gupta and Olugbemi (1988).

Light Interception

Canopy light interception is most important. Crop varieties with more erect leaves, lower extinction coefficients, and hence higher critical leaf area indices, generally have higher photosynthetic rates (crop growth rate, CGR) and in some cases greater economic yield. Tsunoda (1959) is one of the first to indicate an advantage of erect leaves, when he showed that rice varieties with good economic yield had shorter, more erect and thicker leaves. Murata (1961) then related this plant habit to a lower light extinction coefficient, as compared to varieties with taller and more horizontal leaves. Hayashi (1966) attributed the energy utilization of rice variety, Kim, particularly with dense planting, to its better arrangement of leaves. Kim has more erect leaves which result in smaller light extinction coefficient, that is less complete light interception by the uppermost leaves and more interception of incident radiant energy by the lower leaf canopy.

The short, stiff, upright leaf habit of rice is controlled by an apparent single gene (Aquino and Jennings, 1966). This unique gene has many pleiotropic (simultaneous) effects; the recessive allele gives a reduced number of short internodes, shorter, wider and more erect leaves, and a large number of shorter panicles (Morishima et al., 1967). Hsu and Walton (1970) found that the components of leaf area index (LAI) have highly significant additive gene effects and various degrees of dominant effects. A bean variety having a few large leaves was crossed with another having many small leaves (Duarte and Adams, 1963). The F_1 had the leaf number of the many leaved parent and leaf size intermediate between parents indicating that gene action for leaf number was mostly dominant while action for leaf size was mostly additive.

For obtaining maximum yield, leaf area of a crop should expand to reach its optimum as rapidly as possible. The leaves should maintain activity for a prolonged period, and during senescence they should maintain a supply of assimilates to the reproductive and/ or storage organs. Maximum production of dry matter (C max) corresponds to the optimum value or LAI (LAI opt).

C max = Photosynthetic efficiency × LAI opt

With an increase in leaf area and shading of lower leaves, photosynthetic efficiency of the canopy decreases and thus the LAI opt shifts upwards (Fig. 1.1).

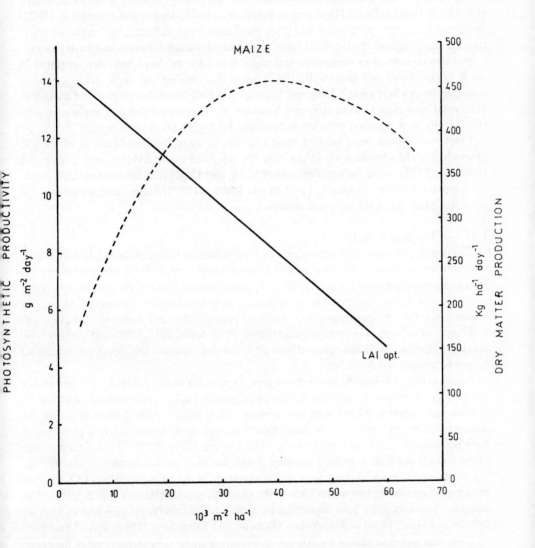

Fig. 1.1.

Leaf Angle

Increasing light penetration into a crop canopy has been suggested as one way of obtaining higher grain yield. Duncan (1971) showed that increased penetration of light into a crop canopy would increase photosynthetic rate and perhaps enhance grain yield. Pearce et al. (1967) found enhanced light penetration in erect leaf canopies, and Angus et al. (1972) found enhanced light penetration and grain yield associated with erect leaf angle in barley. In contrast, Tungland et al. (1987) found no yield advantage for erect leaf barley lines.

Positive contributions of smaller leaf angle (erect leaves) have been demonstrated in small grains. Innes and Blackwell (1983) found that smaller leaf angle made a modest contribution to wheat yield. Chang and Tagumpay (1970) found that smaller leaf angle was associated with high yield in rice, and Yoshida (1972) reported that leaf angle was used successfully as a selection criterion in breeding for higher grain yield in rice.

Erect leaves have been studied most extensively in maize. Pendleton et al. (1968) observed a yield advantage of 41 per cent for erect leaf maize hybrids, and Winter and Ohlrogge (1973) found yield advantages of 9, 14, and 12 per cent for erect maize hybrids at high plant densities. Similarly, Lambert and Johnson (1978) found yield advantages for erect leaf plant types at high plant densities.

LAI — Changing Concept

The concept of larger leaf expansion or canopy close up is fast changing. In fact, plants with smaller and more vertically disposed leaves allowing greater light penetration into the lower canopy strata and thus raising the CO_2 compensation point of the bottom leaves are preferred. Such plants can be sown more densely in order to rapidly obtaining LAI opt and higher yield. Recent developments of 'leafless' peas (Hedley and Ambrose, 1981), 'okra' and "super okra" type of cotton varieties (Kerby, 1977; Kerby et al., 1980) and stemmy type potatoes (Hruska, 1975) have proved beyond doubt that luxuriant leaf development can no longer be considered desirable.

Snoad (1972; 1974) introduced the *st* gene in peas for reduced stipule size and the *af* gene which substitutes tendrils for leaflets, into genetic backgrounds derived from leaved commercial varieties. Plants with this genetic constitution (*afafstst*) have acquired the descriptive name of 'leafless'. The development and reasoning behind this new plant model have been discussed (Snoad and Davies, 1972; Davies, 1976; 1977). The main advantage of the leafless pea is its improved standing ability due to its greater number of tendrils; the risk of lodging is reduced. They have no disadvantage in the efficiency of net CO_2 fixation or in translocation of photoassimilates to the developing pods (Harvey, 1972; 1974). The comparative yield trials have shown leafless peas to be capable of yielding as high as (Snoad and Gent, 1976) or higher than (Debelyi and Knyazkova, 1986) leaved varieties.

Eight near isogenic lines of cv. Alaska, representing all possible phenotypes of the genes *af* (leaflets transformed to tendrils), *st* (reduced stipule area) and *tl* (tendrils transformed to leaflets) were examined (Hobbs, 1986). Carbon exchange rate (CER) was measured on the leaflets (*AfAf*), tendrils (*afafTlTl*) or minute leaflets (*afaftltl*). The total photosynthetic area was significantly reduced by the *st* gene in *AfAf* types with an apparently associated increase in CER. The *st* gene also significantly reduced the total photosynthetic area in *afaf* types but there was no associated increase in CER. Tendrils had a lower CER than normal leaflets and comprised 22 per cent of the total photosynthetic area of the leafless (*afafStStTlTl*) type. Crosses were made between a semileafless pea and four normal leaved types previously selected for high or low CER. The CER means (normal leaflets) of the F_1 progeny showed variability which was related to parental values. This was also true for the CER means (tendrils) of the populations of semileafless F_2 segregates, showing that genetic variability for CER can exist in tendrils. In the F_2, tendril CER was correlated negatively with stomatal resistance and positively with chlorophyll content and final shoot dry weight (biomass). Genetic improvement in CER may be important when a plant ideotype requires substantial reduction in total photosynthetic area.

Leafless canopies do not exceed 30 per cent light interception at the first flowering node level. Radiation interception was measured for crop canopies of leafless cv. Filby, semi-leafless selection BS3 and leafy cv. Birte (Heath and Hebblethwaite, 1985). Tendrils and petioles formed over 60 per cent of the total leaf area for Filby but less than 30 per cent for the others. The photosynthetic area index was related to radiation interception by calculating attenuation coefficients, which indicated that Filby intercepted more radiation per unit photosynthetic area index than either Birte or BS3. Radiation interception was related to dry matter accumulation by calculating photosynthetic efficiencies (Heath and Hebblethwaite, 1985). The photosynthetic structures are produced early in the leafless plants and therefore, may remain above the CO_2 compensation point for a longer period and compensate to some extent for the poor light interception by the canopy above the first fruiting node. The major photosynthetic contribution is derived from the tendrils or leaflets subtending each developing pod (Harvey, 1974). Harvey and Goodwin (1978) have suggested that the photosynthetic potential per unit area of the tendrils of the leafless mutant may be higher than that of the leaflets of the leaved phenotype.

The pod acts as a storage organ for the developing seeds, and also as an efficient organ for trapping and recycling CO_2 respired by the developing seeds. Although the pea pod is capable of a net uptake of CO_2 from the atmosphere during the early stages of pod development (Harvey et al., 1976), its role in refixing and recycling carbon to the seed is substantial and accounts for up to 20 per cent of the pods assimilate requirement (Flinn et al., 1977).

Similar to leafless peas, increase in light penetration has been observed in the canopy of super okra leaf mutant of cotton (Kerby, 1977; Kerby et al., 1980). Kerby (1977) grew three nearly isogenic lines of cotton — normal, okra and super okra, differing in leaf area, under narrow row conditions. At the first week of flowering, LAI was 6.0, 4.7 and 3.9, respectively, in the three lines. Differences were most pronounced in the middle — 30 to 55 cm, and upper - above 55 cm, canopy levels. Rates of $^{14}CO_2$ uptake in the upper canopy were similar (on leaf dry weight basis) for the three lines. Below 55 cm, the carbon uptake rates were 53 per cent greater for super okra and 39 per cent greater for okra than in normal leaf plants. Further to this, Karami and Weaver (1980) reported that cotton plants with the okra leaf shape (conditioned by L^oL^o) had higher values for photosynthetic rate, number of green bolls, number of flowering buds and flowers, dry weight of fruiting plants, fruiting index (dry weight of fruiting parts/biological yield), harvest index (HI), yield and lint percentage and a lower number of vegetative branches, than plants with normal leaves.

Horrocks et al. (1978) conducted narrow row field experiments to investigate the importance of leaves near developing bolls as a source of assimilate in normal and super okra plants of cotton. Plants with super okra leaves provided the primary boll with similar amounts of ^{14}C assimilate as the normal leaved plants. Jones (1982) compared the effects of an open canopy cotton homozygous for okra leaf and comparable broad leaved cultivars on agronomic performance. Okra cotton had less leaf area, which allowed for better penetration of light and insecticides. Landivar et al. (1983a) concluded that okra leaf cottons are very competitive in yield with normal leaf ones under favourable growing conditions, but are likely to be less competitive than normal ones under adverse conditions. Meredith and Wells (1986) concluded that certain populations have the genetic potential of producing okra leaf cottons with higher yielding ability than that of normal leaf ones. Four near isogenic lines differing in degree of leaf lobing (normal, subokra, okra and super okra) and the F_1 and segregating F_2 of the cross normal × okra were field grown (Wells et al., 1986). Wells et al. concluded that intermediately lobed leaves may be advantageous in germ-plasm development. Canopy apparent photosynthesis averaged about 7 per cent higher in subokra lines than in their normal leaf parents (Meredith, 1988). Lint yield was also 3 to 7 per cent higher in the subokra lines.

In a study of near isogenic lines of soybean cv. Clark, plants with lanceolate leaflets consistently maintained a higher CER than the plants with normal leaflets, throughout the reproductive stages of development (Hsieh and Sung, 1986). Increased CER was accompanied by an increase in chlorophyll content, increased RuBPCase activity, as a result of higher leaf soluble protein content, and increased exports of photosynthates from the photosynthetic tissues.

Three years of investigation of 'stemmy' varient of potato (Hruska, 1975) has shown that the stemmy plant habit, with many small leaflets having high chlorophyll content, was the best for tuber yield in all maturity classes except the early ones. The leafy habit suited early varieties better because of the timely attainment of photosynthetic capacity.

Photosynthesis as Related to Yield

Photosynthetic efficiency is the primary component of dry matter productivity. It has, however, been found inconsistently related to economic yield because of several plant factors, like — photorespiration, dark respiration, assimilate transport and partitioning efficiency, filling period duration and sink size — all interacting with the aerial and soil environments. Often, total leaf area per plant has a better relationship with yield (per plant) than leaf carbon exchange rate (Duncan and Hesketh, 1968; Kaplan and Kollar, 1977) and much of the crop improvement in the past has come from an increased rate of leaf area production. But the concept is changing — future production enhancement will increasingly depend upon gains in photosynthetic efficiency and reduction in total leaf area so as to increase plant population and productivity as discussed in the earlier paragraphs.

Leaf area per plant and leaf CER generally vary inversely (Irvine, 1975; Kaplan and Kollar, 1977). Sugarcane cultivars with lower CER have been observed to exhibit as much as two-fold greater leaf area per plant than cultivars with higher CER (Irvine, 1975). Improved varieties of wheat, rice, cotton and other crops have lower rates of photosynthesis than their ancestrol types (Evans and Dunstone, 1970; Evans, 1975; Evans and Wardlaw, 1976). However, yield increases have resulted due to increased partitioning and HI. Enhancement of the plant sink and storage capacity could result in increased agronomic yield. Inadequate supply of photosynthate during reproductive growth is the major limiting factor for grain yield. Indirect evidence supporting this hypothesis includes increased grain yield from CO_2 enrichment of the atmosphere around the plants (Wittwer, 1978) and/or supplemental light to the lower canopy strata (Johnson et al., 1969).

Chloroplasts isolated from the high yielding varieties of wheat have higher rates of electron flow and cyclic and non-cyclic phosphorylation than those from the low yielding varieties (Gavrilenko et al., 1974). The chloroplasts of high yielding wheat variety Kavkaz also has a higher photochemical activity (Volodarskii et al., 1980). Also the two high yielding oat isolines of the early series have higher flag leaf CER, and the low yielding isoline of the midseason series has a lower CER than their respective recurrent parents (Brinkman and Frey, 1978).

Photosynthetic electron transport in high yielding wheat varieties is more rapid and energizes the chloroplast membranes to a greater extent than the lower yielding varieties which show less balance between the light and dark stages of photosynthesis (Volodarskii et al., 1981). A study of the photoreducing activity of chloroplasts at the wax ripeness stage showed that the activity declined faster in the low yielding wheat varieties than in the high yielding ones (Gins et al., 1986). The chloroplasts of high yielding varieties had a high photoreducing activity potential.

Chloroplasts isolated from the leaves of a high yielding cotton mutant, Duplex, had a higher rate of non-cyclic electron transport than the other cotton lines (Khramova et al., 1979). The rate of ATP formation in the photophosphorylation reaction was also highest in Duplex chloroplasts. Landivar et al. (1983b) report that lint yield of cotton could be increased by 54 per cent by 30 per cent increase in the photosynthetic rate.

Kumakov et al. (1983) described a method of assessing the role of different plant organs in yield development during grain filling, based on determining the dynamics of biomass for the organs, respiratory expenditure, rate of photosynthesis, gross and net photosynthesis, reutilization and the "common pool" of metabolites used by the developing grain. In a tall variety, Albidum 43, the contribution of the leaves to common pool of metabolites was 35.4 per cent and that of the straw and the leaf sheaths was 64.6 per cent, while in the short variety Saratovskaya 52, it was respectively 59 and 37.3 per cent, with 3.7 percent contributed by the ear itself. Gej and Posnik (1986) report that on an average, removal of awns three to four days after heading caused a marked fall in photosynthetic activity leading to an average reduction of 10 per cent in grain weight.

Photosynthesis as Related to Photorespiration

In C_3 plants, photosynthetic product losses upon photorespiration vary between 20 and 50 per cent of the gross rate of photosynthetic CO_2 fixation (Zelitch, 1973). Its relative magnitude depends upon species, temperature, light intensity and several other factors. The main source of CO_2 release in the process of photorespiration is decarboxylation of the intermediates of glycolate metabolism. The glycolate pathway is closely connected with the Calvin cycle, particularly with oxygenation of RuBP resulting in the formation of phosphoglycolic acid. It is generally accepted that the key enzyme of the Calvin cycle, RuBPCase plays a bifunctional role catalyzing carboxylation and oxygenation reactions.

Photosynthesis as Related to Dark Respiration

Although dark respiration is an opposite process of photosynthesis where the fixed carbon is released, energy liberated in the process is required for driving the assimilatory process. Hence, a high rate of respiration is generally associated with the high yielding varieties. Seedling vigour is positively correlated with seed and seedling respiration rate and the seedling respiration rate can be used as a criterion for the evaluation of seedling vigour in rice. (Chen et al., 1986).

Respiration plays an important part in attraction by the ear of assimilates from the vegetative organs (Vlasenko, 1982). Babenko and Narsschuck (1983) report that the high yielding wheat varieties have a higher functional activity of mitochondria and greater efficacy of oxidative phosphorylation than the low yielding varieties. Damish (1983) sees respiration rate of the sink organ (ear) as the physiological basis for rapid gain in biomass and associated the high respiration rate in the ear with a rapid gain in dry matter in the ear. Rapid dry matter gain was also associated with high ATP content. The correlation between respiration rate and grain yield in wheat was high ($r = 0.72$).

STOMATAL FREQUENCY AND CONDUCTANCE

Stomatal resistance to diffusion of CO_2 has attracted attention of several scientists. They have suggested that high stomatal resistance in plants would be advantageous in water economy and drought resistance, and have attempted to select plants with high stomatal resistance (Miskin et al., 1972; Wilsom, 1972). The stomatal frequency of mature leaves is influenced by the number of stomata which are differentiated and by the extent of epidermal cell enlargement. Stomatal characters such as size and frequency per unit leaf area have high heritability in a range of species (Heichel, 1971b; Miskin et al., 1972; Walton, 1974; Liang et al., 1975; Tan and Dunn, 1973) and thus breeding for these characters should be successful, even though there are phenotypic variations. Although there is evidence to show that both stomatal conductance and behaviour are reasonably heritable, there are very few quantitative studies on the inheritance or heritability of these characters. There is some evidence to show that low conductance is dominant (Henzell et al., 1976; Roark and Quisenberry, 1977). Dwelle et al. (1978) measured the differences in stomatal conductance and gross photosynthesis among 30 advanced selections and clones of potato. On one day, the stomatal conductance of selections A71617-3, A70369-2 and NDA9268-2 in the afternoon, was 55, 98 and 123 per cent, respectively, and CO_2 assimilation was 79, 107 and 120 per cent, respectively of the morning value. On another day, the stomatal conductance of clone A68587-3 in the afternoon was 123 per cent, whereas the photosynthetic rate was only 88 per cent of the morning value; and on yet another day, selection A7079-24 had values of 121 and 150 per cent, respectively, indicating that changes in stomatal conductance and CO_2 assimilation do not show a direct correlation. However, Yoshida (1976; 1978; 1979) working on seven barley varieties and two near isogenic lines differing in stomatal frequency, established a positive correlation between this character and the rate of photosynthesis. Photosynthetic rate was highest in types with the highest stomatal frequency. In 384 rice varieties, stomatal frequency ranged from 43 to 95 stomata mm^{-2} (Yoshida, 1978). Yoshida (1976) suggested that stomatal frequency can be used for screening for high photosynthetic capacity. Thus, in 1979, Yoshida developed closely related barley populations with either high or low stomatal frequency. In the resulting populations, high stomatal frequency was found to be associated with high photosynthetic rates.

Heichel (1971a) used two inbreds of maize, Wf 9 and Pa 83 having a consistent difference in stomatal frequency. The F_1 generation showed dominance for low epidermal cell frequency and partial dominance for low stomatal frequency. The segregation ratios of F_2 and BC generations, the pattern of segregation, and the correlation between epidermal cell and stomatal frequency, suggest that a simple genetic system controls epidermal cell and stomatal frequency.

In sorghum, Liang et al. (1975), reported a high heritability estimate for stomatal density but heterosis was little. High heritability and little heterosis support the general principle that traits showing little or no heterosis tend to have higher heritability. Total stomatal number per plant was higher in hybrids than in parents; the difference was attributable to more leaf blade area. Stomata on abaxial surfaces were longer than those on adaxial surfaces; hybrids tended to have longer stomata than did the parents. In a greenhouse study with crested wheat grass, Frank (1978) recorded mean photosynthetic rates of 11.6, 21.5 and 12.8 mg CO_2 dm^{-2} h^{-1} and the mean stomatal frequency of 73, 82 and 69 stomata mm^{-2} for 2n, 4n and 6n plants, respectively.

As the stomatal number per unit leaf area varies with leaf expansion, Martin (1970) expressed his data obtained with bean cross M-62 × Redkote as stomatal index (stomatal number/stomatal number plus epidermal cell number). The F_1 showed dominance and overdominance, respectively, for stomatal indicies of upper and lower leaf surfaces. Further, in an evaluation of five soybean parent diallel, a non-additive gene action was noted for adaxial and abaxial stomata density (Santos, 1978).

CHLOROPLAST EFFICIENCY AND PIGMENT CONTENT

Chloroplast Efficiency

Populations of rye grass, selected by Wilson and Cooper (1967) to exhibit a range of photosynthetic activity, were investigated in terms of chloroplast activity (Treharne, 1972). Hill reaction and PMS-mediated phosphorylation of isolated chloroplasts were compared with photosynthetic activity of the leaves. The data (Table 1.1) reveal a wide range of genetic variation in both Hill activity and PMS-phosphorylation in the populations, when expressed in terms of either chlorophyll content or unit leaf area. Photosynthetic rates of the leaves of different populations showed a wide variation (Wilson and Cooper, 1967) but there was no evident relationship between these, in *in vitro* chloroplast activities and leaf photosynthesis. There are reports of differences in the electron transport system of chloroplasts of sun and shade ecotypes of *Solidago virgaurea* (Björkman, 1968), and varietal differences in photophosphorylation in barley (Kleese, 1966) and maize (Miflin and Hageman, 1966) but the later study did not include measurements of leaf photosynthesis.

Higher photosynthetic rates in isolated chloroplasts of hybrids of many crop species have been observed at the seedling stage (Heichel and Musgrave, 1969; Nagy et al., 1972); and there is a concensus for the occurrence of chloroplast heterosis in crop plants (Sinha and Khanna, 1975). Recent electron microscopic studies have provided evidence that hybrids possess more highly developed chloroplast structure than their respective parents (Hraska, 1978), and the increase in the size of lamellae and thylakoid membrane structure in the chloroplasts of the hybrids is directly correlated with their chlorophyll contents (Rakhmankulov et al., 1975). In addition to the enhanced activities of the key enzymes of the Calvin cycle in hybrid chloroplasts, heterosis in chlorophyll content in maize and sorghum has also been reported (Nosberger, 1970; Khanna, 1974; Fleming and Palmer, 1975). Results of chloroplast complementation based on Hill reaction and cyclic phosphorylation in maize show 25 to 60 per cent increase in activity, and the enhanced activities due to chloroplast complementation are closely associated with the degree of grain yield heterosis (Ovchinnikova and Yakovlev, 1978).

In a study of maize crosses, VIR44 × VIR38, VIR26 × VIR27 and Gloriya Yanetskogo × VIR44, the heterotic hybrids had more active photosynthetic apparatus than their parents,

Table 1.1:   Variation in photosynthesis and chloroplast activity of *Lolium* populations (Plants grown at 15°C, 16h 70Wm⁻² irradiance — After Treharne, 1972).

Population No.	Origin	Leaf photosynthesis litres O_2 min⁻¹ cm⁻²	Hill activity mole FeCN h⁻¹		PMS- phosphorylation mole ATP h⁻¹		RuDPCase activity ¹⁴C-dpm min.⁻¹	
			mg⁻¹ chl.	cm⁻² area 10^{-2}	mg⁻¹ chl.	cm⁻² area 10^{-2}	mg⁻¹ protein 10^{-6}	cm⁻² area 10^{-4}
Ba 8283	Spain	2.00	190	45.0	183	43.6	34	68.2
Ba 8408	Spain	1.94	159	39.8	154	38.5	82	67.8
Ba 8319	Macedonia	1.82	198	49.5	180	45.0	84	65.0
Ba 8182	Germany	1.70	191	45.5	305	72.6	93	55.8
Ba 8169	Canada	1.49	180	42.9	367	87.4	83	58.6
Ba 8341/1	Netherlands	1.34	171	46.2	188	50.8	47	48.5
Ba 8341/2	Netherlands	1.31	236	63.8	136	36.8	36	15.5
LSD 5%		0.43	42	10.1	74	19.2	26	9.1

indicated by a higher rate of Hill reaction and a greater rate and duration of CO_2 assimilation during daylight (Fattakhova et al., 1976). The hybrid *Gossypium hirsutum* S3506 × *G. barbadense* 5595V had a 40 to 45 per cent higher photosynthetic rate than its parents, a 30 to 40 per cent higher photochemical activity in the chloroplasts and 20 to 40 per cent higher rate of phosphorylation. The chloroplasts of the interspecific hybrid had a better developed lamella system than those of its parents, the intervarietal hybrid differed little from its parents in this respect (Imamaliev et al., 1975). The heterotic hybrids of cotton had a higher rate of photophosphorylation and a higher chlorophyll content in the leaves than its parents. The non-heterotic hybrid did not differ markedly from its parents in these respects (Azizkhodzhaev et al., 1975).

Kermanskaya and Urazaliev (1979) studied the Hill reaction in chloroplasts of wheat seedlings and leaves of adult plants of heterotic hybrids and their parents. In some hybrids, such as Kavkaz × Lutescens 85, the rate of Hill reaction in the heading and flowering phases reflected a high degree of heterosis for yield. In the same year, four forms of *Triticum monococcum* tested had a higher photochemical activity of chloroplasts than the bread wheat, Zarya (Mogileva et al., 1979). When two of these forms were compared with the bread wheats Svenno and Leningradka, they each had a more rapid rate of non-cyclic and cyclic photophosphorylation than the bread wheats. Further, Volodarskii et al. (1980) compared six high-yielding wheat varieties with the lower yielding variety Novoukrainka 83; a higher photosynthetic rate, especially in the early stages of photosynthesis occurred in the high-yielding varieties.

Bystrykh and Volodarskii (1983) studied the phosphorylation activity in chloroplasts isolated from plants of five wheat cultivars differing in yield. The chloroplasts of high yielding cultivars were characterized by a markedly higher rate of non-cyclic photophosphorylation, higher cyclic photophosphorylation coupled with the phytochrome system and the stability of phosphorylating systems during grain formation and filling period. Shmeleva et al. (1983) found a correlation between yield and photochemical activity in the flag leaf. Flag leaf of the highest yielding variety was able to produce photochemically reduced NADPH twice as effectively. Differences were observed between four wheat cultivars in the number of P_{700+} reaction centres, size of the light harvesting antenna and the rates of $NADP^+$ photoreduction from water per reaction centre (Gins et al., 1986). Gins et al. suggested that photoreduction activity should be used as an index of photosynthetic activity in breeding and genetic studies.

The chloroplasts of high yielding varieties showed greater efficiency in storing light energy in ATP chemical bonds than those of the low yielding rice varieties (Alauddin et al., 1983). The high yielding varieties had higher total chlorophyll contents, not only per unit fresh weight and per unit leaf surface area, but also per chloroplast and per whole leaf, than the low yielding varieties (Khramova, 1986). The chloroplasts of high yielding varieties had more membrane energy-transducing complexes, particularly those involved in electron transport, than low yielding varieties.

With the increase in ploidy level, chloroplast efficiency has been noted to decrease. Abdullaev et al. (1978) noted that an increase in the ploidy of sugar-beet was accompanied by an increase in the number of chloroplasts in the stomatal and palisade cells of the mesophyll. The higher rate of photosynthesis in 4n plants was due to an increase in the palisade number rather than an increase in the activity of individual chloroplasts. Zelenskii et al. (1978) studied the Hill reaction activity in chloroplast suspensions derived from

Triticum monococcum (2n), *T. dicoccum* (4n), *T. durum* (4n), *T. compactum* (6n) and *T. aestivum* (6n). The rate of Hill reaction tended to decrease in species of higher ploidy. Zelenskii et al. (1979) in a study of 64 wheat varieties belonging to six species (2n, 4n and 6n), including 49 bread wheats, showed the highest photochemical activity by chloroplasts of the evolutionarily least advanced species, *T. monococcum*. Photochemical activity decreased with increasing ploidy. Further, Zelenskii et al. (1982) in a study of 70 forms of 2n, 4n and 6n wheat species detected fairly wide phenotypic diversity in photochemical activity of chloroplasts, the range of variation being 16 to 24 per cent, together with satisfactory stability for this character. Zelenskii et al. considered the photochemical activity of chloroplast as a good basis on which to breed for improved photosynthetic capacity. Austin et al. (1986) noted that the flag leaves of wild diploid species had higher rates of photosynthesis than those of the hexaploids, both on the leaf area basis and per unit weight of chlorophyll. Other organs also had higher photosynthetic rates per unit of chlorophyll in the diploids. Austin et al. recommended that gene transfer for higher photosynthetic rate should be attempted from diploids to hexaploids. However, Stasik et al. (1986) observed a higher rate of CO_2 assimilation in the flag leaf of the bread wheats Kharkovskaya 2 and Kharkovskaya 93 than in the *durum* wheats Narodnaya and Kharkovskaya 46. Shannag (1988) reported that the wild diploid wheat with the A genome had the highest net photosynthesis rate and water use efficiency, reflecting its adaptation to a relatively short growing season and considerable sink demands during the vegetative growth. The cultivated tetraploids and hexaploids exhibited intermediate net photosynthesis rates and osmotic adjustments which allowed them to survive a longer growing season in which water stress increased towards the end. Internal resistance to CO_2 was a more important factor in the differences in net photosynthesis rates than stomatal features.

A further study by Timko and Vasconcelos (1981) with haploid, diploid and tetraploid individuals of *Ricinus communis* revealed that photosystem II and O_2 uptake in photosystem I in isolated chloroplasts decreased with increasing ploidy, as did the rates at which O_2 was given off and CO_2 fixed during photosynthesis. In a study of 2n and 4n cotton species and forms, the wild forms had a higher content of chloroplasts per cell and a greater chloroplast activity than the cultivated ones (Khodzhaev and Igamberdieva, 1986). Pehu et al. (1988) studied the effect of genotype and ploidy on RuBPCase activity, chlorophyll content, leaf area, chloroplast ultrastructure and net photosynthesis among monoploid, diploid and tetraploid anther derived plants (four genotypes at each ploidy level) from a single heterozygous *Solanum phureja* genotype. Within the monoploid group, RuBPCase activity and concentration displayed a significant genotype effect. For the diploids, variation among genotypes was significant for total protein content and maximum specific activity of RuBPCase, and among the tetraploids for net photosynthesis and specific leaf weight. Ploidy effect was evident regarding net photosynthesis, leaf area and chlorophyll content.

Pigment Content

Photosynthetic rate of soybean showed a high correlation with chlorophyll content (Buttery and Buzzell, 1977). Among a collection of 48 cultivars, the linear regression of photosynthetic rate on chlorophyll content accounted for 44 per cent of the variation, whereas with a number of genotypes with various mutant chlorophyll genes, the regression accounted for 81 per cent of the variation. Muresan et al. (1975) further observed that the chlorophyll content in maize varied between the seven hybrids studied, between their parental lines and between different stages of growth. Chlorophyll content and accumulation of dry matter were closely related and heterosis was apparent for both in the hybrids. The maize hybrids exceeded their parents by 20 per cent in yield per unit of leaf area and had a slightly higher content of chloroplast pigments than their parents (Vasev, 1977). From the female parent the highly heterotic maize hybrid 33-6-1-1× 248-11-3-1 inherited a high pigment content and from the male parent the ability of the pigments to form more stable complexes with the lipoproteins in the chloroplasts (Filippova and Strelnikova, 1977). As a result, the rate of photosynthesis was higher in the hybrid than in its parents.

Genotypic differences in chlorophyll content of 18 wheat and 20 rice varieties have been observed by Murthy and Singh (1979) and Srivastava and Singh (1979) respectively. Photosynthetic rates of 25 lines of blackgram (*Vigna mungo*) differed significantly among lines with a two-fold difference between the highest and the lowest (Subramanyam and Pandey, 1981). Further, the studies of Augustine et al. (1976) on the cross Ottawa 67 × VF145-7879 of tomato indicated that the high carboxylation efficiency of Ottawa 67 is controlled by a single gene with dominant effects, probably *hp*, which is recessive for high pigment content. Carboxylation efficiency and chlorophyll content in Ottawa 67 × LA959 appeared to be controlled by the genes *hp* and *lur* (lurida). The results indicated that *lur* is epistatic to *hp* for carboxylation efficiency. F_2 segregation ratios differed for carboxylation efficiency and chlorophyll content.

PHOTOSYNTHESIS

C_3 Mechanism

The C_3 mechanism operating through Calvin cycle where the first stable product is a three carbon compound (PGA), is a commonplace knowledge and will not be discussed here except re-inforcing the fact that RuBPCase is the key enzyme in this mechanism. Measuring photosynthesis in the field of the breeding lines has been difficult due to the requirement of extensive equipments or radio isotopes which are costly and time consuming. So far the most practical field method has been the determination of specific leaf weight (SLW). SLW has frequently been correlated with net photosynthesis in several crops and can be measured rapidly with a leaf punch with little damage to the plant. There is a significant positive correlation between SLW and NAR (Buttery and Buzzell, 1972). SLW is a heritable character (Ojima and Kawashima, 1970; Buttery and Buzzell, 1972). Lugg and Sinclair (1979), however, found an increase in SLW from flowering to after mid-pod-fill, and thereafter a sharp decline due to the transport of assimilates to the developing grains.

Sullivan et al. (1976) and Clegg and Sullivan (1976) developed a simplified portable photosynthesis chamber and technique for measuring CO_2 concentrations which made the measurement of photosynthesis of many plants within a reasonable time span, possible. The technique involves the use of a small battery-driven fan inside for air circulation. When the chamber is clamped on a plant leaf in light for a few seconds, CO_2 is utilized from air within the chamber if the leaf is photosynthesizing. A small sample of air is taken with a syringe and needle from the chamber immediately after the chamber is clamped on the leaf and a second sample is taken after 10 to 15 seconds. The sample of air in the syringes are then injected into a precalibrated infrared CO_2-analyzer and the concentration of CO_2 determined in 5 to 10 seconds. The rate of photosynthesis can then be calculated very rapidly. The infrared analyzer may be kept in a laboratory many miles from the field where the plants are sampled. The method is non-destructive and repeated samples can be taken from a single plant, if desired.

By darkening the plants, respiration rate can also be measured. This system permits rapid sampling of many plants at different locations in fields, with a minimum amount of equipment to move. This has greatly improved the feasibility of using photosynthesis and respiration as selection criteria.

Further to this, Ito et al. (1981) designed an apparatus used for measuring photosynthetic O_2 production by small pieces of leaves of 17 rice varieties. The rate ranged from 302 to 580 μ moles O_2 $cm^{-2}h^{-1}$. Ito et al. concluded this as a useful method for detecting varietal differences in photosynthetic capacity in rice. Then Nagamine et al. (1987) developed a simple, rapid and stable method of measuring photosynthetic activity in rice leaves using multi-channel O_2 electrodes. Changes in O_2 evolution activity associated with sampling time were eliminated by punching leaf discs immediately after sampling the leaves and incubating the discs on distilled water for 24 hours before measurements were made. Oxygen evolution of the leaf discs was stable irrespective of the field weather conditions at sampling. Maximum O_2 evolution was obtained at a light intensity of 70 K lux and 25 °C, using N-2-hydroxyethyl piperazine-N"-ethanesulphonic acid buffer solution at pH 7.2. Oxygen evolution could be measured in 150 samples per day by this method. The method was developed for analysis of genetic variation in photosynthetic capacity and is particularly recommended for this purpose.

An apparatus for the simultaneous measurement of photosynthetic $^{14}CO_2$ fixation by a large number of leaf samples was described (Albergoni and Basso, 1985). The apparatus offers the possibility of simultaneous measurements of up to 100 samples under constant environmental conditions. Uniform illumination of all samples is obtained by continuous rotation of the sample bearing disc during exposure to $^{14}CO_2$. The use of apparatus is illustrated by experiments with maize, beans, tomato, cucumber and tobacco, the results of which were generally in good agreement with those obtained by other methods.

Variability : About 20 per cent differences in the rate of photosynthesis of soybean varieties have been reported (Ojima and Kawashima, 1970). Ojima and Kawashima presented data showing genetic segregation in crosses between parents with high and low CER. Differences in soybean parents and derived F_3 progenies were highly significant. In 55 soybean genotypes studied by Ojima (1974), the coefficient of variation in the rate of photosynthesis was up to 25 per cent, while Bhagsari et al. (1977) observed that in 16 cultivars of soybean, the rate of photosynthesis ranged from 23 to 37 mg CO_2 dm^{-2} h^{-1}. Sinclair (1980) also observed over 60 per cent differences in CER among the 21 field-grown soybean cultivars. Among the indeterminate soybean cultivars, as great differences in canopy CER as 20 mg CO_2 dm^{-2} h^{-1} occur (Hansen, 1972).

The variability and heritability in CER of peas were examined by Mahon and Hobbs (1981). The CER of 25 genotypes ranged from 18.9 to 35.5 mg CO_2 dm^{-2} h^{-1}. Three genotypes with high and three with low CER were selected and retested in the field. Despite a significant genotype × environment (G × E) interaction, the high and low groups were consistently separable.

The apparent photosynthetic rate of peanut lines showed an approximate variation of over 50 per cent (Bhagsari, 1975; Rao and Ram Das, 1981). A very rapid rate of photosynthesis was observed in the Virginia type between these genotypes (Pallas and Samish, 1974). The net photosynthetic rate of several genotypes exceeded 50 mg CO_2 dm^{-2} h^{-1}.

The tall *indica* rice varieties are more efficient in photosynthesis than the high yielding semidwarf varieties (Murty et al., 1976). About 15 per cent of the varieties (out of 66) showed higher NAR values than the control variety Kalinga I (Murty and Pattanaik, 1986). Further, among the 24 varieties studied, Cauvery (early), Swarnaprabha (medium) and Vikram (late) showed the highest photosynthetic rate at flowering in their respective maturity groups (Swain et al., 1986). Swarnaprabha combined high photosynthetic rate with optimum LAI and grain yield. Photosynthetic rate at flowering was lowest in the late varieties. LAI at flowering was associated negatively with photosynthetic rate but was associated positively with dry matter content and yield. The photosynthetic rate at flowering was negatively associated with yield. Swain et al. suggested that varieties should be selected for the best combination of photosynthetic rate and LAI, as in Swarnaprabha. Watanabe (1987) measured photosynthetic oxygen evolution by means of an oxygen electrode in 15 *Oryza* species. Cultivated species had a higher photosynthetic oxygen evolution than did wild species. Among 15 varieties of *O. sativa* (three subspecies), autotetraploids had a higher chlorophyll content per unit leaf area than diploids. Photosynthetic oxygen evolution of diploids was around half that of autotetraploids. Kawamitsu and Agata (1987) further compared 50 rice varieties under controlled conditions at three stages between maximum tillering and panicle heading. The maximum photosynthetic rate observed was 51 mg CO_2 dm^{-2} h^{-1} in Century Patna 231 and the minimum was 22 mg CO_2 dm^{-2} h^{-1} in Senbon Asahi.

Differences in photosynthetic efficiency of wheat genotypes have been observed at the late tillering and flowering stages (Zaigraev, 1978; Aliev and Kazibekova, 1979). Significant differences in photosynthetic rates of oat genotypes have also been observed (Henshaw and McDaniel, 1972).

In sweet potato, photosynthetic rates have been reported to range from 19.1 to 32.4 mg CO_2 dm^{-2} h^{-1} (Bhagsari, 1980) and 28.5 to 38.9 mg CO_2 dm^{-2} h^{-1} (Bhagsari and Harmon, 1980). Photosynthetic rates of nine wild species and 10 cultivars of cassava were measured (Mahon et al., 1977). The rates ranged from 15 to 29 mg CO_2 dm^{-2} h^{-1} in *Manihot quinquipartita* and *M. dichotoma*, respectively. Varieties of sugar-beet have been shown to differ in photosynthetic potential up to two-fold (Lutkov, 1977).

Heritability and Heterosis : Heritability estimates for photosynthesis have been made for several crops, for example, soybean (Ojima, 1974; Wiebold et al., 1981), peas (Mahon and Hobbs, 1981), cotton (Rakhmankulov and Gaziyants, 1980) and tobacco (Matsuda, 1978) etc. Of the 22 soybean varieties bred from Tokachinagaha (high rate of photosynthesis), Mandarin (high rate of photosynthesis) and Ani (low rate of photosynthesis), 77 per cent were equal to or better than the better parent. Crosses with Harosoy indicated that the low photosynthetic rate of Norin 1 and Manshu is dominant and is inherited quantitatively. A number of F_3 lines with high photosynthetic capacity were selected from F_2 lines with high rates of photosynthesis. In the F_4 the highest yielding lines were obtained from the F_3 lines with high photosynthetic rates. Harrison et al. (1980) grew 34 soybean lines from the F_3 of the crosses Dare × Forrest and Tracy × Davis. Heritability estimates for canopy apparent photosynthesis during pod-filling period were 36 to 66 per cent. Significant genotypic variations and transgressive segregation for canopy apparent photosynthesis occurred at each measurement of both crosses (Harrison et al., 1981). Heritabilities based on variance component estimates were 41 per cent for canopy apparent photosynthesis and 28 per cent for seed yield in the former and 65 and 14 per cent for the latter cross, respectively. Wiebold et al. (1981) selected the soybean parental materials consisting of cultivars with high (Amsoy and Corsoy) and low (Ford and Hawkeye) leaf CER and made the crosses; Amsoy × Ford and Corsoy × Hawkeye. CER, SLW and leaf thickness and leaf density were measured in parents, F_1s and F_2 s. CER was measured in selected F_3 and F_4 lines of Amsoy × Ford, and the other variables were measured in all F_3 and F_4 lines. All traits showed quantitative inheritance without dominant effects. Broad sense heritability, based on parental, F_1 and F_2 variances, was low to moderate for CER.

The *Gossypium hirsutum* varieties Kzyl Ravat and S4727 have a higher photosynthetic rate than the *G. barbadense* varieties S6030 and S6037 (Rakhmankulov and Gaziyants, 1980). In the interspecific hybrids between these varieties, photosynthetic rate and photochemical activity of the chloroplasts depended on the direction of the cross; reciprocal differences revealed the major role played by cytoplasmic inheritance in photosynthesis. Photosynthetic rate, photochemical activity of the chloroplasts and total photosynthesizing surface area in cotton are all inherited as quantitative characters controlled by additive and dominant genes (Gaziyants, 1983). Overdominance is important in the inheritance of these characters and dominance of low photosynthetic activity was observed. The hybrids with a high photosynthetic rate did not show significant heterosis for leaf surface area. When high yield was due to well-developed leaf surface, photosynthetic rate was lower than the parent (Gaziyants and Laiskhram, 1986). Krasichkova et al. (1988) observed heterosis for chlorophyll content in the hybrids at four-leaf stage. Heterosis for photosynthetic activity was not observed, with hybrids being intermediate between their parents or inferior to their parents. Even at the early developmental stages studied, high yielding varieties exceeded low yielding varieties in pigment content and photosynthetic activity. Krasichkova et al. recommended that high yielding varieties be used as maternal parents in crossing to give hybrids with useful photosynthetic characters. Apparent photosynthetic rate of tobacco had low heritability in F_2 and F_3 populations (Matsuda, 1978) but high heritability in the F_4, indicating that selection would be effective in the early generations.

Heterosis for high photosynthesis has been observed in rice (McDonald et al., 1974) and wheat (Kermanskaya et al., 1979) but not in dry bean (Izhar and Wallace, 1967) or soyabean (Wiebold, 1975). Rather, dominance for low photosynthetic rate has been reported in dry bean (Izhar and Wallace, 1967) and in certain soybean crosses (Wiebold, 1975). Although the photosynthetic mechanism is a component of cytoplasm, cytoplasmic inheritance of photosynthesis has not been detected in soybean (Wiebold, 1975) and dry bean (Izhar and Wallace, 1967).

McDonald et al. (1974) recorded the highest (44 per cent) heterosis in photosynthetic rate of the rice cross Kulu × Taichung Native 1 on leaf area basis. However, when heterosis was determined on dry weight basis, the relative ranking of lines changed. At high light intensity, heterosis at 40 °C was slightly higher than at 30 °C suggesting a greater degree of homeostasis in the hybrids than in their parents. Broad sense heritability estimates for photosynthetic rate at 30 ° and 40 °C were 0.75 and 0.78, respectively. In crosses among seven rice cultivars, heterosis for photosynthetic rate under-controlled conditions was the highest in the F_1 from Suzunari × Zenith which exceeded the parental mean by 57 per cent and the better parent by 51 per cent (Murayama et al., 1987). Four of the seven hybrids surpassed the better parent in photosynthetic rate.

In a study of 132 hybrids obtained by crossing 27 wheat varieties, those with the highest rate of photosynthesis were obtained by crossing varieties contrasting in size of photosynthesizing surface area (Orlyuk, 1974). Heterosis for this character occurred most often in hybrids from crosses including Kavkaz × Avrora, Odessa 51, Dnepr 537 and Mironovskaya yubileinya. The highly heterotic wheat hybrids Siete Cerros × Hostianum 8065 and Erythrospermum 8086 × Dneper 521 exceeded their parents in leaf area and photosynthetic rate during the reproductive phase of development (Kermanskaya et al., 1979). It is thought that the area of two to three upper most leaves and the photosynthetic rate of the flag leaf during the reproductive phase can serve as a test for heterosis. While heterosis would appear to affect canopy photosynthesis indirectly by leaf area changes, there have been conflicting reports about whether heterosis affects cotton leaf photosynthetic efficiency *per se*. While Bhatt and Rao (1981) found considerable increases in hybrid leaf photosynthesis when hybrids were compared to their parents, Muramoto et al. (1965) and Wells et al. (1988) found no increases due to heterosis.

The intraspecific (*G. hirsutum*) and two interspecific (*G. hirsutum* × *G. barbadense*) F_1 hybrids exhibited heterosis in photosynthetic rates (Bhatt and Rao, 1981). The developing bolls of the hybrids and significantly higher weights than those of their parents until the twentieth day after anthesis. Patterns of leaf area development among interspecific hybrids were different from those of the parent plants.

In five heterotic cucumber hybrids, the rate of photosynthesis was 7.2 to 33 per cent higher than in their parents (Brezhnev and Tagmazyan, 1973). VIR501 and VIR505 had a more intense photosynthesis and greater productivity than their parents. Compared with their initial forms, the tomato F_1 hybrids studied had a higher rate of CO_2 absorption and a lower rate of respiration from the budding stage to the beginning of ripening (Bontar et al., 1974; Erina, 1975) and were intermediate in chlorophyll content (Erina, 1975). A further study of five tomato hybrids and their parents showed that hybrids between varieties with high photosynthetic activity had high photosynthetic activity and heterosis of yield. The hybrids exceeded their parents in net photosynthetic production and chlorophyll content (Tagmazyan, 1980). Heterosis for photosynthetic rate was found in reciprocal hybrids between Ideal and XXIV 2 and was higher when Ideal was the maternal parent (Stanev et al., 1984).

Genetics, Breeding and Selection: Photosynthetic rate of rice for which parents differed were measured in the parents, F_1 and F_2 of the cross Nakate Shinsenbon × CP-SLO (Hayashi et al., 1977). Photosynthetic rate in the F_1 was more similar to that of Nakate Shinsenbon than CP-SLO, and in the F_2 its distribution was bimodal, suggesting that a single locus was involved and that low photosynthetic rate was dominant over high rate. In the F_2, photosynthetic rate was negatively correlated with the leaf area and positively correlated with N content per unit leaf area.

In general, it appears that the rate of photosynthesis decreases with an increase in the ploidy level. Rates of net photosynthesis of the flag leaves of five hexaploid wheat cultivars, three tetraploid species and seven diploid species were measured (Austin et al., 1982). At the stage when rates were maximal, they were in general, highest for the diploids, intermediate for the tetraploids and lowest for the hexaploids — their mean rates being 38, 32 and 28 mg CO_2 dm^{-2} h^{-1}. Singh and Tsunoda (1978) also recorded a higher rate of photosynthesis in *Triticum aegilopoides* var. *boecticum* (2n) than in *T . aestivum* var. *erythrospermum* (6n). Photosynthetic rate of tobacco leaves was higher in diploid plants than in their haploid derivatives (Avratovscukova and Rerabek, 1973). In onion and potato, tetraploidy is the optimum ploidy in relation to photosynthetic rate (Bykov et al., 1977; Vodyanova, 1979).

Mahon (1979) selected field peas for high leaf photosynthesis. Diallel crosses were made among the three varieties with the highest rates and three with the lowest rates, and after three generations of selfing the progeny from six plants from each combination were assayed. Recombinants from the diallel of cultivars with the highest CO_2 uptake showed a 30 per cent increase in CO_2 uptake, while those from the diallel of cultivars with the lowest CO_2 uptake showed a 30 per cent decrease. Further to this, Mahon and Hobbs (1981) selected peas for CER under field conditions. They made crosses within the high and low groups; the F_4 lines showed significant segregation, and had higher values than the parents. Heritability of CER estimated by several methods averaged 0.7. The results demonstrate the theoretical feasibility of using CER as a selection criterion.

Breeding needs to concentrate on developing plants morphologically adapted for high light-energy utilization (Yang et al., 1984). Nasyrov et al. (1983) recommended that the maternal variety selected should have high photosynthetic rate. The optimum number of functionally active chloroplasts per cell was seen as an important criterion in breeding. Baratov et al. (1985) also observed that in cotton photochemical activity was inherited through the maternal parent. In comparison with the mutant 374, the variety S6040 had a more developed photosynthetic apparatus and a more active photosynthetic enzyme system. Hybrids involving S6040 as maternal parent were recommended as maternal parents when breeding high yielding varieties. For species that cannot be sexually hybridized, namely, tomato and potato, Wettstein (1983) has suggested isolation and fusion of protoplasts for achieving somatic hybrids.

Genetic variability for CER in peas is positively correlated with chlorophyll content and RuBPCase activity, and negatively correlated with stomatal resistance. To examine the genetic control of these characters, six pea genotypes selected for high or low expression of CER were crossed in all possible combinations to produce a complete diallel (Hobbs and Mahon, 1985). Means ranged from 30.8 to 46.0 µg cm^{-2} for chlorophyll content, from 12.4 to 22.5 nmol s^{-1} cm^{-2} for RuBPCase activity, and from 0.44 to 1.25 s cm^{-1} for stomatal resistance. General combining ability (GCA) was significant for all characters and specific combining ability (SCA) was also significant for chlorophyll content and RuBPCase activity. The high proportion of additive gene action should allow selection for improvement of these characters, using as parents those genotypes with high positive (or negative in the case of stomatal resistance) GCA effects.

Bjornstad (1986) presented a number of examples from the literature on cereal and pulse breeding in terms of effectiveness of indirect selection for yield via physiological characters such as photosynthesis, respiration and related morphological sink and source characters. Bjornstad concluded that physiological selection has rarely resulted in yield increases in practice but might do so at higher yield levels. Harrison et al. (1981) have presented evidence that in soybean, canopy apparent photosynthesis (CAP) may serve as a reliable criterion of selection for seed yield. Thirty-four F_3-derived soybean lines from each of the crosses, Dare × Forrest and Tracy × Davis were used for determining CAP per plot during the period of pod and seed development. CAP and seed yield were positively correlated.

Aslam and Hunt (1978) reported that the CER began to decrease rapidly two weeks after ear emergence in wheat variety Glenlea, Neepawa and Kolibri but only after four weeks in Opal. Decline in CER was continuous throughout the period of grain development in all the varieties except Kolibri in which CER was maintained at a constant level between the fourth and the sixth week after ear emergence. In wheat, leaf thickness and SLW were correlated with mean photosynthetic rate (Migus, 1983). Migus suggested that these traits can be used as selection criteria for field screening to identify genotypes possessing high mean photosynthetic rates.

Su et al. (1985) screened rice cultivars with high photosynthetic rates by growing seedlings in an illuminated chamber flushed with low CO_2 air or high O_2 air. Absorbance of photosynthetic pigments at 670 nm was also used as a parameter in evaluation. High O_2 air screening was more effective because it increased direct O_2 inhibition of photosynthesis and enhanced photorespiration. Of 1440 cultivars screened, 100 cultivars with high photosynthetic rates were selected. Based on other desirable agronomic traits, 28 cultivars were recommended as source material for breeding high yielding cultivars with high photosynthetic rates. In the same year, Jiao et al. (1985) used a variant method of screening wheat seedlings at lowered CO_2 concentration created by growing *Pennisetum purpurium* for five to seven days, followed by further screening at lower light intensity (one-third of full sunshine) to select plants with desirable morphological and physiological characteristics. In this way, two ecotypes were distinguished : shade avoiding and shade tolerant (could be used at high plant population). The former was characterized by elongation of leaves competing for light, quicker stomatal responses and rapid decrease in photosynthetic activity at low light intensity while the latter was characterized by an apparent decrease in SLW, slow stomatal responses and relative stability of photosynthetic activity.

In tobacco, populations with high photosynthetic efficiency were selected by exposure to atmospheric CO_2 concentration close to the compensation point in a hydroponic culture chamber for 45 days (Medrano and Primo-Millo, 1985). Selected plants were propagated by *in vitro* culture of buds and then diploidized by treatment with colchicine to obtain seeds. In a field test, plants grown from these seeds were significantly better than controls for growth characters, dry weight, leaf area and net photosynthesis.

The results of Mauney et al. (1978) indicate that selection for CER is effective in increasing yield of indeterminate cotton and soybean varieties. However, Sinclair (1980) reported that most of the soybean cultivars with high CER were selected by screening for high SLW but not all cultivars with a high SLW had high CER.

Physiological Basis of Improvement : RuBPCase has a direct role in C_3 photosynthesis and thus the genotypes possessing higher activity of this enzyme have a higher rate of photosynthesis. A close correlation between photosynthetic rate and RuBPCase activity at the early pod-set and maturity stages of beans and peas has been reported (Peet, 1976; Peet et al., 1977; Hobbs and Mahon, 1985). The peanut genotypes having high net photosynthesis (> 40 mg CO_2 dm^{-2} h^{-1}) were also found to have higher specific activity and content of RuBPCase (Reger et al., 1981). In a study of 20 varieties of green gram, significant differences in RuBPCase activity were recorded at the early reproductive stage (Kuo et al., 1980). The high yielding varieties, PHLV 18 and KJ 5, also possess higher RuBPCase activity. Significant positive correlations were found among RuBPCase, seed yield and HI. Further, Subramanyam and Pandey (1981) examined 25 lines of black gram which showed significant differences in photosynthesis and RuBPCase activity.

Among the haploids of *Solanum phureja*, RuBPCase activity and concentration displayed a significant genotypic effect (Pehu, 1986). Pehu identified two exceptional genotypes : (i) a haploid with 28 per cent greater RuBPCase activity than the donor plant, and (ii) a tetraploid with 30 per cent net greater photosynthesis. Pehu et al. (1988) identified a monoploid genotype (A205) with an increase of 9 per cent in maximum activity of

RuBPCase over another donor plant. Segregation of traits and differential gene expression, together with possible mutations during anther culture, are thought to have possibly produced the variation detected.

Preparations of partially purified RuBPCase from seedling leaves of 2n, 4n and 6n wheat species were assayed for simultaneous carboxylase and oxygenase activities (Holbrook et al., 1984). The mean ratio of carboxylase to oxygenase activities was 6.11 ± 0.16; species differences were not significant.

Growth Behaviour : Early and dwarf varieties appear to possess a higher rate of photosynthesis. During the early stages of development, all plants have a high photosynthetic rate, but the early maturing cotton varieties generally have a higher rate throughout the growing period (Vodogreeva, 1976). A short duration soybean variety, Parana, has a greater photosynthetic efficiency than the long duration variety UFV 1 (Santos et al., 1977).

A comparative analysis of eight wheat cultivars classified by height as tall, semidwarf, dwarf or very dwarf revealed that cultivars carrying Norin 10 dwarfing genes have a significantly higher photosynthetic rate per unit leaf area than the tall cultivars that lack these genes. Among the groups carrying Norin 10 genes, dwarf had the highest photosynthetic and respiratory rates, followed by very dwarf and semidwarf (Kulshrestha and Tsunoda, 1981). Breeding dwarf genomes in several crops for higher productivity have a definite photosynthetic advantage (Gupta, 1978).

C_4 Mechanism

As described briefly but beautifully by Nasyrov (1978), C_4 pathways are an autocatalytic cycle of CO_2 assimilation through transcarboxylation of C_4 acids, and is an auxiliary channel of primary CO_2 fixation via the Calvin cycle. In all higher plants, CO_2 reduction to carbohydrates occurs in the pentosephosphate cycle. However, C_4 plants have the advantage of greater resistance to high temperature, greater water use efficiency, a very low CO_2 compensation point, and a very high rate of CO_2 fixation; for instance, 85 mg CO_2 $dm^{-2}h^{-1}$ being fixed by maize under natural conditions (Heichel and Musgrave, 1969).

The most important feature of C_4 plants is the presence in their leaves of two types of chloroplasts which differ in their structure and functions. In pallisade mesophyll cells, chloroplasts have a pronounced granal structure while in parenchymatous bundle sheath cells they are usually paucigranal. These chloroplasts differ greatly in their photochemical properties : photosystem II is active in mesophyll cells while photosystem I is active in bundle sheath chloroplasts. The key enzyme of the Calvin cycle, RuBPCase is absent in the mesophyll cells. The primary CO_2 fixation in C_4 plants occurs by phosphoenol pyruvate (PEP) carboxylation which leads to the formation of oxalic acid. This process proceeds in the cytoplasm since PEPCase is a purely cytoplasmic enzyme. In chloroplasts oxalacetic acid is transformed into malate and aspartate; depending on species. Malate or aspartate is transferred to the bundle sheath cells where due to decarboxylation of the acid, CO_2 is released and refixed in the Calvin cycle.

Cooperation of the two types of plastids in C_4 plants leads to a concentration of CO_2 in the bundle sheath cells where it decreases to a certain extent the limitation of photosynthesis by the diffusive CO_2 resistance and thus aids in increasing the efficiency of the process. On the other hand, all this prevents the CO_2 burst due to photorespiration. Probably the primary function of C_4 pathway is the reassimilation of CO_2 released as a result of this wasteful process.

Variability : Irvine (1967) reported a 2.5-fold range in $^{14}CO_2$ uptake of 10 sugarcane varieties. Heichel and Musgrave (1969) reported 3-fold differences among 27 corn inbreds and varieties. Among the corn hybrids, Gaskel and Pearce (1981) identified high CER and low CER types. The high CER hybrids produced more tillers and had greater overall leaf areas than the low CER types. Wu (1982) from a variability study of 100 maize genotypes suggested that it is possible to select materials with high photosynthetic capacity and tolerance of shade for high density planting. A direct positive correlation has been noted between photosynthetic rate and grain yield in maize (Han, 1982). While photosynthetic rate and leaf area both influence dry matter production and grain yield significantly, the influence of leaf area is greater.

Among the *Pennisetum* species, Lavergne et al. (1979) noted that *P. mollissimum* had a higher CO_2 fixation rate and a greater accumulation of starch in bundle sheath chloroplasts than in *P. americanum* 23DB.

Productivity advantage of hybrids depends on the assimilatory area of the leaves and the interaction between photosynthesis and respiration (Zhebin, 1970). Leaf area primarily affects total plant production, while increased photosynthetic activity affects grain yield per year. From a computer simulation model, Kaiser (1979) suggested that the grain yield potential of maize cultivars is limited by the rate and duration of photosynthesis during the reproductive growth period. The genetically controlled characters that could increase the rate and duration of photosynthesis should be included in breeding programmes.

Heritability and Heterosis : Crosbie et al. (1977a) working on the CER of maize inbred lines reported that the actual advance for one cycle of selection among 10 inbred lines was 2.5 (4.7 per cent) and 2.3 (5.7 per cent) mg CO_2 dm^{-2} h^{-1} during vegetative and grain filling stages, respectively. Realized heritabilities for CER were 0.33 and 0.26 for the reproductive growth stages. Further, Crosbie et al. (1977 b) measured CER during vegetative (CER_1) and grain filling (CER_2) stages of maize to estimate genotypic variances and heritabilities and to calculate selection advances at these stages. Estimates of additive variance were 1.7 and 4.9 times larger than the estimates of $G \times E$ variance at CER_1 and CER_2 respectively. Narrow sense heritabilities were 0.58 for CER_1 and 0.80 for CER_2 and realized heritability values were 0.72 (CER_1) and 0.66 (CER_2). A high genotypic correlation between CER and CER_1 and CER_2 indicated that selection at either developmental stage would produce nearly identical results.

Significant amounts of heterosis for high photosynthesis in F_1 hybrids of maize have been observed (Heichel and Musgrave, 1969; Wu, 1982; Albergoni et al., 1983; Markowski and Grzesiak, 1984; Akita et al., 1986; Iordanov and Pok, 1987). The actual amount of heterosis depends upon the genetic diversity of plants. Heichel and Musgrave (1969) reported midparent heterosis values ranging from 2 to 97 per cent. Bykov (1975), noted single interline maize hybrids to show heterosis for photosynthetic activity, for example, Iskra exceeded the midparental value for CER by 11 per cent and Slava by 15 per cent. The double interline hybrid VIR 359 exceeded the midparental value by 12 per cent in this respect.

Crosbie et al. (1977a) recorded heterosis in CER of maize during vegetative growth after analysis of a diallel cross involving lines with high or low CER. Overdominance for high CER was shown by two crosses of high CER × high CER lines. General combining ability was the largest component of differences among crosses at both vegetative and grain filling stages, and was positive in high CER lines and negative in low CER lines. Further, Crosbie et al. (1978a) measured CER of a complete diallel (parent, F_1 crosses and their reciprocals) among eight inbred lines of maize (four with low and four with high CER). Measurements were made during vegetative (CER_1) and grain filling (CER_2) stages of growth, in two consecutive years. They recorded large differences among crosses and significant heterosis for high CER at both stages. Several crosses exhibited overdominant phenotypes for high CER at both growth stages but one cross showed significant overdominance for low CER at CER_1.

Wu (1982) recorded 24 to 174 per cent higher photosynthetic rates in the hybrids than those recorded in the parental inbreds, while Markowski and Crzesiak (1984) observed an increase of 23 per cent at the milk ripe stage. The maize line NI72 had the highest CER and produced hybrids with long ears and high CER (Albergoni et al., 1983). High heterosis values for photosynthetic activity were found in progenies of lines H55 and Mo 17. Akita et al. (1986) measured the rate of photosynthesis in four hybrids and four of their parents at five growth stages. All hybrids showed heterosis on the first two stages but heterosis tended to disappear at the later developmental stages. Iordanov and Pok (1987) observed low heterosis in one hybrid but high in the other. However, overdominance was apparent in the inheritance of photosynthetic rate.

In a study of 15 sorghum hybrids and their parents, leaf area showed higher heterosis (30 to 45 per cent) than photosynthetic rate (11 per cent) (Zhao et al., 1983). Positive correlations were also noted between 1000-grain weight and photosynthetic rate at flowering. Heterosis was noted in PEPCase activity in one pearl millet hybrid but not in the other two (Joshi et al., 1986).

Genetics: The heritability studies on photosynthesis in maize confirmed that it is under genetic control and that selection for high net photosynthesis can be accomplished during the early growth stages (Ariyanayagam, 1974). The female parent has a greater effect on photosynthetic activity than the male, that is hybrids with high photosynthetic activity are obtained most often when the female parent has a greater activity than the male (Filatov et al., 1978).

Maize lines with high CER show positive general combining ability (GCA) effects, and low CER lines show negative GCA effects (Crosbie et al., 1978b). Maternal and reciprocal effects are not significant which indicate that CER is controlled largely by additive effects of nuclear genes. A positive genotypic correlation between CER_1 and CER_2 suggests that selection at either growth stage would improve CER throughout the growing season. Nucleus controls formation of the photosynthetic enzymes and proteins of the electron transport chain while DNA of the chloroplasts is responsible for the synthesis of membrane proteins of plastids (Suponina et al., 1975).

In tests on sorghum parental lines and their hybrids, Krieg et al. (1978) observed photosynthetic rate reductions of 20 to 85 per cent among the genotypes at comparable leaf water potentials. Under stress condition, the photosynthetic ability of the hybrids tended to follow the pattern of the female parent and that of the male parent under non-stressed condition.

In 2n and spontaneous 4n plants of the inbred pearl millet line Tift 239DB, the tetraploid had half as many photosynthetic cells per unit leaf area as the diploid, and the mesophyll and bundle sheath cells were 16 μm longer in the tetraploid (Warner and Edwards, 1988). The volume of the bundle sheath cells was twice as great in the tetraploid as in the diploid, but the mesophyll cell volume was only 45 per cent greater in the tetraploid. There were twice as many chloroplasts per cell in the tetraploid but the number of chloroplasts per unit leaf area was the same at each ploidy, as the rate of CO_2 uptake when calculated on a leaf area basis (though it was twice as high in the tetraploid on a per cell basis). Polyploidy-related changes in content of chlorophyll and soluble protein in the leaves, and in photosynthetic enzyme activity on a leaf area and per cell basis, corresponded to changes in photosynthetic rates.

Physiological Basis of Improvement: PEPCase operates in the cytoplasm of C_4 plants (mainly) and its synthesis is controlled by the nucleus. It has a higher affinity for HCO_3^- than for CO_2 (Filmer and Cooper, 1970), therefore, it can bind the hydrated CO_2 more efficiently and transfer it to the chloroplasts, thus decreasing the limitation of photosynthesis by carboxylation resistance. The activity of PEPCase in C_3 plants is 60 times lower than that in C_4 plants (Slack and Hatch, 1967). Induction of mutants with a high PEPCase activity, stimulating CO_2 transport to the chloroplasts could be one of the methods of improving photosynthetic efficiency (Nasyrov, 1979). On the basis of a study of chloroplast DNA in maize and the genetic control of C-metabolism during photosynthesis, Nasyrov (1981) presented a model for the activation of CO_2 transport to the chloroplasts. This involves induction of PEPCase and PEP carboxykinase, which are responsible for regeneration of the CO_2 acceptor, PEP, in the cytoplasm.

CAN WE BREED FOR C$_4$ MECHANISM ?

The C_4 pathway is connected with leaf Krantz anatomy, and therefore, it seems rather difficult to impart properties of cooperative photosynthesis to C_3 plants. However, in nature there are examples where the more highly evolved species in the subgenera *Euatriplex* and *Obione* possess C_4 photosynthetic pathway while the more primitive species possess C_3 pathway. Hybridization between *Atriplex rosea* (C_4) and *A. glabriuscula* or *A. hortensis* or *A. patula* (all C_3) of *Euatriplex* showed that: (i) genetic differences between them were small, and (ii) inheritance of the photosynthetic pathway is determined largely by nuclear genes (Björkman et al., 1974). Wide segregation in F_2 and F_3 progeny suggested that a smaller number of genes determine the inheritance of Krantz anatomy and PEPCase activity. Björkman et al. concluded that it may be possible to incorporate the C_4 pathway into economically important plants in which it is lacking, provided only a small number of genes is involved.

Several genera of plants (for example, *Panicum*) possess what is termed intermediate, or C_3-C_4 photosynthesis. These plants are characterized by reduced photorespiration compared to C_3 species, a CO_2 compensation point that is intermediate to C_3 and C_4 plants (Brown and Brown, 1975), and leaf anatomy that closely resembles the Krantz anatomy (Brown et al., 1983). Brown et al. (1985) developed hybrids between C_3 and C_3-C_4 *Panicum* species and characterized for photosynthesis and leaf anatomy. The F_1 progenies were intermediate to their parents for anatomical characteristics, CO_2 compensation concentration, and photorespiration.

Other experimental approaches to improving photosynthetic efficiency will include: (i) enhancement of the auxiliary CO_2 fixation through PEPCase — the observation of Mehrotra et al. (1981) of a close correlation between net photosynthesis and PEPCase activity in chickpea (a C_3 plant) varieties at flowering is significant, (ii) reconstructing the genome and breeding new promising forms by use of mutagens. Though the overwhelming majority of viable mutants have a reduced photosynthetic capacity, at least some mutants (1 to 2 per cent) may belong to 'plus' class as measured by the photosynthetic efficiency. Photosynthetically positive mutants have been obtained in pea (Highkin et al., 1969) and cotton (Usmanov et al., 1975). With γ -radiation, 60 cotton mutants were obtained among which there was one promising mutant, 'Duplex' with a compact type of branching, high frequency of double sympodial bolls, fast ripening, and high agronomic yield. Its harvest index (HI) amounts to 50 per cent, which points to rational partition and utilization of assimilates. The economic yield of this mutant is at the rate of 1 t ha^{-1} of seed cotton higher than that of the initial variety 108F, and (iii) incorporation of C_4 dimorphic chloroplasts into protoplasts by somatic cell hybridization. The protoplasts from leaf mesophyll cells can be stimulated to fuse by defined experimental manipulations and a hybrid cell obtained. Fusion of protoplasts isolated from two different species may give rise to a somatically produced hybrid plant. A somatic hybrid plant of potato and tomato (both C_3 species) from fused protoplast has been successfully produced (Melchers et al., 1978). It is hoped that in future, the possibility of genome mapping and genotype manipulation by means of somatic hybridization and gene engineering will allow breeders to create new, synthetic plant varieties with a desirable combination of the valuable traits, high photosynthetic efficiency and maximal expression of economic yield (Nasyrov, 1978).

PHOTORESPIRATION

The rate of photorespiration is three to five times greater than dark respiration in C_3 species. Photorespiration results from the oxidation of compounds produced during photosynthesis. By blocking photorespiration, net photosynthetic CO_2 incorporation in many species is increased by at least 50 per cent. By a number of different methods of assay, it has been observed that the release of CO_2 by this process consumes about 50 per cent of the net CO_2 assimilated in C_3 species, while it is barely detectable in C_4 species (Zelitch, 1980). At higher temperatures, photorespiration in C_3 species increases with a Q_{10} value of about three (Zelitch, 1971). The C_4 crop species have a very low rate of photorespiration and hence their rates of photosynthesis are about twice of C_3 species. Average yields and average seasonal growth rates are two to three times higher for C_4 than for C_3 species. Increasing the intracellular concentration of aspartate, glutamate, phosphoenol pyruvate or glyoxylate by floating tobacco leaf discs on solutions of these compounds in light, causes glycolate synthesis and photorespiration to be inhibited (Oliver and Zelitch, 1977 a, b). The $^{14}CO_2$ uptake in light increases as much as two-fold in the presence of glyoxylate. The results obtained on the regulation of glycolate synthesis and photorespiration in leaf tissue provide evidence that chemical or genetic regulation of the formation and metabolism of some commonly occurring compounds should produce plants with decreased photorespiration and higher rates of photosynthesis (Zelitch, 1980).

Variability : Since elimination of photorespiration from C_3 plants is difficult, we must look for the natural variability and heritability of low photorespiration and try to improve upon the commercial varieties by reducing this unwanted process. A 50 per cent reduction in the rate of photorespiration led to a 38 per cent increase in photosynthesis of tobacco, var. Havana Seed (Zelitch, 1976). Further, in sunflower (a C_3 plant) which has a faster growth is because it was possible to select types with only 14.3 per cent photorespiration of true photosynthesis or 16.7 per cent of apparent photosynthesis (Lloyd and Canvin, 1977). But even in species like *Panicum* in which rate of photorespiration is high, Brown (1978) found in a South American species, *P. milioides*, that photorespiration resulted in only 20 per cent loss of the photosynthetic products, and recommended that this species may be used in breeding for increased photosynthetic efficiency.

Zelitch (1975) distinguished a tobacco population from the parent variety "Havana Seed" in which 25 per cent of the progeny had low photorespiration. Some promising varieties with low photorespiration have been obtained through exhaustive screening in populations of C_3 species, *Lolium* and *Paspalum* (Wilson, 1972).

In a comparison of barley varieties Bankuti Korai (Bankut Early) with short day winter barley Horpacs and the long day spring barley Taplan, the first had the lowest rate of photorespiration while the other two varieties were similar to each other in terms of both CO_2 production and O_2 consumption (Pozsar, 1974). In wheat, photorespiration rate varied widely according to developmental phase, variety and growth conditions (Safarov and Akhmedov, 1984; Akhmedov, 1985). The low yielding varieties generally had higher values of photorespiration than of the high yielding varieties. Similar differences were also found in the ratio of photorespiration to photosynthesis. Glycolate oxidase activity rose by 15 times in low yielding and five to six times in the high yielding wheat varieties at the end of tillering and the beginning of stem extension (Gins et al., 1985). From 40 to 75 per cent ^{14}C incorporated in sucrose is provided by the glycolate pathway (Champigny and Moyse, 1979). The RuBPOase/RuBPCase ratio was virtually constant during the growth period. A slow ratio was, however, observed at the active tillering stage (Saka, 1985). Stasik (1987), however, observed a higher photorespiration rate in the flag leaf of high yielding 6n bread wheat varieties as against the low yielding 4n semiwild species *Triticum dicocum*. The earlier reports (Evans and Dunstone, 1970; Evans, 1975; Evans and Wardlaw, 1976) that 6n wheats have lower photosynthesis rate than 4n or 2n wheat varieties, are in conformity to the above observation.

In soybean, Bhagsari et al. (1977) observed significant differences in photorespiration of different cultivars. The soybean variety Ha 79-9440 has a high photosynthetic and a low photorespiration rate and can be classed as a variety with high photosynthetic efficiency (Hao et al., 1983). Further, Latche et al. (1978) made a comparative study of the photorespiratory metabolism of nine soybean varieties using ^{14}C-glycolate. After 30 minutes, all but 8 per cent of the glycolate was metabolized in VNIIMK 1 and Hodgson, whereas 11 to 19 per cent remained unmetabolized in wells, GSZ 3, Amsoy 71 and Hei-long 3. The highest incorporation of glycolate was in Swift and Hodgson and the lowest in Adepta and Wells.

Zelitch and Day (1973) noted that the normal appearing tobacco plants with lower than usual rates of photorespiration and rapid rates of CO_2 uptake had a greater tendency to transmit their traits to their offspring following self-pollination and selection than did plants with high photorespiratory rates. They also showed that diminishing photorespiration resulted in large increases in net photosynthesis and plant productivity.

Genetics : Schmid et al. (1981) studied the dependence of photorespiration and photosynthetic unit sizes on two interdependent nuclear gene factors in tobacco. The seedlings obtained after selfing a variegated individual derived from vari Consolation fell into four mutant categories : variegated, yellow, yellow-green and green. The yellow-green plants were homozygous (aabb). Both green and yellow plants were heterozygous (AaBb). The yellow phenotype was controlled by a labile gene, Cc. The heterozygous green phenotype exhibited an abnormally high rate of photorespiration. The chlorophyll content and photo-synthetic unit size of the yellow phenotype were considerably less than in the green phenotype. Incorporation of Cc into the green phenotype greatly reduced its rate of photorespiration.

Mc Hall et al. (1988) selected a photorespiration mutant of *Nicotiana sylvestris* lacking serine : glyoxylate amino-transferase (SGAT) activity in the M_2 following seed treatment with ethyl methane sulphonate. Mutant strain NS 349 displayed a nine-fold increase in serine accumulation relative to the wild type controls. Enzyme assays revealed an absence of SGAT activity in NS 349. Heterozygous siblings of NS 349 segregating air-sensitive M_3 progeny in a 3:1 ratio were shown to contain half the normal level of SGAT activity, indicating that air sensitivity in NS 349 results from a single nuclear recessive mutation eliminating SGAT activity. Lea et al. (1984) developed a barley mutant that is unable to photorespire. After 10 minutes exposure of the mutant leaves to $^{14}CO_2$, 65 per cent of radioactivity was detected in glycine and only 6 per cent in sucrose while in the wild type, 13 per cent occurred in glycine and 44 per cent in sucrose. The mutant was unable to metabolize $^{14}CO_2$ glycine whilst the wild type converted it to serine and sucrose. Hall et al. (1987) developed a barley mutant deficient in phosphoglycollate phosphatase. The enzyme activity in the mutant was only one-tenth of the wild type leaves. Genetic studies revealed that the mutant trait is controlled by a single recessive gene.

Physiological Methods of Suppression: By using very high concentrations of potassium glycidate (a by-product of glycolate metabolism), it was possible to induce mutation in tobacco leaf single-cell cultures, resulting in cells with low rates of photorespiration (Zelitch, 1976). Such single cells were cultured and grown into complete plants having low rates of photorespiration. Wildner and Henkel (1976) further demonstrated that the pho-torespiration inhibitor glycidate irreversibly inactivated oxygenation, the carboxylase activity of the enzyme not being affected. Apparently oxygenation specific inhibition is connected with blocking of enzyme SH groups located in its active centre.

Oliver and Zelitch (1977b) worked on the possibility of increasing photosynthesis of tobacco leaf discs by inhibiting photorespiration with glyoxylate. Glyoxylate treatment doubled net photosynthetic CO_2 fixation as glyoxylate resulted in decreased photorespiration. These observations showed that photorespiration can be metabolically regulated and suggest that genetic or chemical alteration of pool sizes of metabolites can produce plants with increased photosynthesis. Further to this, Oliver (1978) demonstrated inhibition of photorespiration and increase in net photosynthesis in isolated maize (a C_4 plant[*]) bundle sheath cells treated with glutamate or aspartate. Treatment of the bundle sheath strands with glutamate inhibited glycolate synthesis by 59 per cent. Photorespiration in this tissue, measured as O_2 inhibition of CO_2 fixation was inhibited by treatment with glutamate. Stimulation in net photosynthetic CO_2 fixation probably results from decrease in pho-torespiratory CO_2 loss. This metabolic regulation of the rate of glycolate synthesis and photorespiration observed with isolated bundle sheath strands could account for the inability to detect rapid photorespiration in the mature intact maize leaf.

The possibility of reducing photorespiration by genetic manipulation with the aim of

---

[*] Photorespiration in maize is about 2 per cent of that in tobacco tissue (Zelitch, 1971).

increasing crop yield has been investigated by studying a range of induced *Arabidopsis thaliana* mutants which carried defects in various enzymes associated with the photorespiratory pathway (Somerville, 1982). Somerville concluded that selection for reduced RuBPCase activity, rather than reduction of photorespiration by manipulation of the pathway *per se*, is the most important way of achieving this. Manipulation of CO_2 concentration of the atmosphere allows selection of photorespiratory mutants from the populations of seeds treated with mutagens. Barley lines deficient in activity of phosphoglycollate phosphatase, catalase, the glycine to serine conversion, glutamic synthetase, glutamate synthetase, 2-oxo-glutarate uptake and serine : glyoxylate aminotransferase have been isolated (Blackwell et al., 1988). Also a line of *Pisum sativum* lacking glutamate synthetase was discovered.

Muszynski and Schmid (1986) studied the Warburg effect (the stimulatory effect of low O_2 partial pressure on photosynthetic CO_2 fixation) as a measure of photorespiration in 10 pea mutants and four commercial soybean cultivars. Up to 50 per cent of freshly fixed CO_2 was lost by photorespiration in some cultivars. Muszynski and Schmid concluded that there is a potential for increasing yields by breeding for reduced photorespiration. After studying the photosynthetic characteristics of 11 barley mutant lines deficient in enzymes of the photorespiratory nitrogen cycle in atmosphere containing 1, 21 and 50 per cent O_2, Sivak et al. (1988) separated the lines into two classes : (i) those that displayed normal rates of photosynthetic CO_2 assimilation when compared to wild-type rates, and (ii) those that exhibited low rates of CO_2 assimilation at high O_2 concentrations and where there was no restoration of the original rate on return to 1 per cent O_2.

CONTRIBUTION OF DARK RESPIRATION

Photosynthesis has a direct bearing on dry matter production, but respiration has an indirect effect — a negative effect by way of decarboxylating organic substances and a positive effect by way of generating the much needed ATP' energy. Thus a balanced rate of respiration is essential which is influenced greatly by temperature, moisture regime, age of the plant, etc. Bogdan (1976) noted that the higher yielding wheat variety Polese '70 differed from Polese '71 in having a higher respiration rate, a greater degree of coupling between oxidation and photorespiration, and a higher photochemical activity in the leaves. The contribution of grain respiration to the respiration of the whole ear varies from 30 to 50 per cent in different genotypes (Apel and Tschape, 1973). A high rate of respiration in the growing grain releases energy for starch synthesis, and this results in removal of assimilates, which in turn, causes an increase in photosynthetic rate, provided that CO_2, light, etc are adequate (Damisch, 1974a). The 'magnitude of attraction' (or 'sink capacity') thus depends upon the energy available for starch synthesis. It is shown that the 'total attraction respiration' which is the sum of attraction respiration throughout the grain-filling period; and also the maximum attraction respiration, are both directly related to yield. Further, Damisch (1974b) measured the attraction respiration or respiration of the ear as a storage organ in relation to starch increment and a direct relation between grain yield and respiration rate was established (Damisch, 1974c). Damisch (1974d) argued that the magnitude of attraction depends upon the amount of energy available for starch synthesis, and this energy is made available by means of respiration; there is a feedback relationship between photosynthetic rate and the respiration rate, such that the photosynthetic rate depends upon the control of assimilate consumption by the respiration rate. Thus the high yielding modern varieties have a higher level of energy exchange combined with a lower basic photosynthetic rate than older varieties.

In soybean on the other hand, Dobrunova (1971) reports that the high yielding varieties Vysokostebelnaya 1 (Tall stemmed 1) and Lincoln 8 have low rates of respiration and the low yielding variety VIR 29 has a high rate of respiration and a low rate of photosynthesis.

Since respiration has an indirect effect on photosynthesis and its contribution in increasing photosynthetic efficiency and yield is disputable, the aspects of variability, heritability, genetics and breeding have not been discussed.

REFERENCES

Abdullaev, A.Kh, Kh. Khadzhibaev, R. Khodzhaeva, V.N. Kovalenko and Yu.S. Nasyrov. (Genome control of chloroplast number and function). Dokl. Akad. Fanhoi RSS Tocikiston.21 (9): 48–51, 1978.

Akhmedov, G.A. (Photorespiration in different winter wheat varieties). *In*: Svyaz' metabolizma ugleroda i azota prifotosinteze. Vses. simp., Pushchino, 24–27 iyunya, 1985. Tez. dokh. Pushchino, USSR. pp. 71–72, 1985.

Akita, S., N. Mochizuki, M.Yamada and I. Tanaka. Variations of heterosis in leaf photosynthetic activity of maize (*Zea mays* L) with growth stages. *Jap J. Crop Sci.* 55: 404–07, 1986.

Alauddin, M., T.E. Krendeleva, N.V. Nizovskaya and G.V. Tulbu. Primary processes of photosynthesis in seedlings of rice varieties differing in yield). Fiziol.i Biokhim. Kulturnykh Rast. 15: 327–32, 1983.

Albergoni, F.G. and B. Basso. An apparatus for the simultaneous measurement of photosynthetic $^{14}CO_2$ fixation by a large number of leaf samples. *Maydica*. 30: 75–84, 1985.

Albergoni, F.G., B. Basso, E.Pe and E. Ottaviano. Photosynthetic rate in maize. Inheritance and correlation with morphological traits. *Maydica* 28: 439–48, 1983.

Aliev, D.A. and E.G. Kazibekova. (Features of photosynthetic rate in extensive and intensive wheat varieties). Az SSR Elmlar habarlari, elmlari ser No. 3: 10–5, 1979.

Angus, J.F.; R. Jones and J.H. Wilson. A comparison of barley cultivars with different leaf inclination. *Aust.J. Agric. Res.* 23: 945–57, 1972.

Apel, P. and M. Tschape. Respiration of the ear in the grain filling phase of wheat (*Triticum aestivum* L.). *Biochem. Physiol. Pflanzen.* 164: 266–75, 1973.

Aquino, R.C. and P.R. Jennings. Inheritance and significance of dwarfism in an Indian rice variety. *Crop Sci.* 6: 551–54, 1966.

Ariyanayagam, R.P. Some genetic and physiological features of photosynthesis and its relationship to yield of corn (*Zea mays* L.). *Diss. Abstr. Int. B.* 34 (12): 5763B, 1974.

Aslam, M. and L.A. Hunt. Photosynthesis and transpiration of the flag leaf in four spring wheat cultivars. *Planta* 141: 23–28, 1978.

Augustine, J.J., M.A. Stevens and R.W. Breidenbach. Inheritance of carboxylation efficiency in the tomato. *J.Am. Soc. hort. Sci.* 101: 456–60, 1976.

Austin, R.B., C.L. Morgan and M.A. Ford. Dry matter yields and photosynthetic rates of diploid and hexaploid *Triticum* species. *Ann. Bot.* 57: 847–57, 1986.

Austin, R.B., C.L. Morgan, M.A. Ford and S.G. Bhagwat, Flag leaf photosynthesis of *Triticum aestivum* and diploid and tetraploid species. *Ann. Bot.* 49: 177–89, 1982.

Avratovscukova, N. and J. Rerabek. (A comparison of the photosynthetic activity of haploid and diploid plants *Nicotiana tabacum* and the intensity of growth of tissue cultures derived from then). *In*: First Colloquium on the use of plant tissue cultures *in vitro* in genetics and breeding,

Olomooc, 15–16 November, 1972). Z. Landa, F.J. Novak and Z. Opatrny (eds), Prague, Czechoslovakia, Ustav experimental in botaniky, Cesloslovenska akademie ved. p. 326; pp. 241–47, 1973.

Azizkhodzhaev, A., S.A. Rakhmankulov and A.I. Imamaliev. (Photochemical reactions of the chloroplasts of cotton in relation to heterosis). *Uzbekiston Biologija Zurnali*. 5: 19–21, 1975.

Babenko, V.I. and F.D. Naruchuk. (Functional activity of the mitochondria of the developing ear in winter wheat genotypes differing in yield). Nauchno-tekhnicheskii Byulleten vsesoyuznogo Selektsionno-geneticheskogo Instituta. 4: 52–6, 1983.

Baratov, L., I. Sh. Muratova and P.D. Kolesnikova. (Photo-chemical activity of the chloroplasts from the leaves of *Gossypium barbadense* hybrids S6040 × 5595V and 374 × 5595V). *In*: Materialy Resp. nauchteor. Komf. mol. uchenykh i spets. Tadzh. SSR. Sekts. biol., Dushanbe, Tajik SSK. p. 23, 1985.

Bhagsari, A.S. Photosynthesis in peanut (*Arachis*) genotypes. *Diss. Abstr. Int. B*. 35(10): 4747B, 1975.

Bhagsari, A.S. Photosynthesis of sweet potato genotypes (Abstr.). *Hort.Science*. 15: 279, 1980.

Bhagsari, A.S., D.A. Ashley, R.H. Brown and H.R. Boerma. Leaf photosynthetic characteristics of determinate soybean cultivars. *Crop Sci*. 17: 929–32, 1977.

Bhagsari, A.S. and S.A. Harmon. Screening sweet potato genotypes for photosynthesis. *Agron. Abstr*. Madison, U.S.A., Am. Soc. Agron. 76–77, 1980.

Bhatt, J.G. and M.R.K. Rao. Heterosis in growth and photosynthetic rate in hybrids of cotton. *Euphytica* 30: 129–33, 1981.

Bjorkman, O. Further studies on differentiation of photosynthetic properties in sun and shade ecotypes of *Solidago virgaurea*. *Physiol. Plant*. 19: 618–33, 1968.

Bjorkman, O., J. Troughton and M. Nobs. Photosynthesis in relation to leaf structure. *In*: *Basic Mechanisms in Plant Morphogenesis*. Brookhaven Symposia in Biology, 25: p. 429; pp. 206–226, 1974.

Bjornstad, A. (Physiological principles and methods in plant breeding). Forskning og Forssk i Landbruket. 37: 253–61, 1986.

Blackwell, R.D., A.J.S. Murray, P.J. Lea, A.C. Kendall, N.P. Hall, J.C. Turner and R.M. Wallsgrove. The value of mutants unable to carry out photorespiration. *Photosynthesis Res*. 16: 155–76, 1988.

Bogdan, I.K. (The comparative physiological and biochemical characteristics of the winter wheats Polese 70 and Polese 71). Nauch. tr. Ukr. s.kh. akad. 183: 20–23, 1976.

Bontar, L.S., V.E. Net and L.A. Soboleva. (Features of photosynthesis and respiration in intervarietal tomato hybrids). *In*: *Obmen veshchestv i produktivnost rast*. Kishinev, Moldavian SSR; Stunea. pp. 64–75, 1974.

Brezhnev, D.D. and I.A. Tagmazyan. (A study of photosynthesis and gibberellin content in cucumbers in relation to the phenomenon of heterosis). Byulleten Vsesoyuznogo Ordena Lenina Instituta Rastenievodstva Imeni N.I. Vavilova, 36: 54–57, 1973.

Brinkman, M.A. and K.J. Frey. Flag leaf physiological analysis of oat isolines that differ in grain yield from their recurrent parents. *Crop Sci*. 18: 69–73, 1978.

Brown, R.H. and W.V. Brown. Photosynthetic characteristics of *Panicum milioides*, a species with reduced photorespiration. *Crop Sci*. 15: 681–85, 1975.

Brown, H.R., J.H. Bouton, P.T. Evans, H.E. Malter and L.L. Rigsby. Photosynthesis, morphology, and leaf anatomy, and cytogenetics of hybrids between C_3 and C_3/C_4 *Panicum* species. *Pl. Physiol*. 77: 653–58, 1985.

Brown, R.H., J.H. Bouton, L.L. Rigsby and M. Rigler. Photosynthesis of grass species differing in carbon dioxide fixation pathways. VIII. Ultra-structural characteristics of *Ricinus* species in the *Laxa* group. *Pl. Physiol*. 71: 425–31, 1983.

Buttery, B.R. and R.I. Buzzell. Some differences between soybean cultivars observed by growth analysis. *Can. J. Pl. Sci*. 52: 13–20, 1972.

Buttery, B.R. and R.I. Buzzell. The relationship between chlorophyll content and rate of photosynthesis in soybeans. *Can. J. Pl. Sci.* 57: 1–5, 1977.

Bykov, O.D. (Physiological research in the field of photosynthesis (1965-74). Trudy po Prikladnoi Botanike, Genetike i Selektsii. 56: 162–72, 1975.

Bykov, O.D., V.I. Galkin, N.A. Zhitlova and V.A. Koshkin. The effect of temperature on changes in the rate of photosynthesis and respiration in potato diploids and polyploids. *In*: Genet. fotosinteza o Dushanbe, Tajik SSR; Douis. 241–49, 1977.

Bykov, O.D. and M.I. Zelenskii. (Possibility of breeding for improvement of photosynthetic characters in crop plants). Fiziol. fotosinteza. Moscow, USSR (1982): 294–310, 1982.

Bystrykh, E.E. and N.I. Volodarskii. (Photophosphorylation and yield of spring wheat). Selskokhozyaistvennaya Biologiya, 9: 27–30, 1983.

Champigny, M.L. and A. Moyse. Photosynthetic carbon metabolism in wild, primitive and cultivated forms of wheat at three levels of ploidy : role of glycolate pathway. *Pl. Cell Physiol.* 20: 1167–78, 1979.

Chang, T.T. and O. Tagumpay. Genetic association between grain yield and six agronomic traits in a cross between rice varieties of contrasting plant types. *Euphytica.* 19: 356–63, 1970.

Chen, C.L., F.J.M. Sung and C.C. Li. Physiological and genetic studies on seedling vigour in rice (*Oryza sativa* L.). I. Relations between seed weight, seedling respiration rate, and growth of rice seedlings. *J. Agric. Assoc. China.* 135: 10–16, 1986.

Clegg, M.D. and C.Y. Sullivan. *Agron. Abstr.* Madison, Am. Soc. Agron., U.S.A. p. 70, 1976.

Crosbie, T.M.; J.J. Mock and R.B. Pearce. Inheritance of photosynthesis in a diallel analysis of maize inbred lines from Iowa Stiff Stalk Synthetic (Abstr.) *In*: *Agron. Abstr.*, Madison, U.S.A., Am. Soc. Agron. p. 52, 1977a.

Crosbie, T.M., J.J. Mock and R.B. Pearce. Variability and selection advance for photosynthesis in Iowa Stiff Stalk synthetic maize population. *Crop Sci.* 17: 511–14, 1977b.

Crosbie, T.M., J.J. Mock and R.B. Pearce. Inheritance of photosynthesis in a diallel among eight maize inbred lines from Iowa Stiff.Stalk synthetics. *Euphytica*, 27: 657–64, 1978a.

Crosbie, T.M., R.B. Pearce and J.J. Mock. Relationships among CO_2 exchange rate and plant traits in Iowa Stiff Stalk synthetic maize population. *Crop Sci.* 18: 87–90, 1978b.

Damish, V. (Studies of physiological indices in a high yielding varietal type of winter wheat and proposals for their utilization in the breeding process). *In*: *Voprosy selektsii i genetiki zernovykh Kultur.* L.Sechnyak (ed), Moscow, USSR. pp. 199–211, 1983.

Damisch, W. (On the yield physiology of cereals. 3. Studies on CO_2 gas exchange during the grain filling period in different winter wheat genotypes). *Archiv für Züchtungsforschung.* 4(2): 81–96, 1974a.

Damisch, W. (On the yield physiology of cereals. 4. Respiration rate and ear increment during the grain filling period in winter wheat). *Archiv für Züchtungsforschung.* 4(3): 161–68, 1974b.

Damisch, W. (On the yield physiology of cereals. 5. Photosynthesis and respiration in young plants of different winter wheat genotypes). *Archiv für Züchtungsforschung.* 4: 219–32, 1974c.

Damisch, W. (The relationship between photosynthesis and yield — a problem in modern breeding research). Tagungsbericht, Akademie der Landwirtschaftswissenschaften der Deutschen Demokratischen Republic. 127: 143–51, 1974d.

Davies, D.R. Creation of new models for crop plants and their use in plant breeding. *Appl. Biol.* 2: 87–127, 1976.

Davies, D.R. Restructuring the pea plant. *Sci. Prog.* (Oxford) 64: 201–14, 1977.

Debelyi, G.A. and S.R. Knyazkova. (Features of photosynthesis in pea forms with different leaf and stem types). Doklady Vsesoyuznoi Ordena Lenina i Akademii Selskokhozyaistvennykh Nauk Imeni V.I. Lenina, 12: 14–17, 1986.

Dobrunova, N.L. (Physiological differences between certain varieties of soybean in the foothill zone of Alma-Ata Province). Anyl Saruaslyk Sylymynyn Habarsysy. 3: 34–39, 1971.

Duarte, R. and M.W. Adams. Component interaction in relation to expression of a complex trait in a field bean cross. *Crop Sci.* 3: 185–86, 1963.

Duncan, W.G. Leaf angle, leaf area and canopy photosynthesis. *Crop Sci.* 11: 482–85, 1971.

Duncan, W.G. and J.D. Hesketh. Net photosynthetic rates, relative leaf growth rates and leaf numbers of 22 races of maize grown at eight temperatures. *Crop Sci.* 8: 670–74, 1968.

Dwelle, R.B., G.E. Kleinkopf and J.J. Pavek. Comparative measurements of stomatal conductance and gross photosynthesis among varied clones of potato (*Solanum tuberosum* L). (Abstr.) *In: Agron. Abstr.*, Mafison, U.S.A., *Am.Soc. Agron.* p. 73, 1978.

Erina, O.I. (Study of physiological and biochemical characteristics of tomatoes in relation to the prediction of heterosis). *In*: Geterosis s-kh. rast., ego fiziol.-biokhim. i biofiz. osnovy., Moscow, USSR; Kolos. pp. 75–81, 1975.

Evans, L.T. The physiological basis of crop yield. *In: Crop Physiology : Some Case Histories.* L.T. Evans, (ed) Cambridge Univ. Press, London. 1975.

Evans, L.T. and R.L. Dunstone. Some physiological aspects of evolution in wheat. *Aust. J. Biol. Sci.* 23: 725–41, 1970

Evans, L.T. and I.F. Wardlaw. Aspects of the comparative physiology of grain yields in cereals. *Adv. Agron.* 28: 301–50, 1976.

Fattakhova, F.Z., G.I. Chernikova, S.G. Akhmetshina, F.T. Akhmetkhanova and L.M. Otzhigova. (Study of photosynthetic rate and productivity in maize in relation to heterosis). Dep. 484–77. p. 16, 1976.

Filatov, G.V., S.L. Suponina and G.P. Kotova. (Photosynthetic activity of maize hybrids in relation to the maternal form.) *Selskokhozyaistvennya Biologiya.* 13: 201–04, 1978.

Filippova, R.I. and T.R. Strelnikova. (Characteristics of the pigment system of sweet corn plants in relation to the phenomenon of heterosis). *In*: Genet. fotosinteza, Dushanbe, Tjik SSR; Donis, 257–60, 1977.

Filmer, D.L. and T.A. Cooper. *J. Theor. Biol.* 29: 131–45, 1970.

Fleming, A.A. and J.H. Palmer. Variation in chlorophyll content of maize lines and hybrids. *Crop Sci.* 15: 617-20, 1975.

Flinn, A.M., C.A. Atkins and J.S. Pate. Significance of photosynthetic and respiratory exchanges in the carbon economy of the developing pea fruit. *Pl. Physiol.* 60: 412–18, 1977.

Frank, A.B. Photosynthesis and water relations for three ploidy levels of crested wheat grass (*Agropyron cristatum*) (Abstr.) *In: Agron. Abstr.* Madison, U.S.A., Am. Soc. Agron. p. 74, 1978.

Gaskel, M.L. and R.B. Pearce. Growth analysis of maize hybrids differing in photosynthetic capability. *Agron. J.* 73: 817–21, 1981.

Gavrilenko, V.F., B.A. Rubin and T.V. Zhigalova. (Photosynthetic phosphorylation in isolated chloroplasts and the rate of photosynthesis in wheat seedlings varying in productivity). *Selskokhozyaistvennaya Biologiya.* 9: 345–51, 1974.

Gaziyants, S.M. (Diallel analysis of photosynthetic activity in cotton). *Genetica*, USSR. 19: 1720–26, 1983.

Gaziyants, S.M. and D.M. Laiskhram. (Genetic analysis of photosynthetic processes following interspecific hybridization in cotton). *Selskokhozyaistvennaya Biologiya.* 5: 72–75, 1986.

Gej, B. and J. Posnik. Ear and awn contribution to the photosynthetic productivity of some rye cultivars. Ann. Warsaw Agric. Univ., Agriculture. No. 19: 13–17, 1986.

Ghose, R., Y.P. Abrol and S.K. Sinha. Amylase heterosis and complementation in sorghum. *Pl. Sci. Letters* 2: 173–76, 1974.

Gins, V.K., N.P. Piskunova, G.B. Khomutov, A.N. Tikhonov, E.N. Mukhin and V.A. Pukhalskii.

(The content of photosystem I reaction centres and their photoreduction activity during vegetative growth of spring wheat cultivars). *Fiziol. Rast.* 33: 905–12, 1986.

Gins, V.K., N.P. Piskunova; A.A. Mutuskin, V.A. Pukhalskii and E.N. Mukhin. (Light-dependent NADP reduction in chloroplasts and glycolate oxidase activity in spring wheats of different yield during ontogeny). *Fiziol. Rast.* 32: 53–60, 1985.

Gins, V.K., E.T. Varenitsa and S.G. Chaplya. (The photoreducing activity of chloroplasts and the activity of RuBPCase during the development of winter wheat plants differing in productivity). Doklady Vsesoyuznoi Ordena Lenina i Ordena. Trudovogo Krasnogo Znameni Akademii Selskokhozyaistvennykh Nauk Imeni V.I. Lenina. 9: 2–4, 1986.

Gupta, U.S. Production potential of dwarf genomes. *In: Crop Physiology* (ed.) U.S. Gupta, Oxford & IBH Pub. Co., New Delhi. pp. 374–407, 1978.

Gupta, U.S. and L.B. Olugbemi. Improving photosynthetic efficiency of crops. *In*: Progress in Crop Physiology, (ed) U.S. Gupta, Oxford & IBH Pub. Co., New Delhi. pp. 189–238, 1988.

Hall, N.P., A.C. Kendall, P.J. Lea, J.C. Turner and R.M. Wallsgrove. Characteristics of a respiratory mutant of barley (*Hordium vulgare* L.) deficient in phosphoglycollate phosphatase. *Photosyn. Res.* 11: 89–96, 1987.

Han, G.C. (Correlations between major photosynthetic traits and yield in maize and the analysis of genetic effects). *Acta Agronomica Sinica.* 8: 237–44, 1982.

Hansen, W.R. "Net photosynthesis and evapotranspiration of field grown soybean communities." Ph.D. Thesis, Iowa State Univ. 1972.

Hao, N.B., K.H. Tan, Y.Z. Zhang, W.G. Du and Y.M. Wang. (Study of the photosynthetic characteristics of the soybean strain Ha79-9440, with high photosynthetic efficiency). *Scientia Agricultura Sinica.* 1: 42–48, 1983.

Harrison, S.A., H.R. Boerma and D.A. Ashley. Heritability of canopy photosynthetic capacity and its relationship to seed yield (Abstr.) *In*: World Soybean Res. Conf. II, 1979 (Abstr.), F.T. Cortoin, (ed) Boulder, Colorado, U.S.A., Westview Press. p. 49, 1980.

Harrison, S.A., H.R. Boerma and D.A. Ashley. Heritability of canopy apparent photosynthesis and its relationship to seed yield in soybeans. *Crop Sci.* 21: 222–26, 1981.

Harvey, D.M. Carbon dioxide photoassimilation in normal leaved and mutant forms of *Pisum sativum* L. *Ann. Bot.* 36: 981–91, 1972.

Harvey, D.M. The translocation of ^{14}C-photoassimilate from normal and mutant leaves to the pods of *Pisum sativum* L. *Ann. Bot.* 38: 327–35, 1974.

Harvey, D.M. and J. Goodwin. The photosynthetic net carbon dioxide exchange potential in conventional and 'leafless' phenotypes of *Pisum sativum* L. in relation to foliage area, dry matter production and seed yield. *Ann. Bot.* 42: 1091–98, 1978.

Harvey, D.M., C.L. Hedley and R. Keely. Photosynthetic and respiratory studies during pod and seed development in *Pisum sativum* L. *Ann. Bot.* 40: 993–1001, 1976.

Hayashi, K. Efficiencies of solar energy conversion in rice varieties as affected by planting density. *Proc. Crop Sci. Soc.* Japan, 35: 205–11, 1966.

Hayashi, K., T. Yamamoto and M. Nakagahra. Genetic control for leaf photosynthesis in rice, *Oryza sativa* L. *Jap. J. Breed.* 27: 49–56, 1977.

Heath, M.C. and P.D. Hebblethwaite. Solar radiation interception by leafless, semiless and leafed peas (*Pisum sativum*) under contrasting field conditions. *Ann. appl. Biol.* 107: 309–18, 1985.

Hedley, C.L. and M.J. Ambrose. Designing leafless plants for improving yields of the dried pea crop. *Adv. Agron.* 34: 225–77, 1981.

Heichel, G.H. Genetic control of epidermal cell and stomatal frequency in maize. *Crop Sci.* 11: 830–32, 1971a.

Heichel, G.H. Stomatal movements, frequencies and resistances in two maize varieties differing in photosynthetic capacity. *J. exp. Bot.* 22: 644–49, 1971b.

Heichel, G.H. and R.B. Musgrave. Varietal differences in net photosynthesis of *Zea mays* L. *Crop Sci.* 9: 483–86, 1969.

Henshaw, J.N. and M.E. Mc Daniel. Genetic variation and heterosis for rate of photosynthesis in oats. *In: Agron. Abstr.*, Madison, U.S.A., Am. Soc. Agron. p. 54, 1972.

Henzel, R.G., K.J. Mc Cree, C.H.M. van Bavel and K.F. Schertz. Sorghum genotype variation in stomatal sensitivity to leaf water deficit. *Crop Sci.* 16: 660–62, 1976.

Highkin, H.R., N.K. Broandman and D.J. Goodchild. Photosynthetic studies on a pea-mutant deficient in chlorophyll. *Pl. Physiol.* 44: 1310–20, 1969.

Hobbs, S.L.A. Relationships between carbon dioxide exchange rate, photosynthetic area and biomass in pea. *Can. J. Pl. Sci.* 66: 465–72, 1986.

Hobbs, S.L.A. and J.D. Mahon. Inheritance of chlorophyll content, ribulose-1, 5-biphosphate carboxylase activity, and stomatal resistance in peas. *Crop Sci.* 25: 1031–34, 1985a.

Hobbs, S.L.A. and J.D. Mahon. Genetic, environmental and interactive components of photosynthesis in peas. *In: The Pea Crop*, P.D. Heath and T.C.K. Dawkins (eds.), Butterworths, London. p. 307–15, 1985b.

Holbrook, G.P., A.J. Keys and R.M. Leech. Biochemistry of photosynthesis in species of *Triticum* of differing ploidy. *Pl. Physiol.* 74: 12–15, 1984.

Horrocks, R.D., T.A. Kerby and D.R. Buxton. Carbon source for developing bolls in normal and super okra leaf cotton. *New Phytol.* 80: 335–40, 1978.

Hraska, S. (The relationship between number of chloroplasts and number of lamellae in the grana of the chloroplasts of winter wheat hybrids). *Polnohospodarstvo.* 24: 22–27, 1978.

Hraska, L. (Inheritance of a model-analysis of the growth of selected potato varieties). Thesis, Vysoka skola zemedelska, Brno, Czechoslovakia. p. 48, 1975.

Hsieh, W.L. and F.J.M. Sung. CO_2 exchange rate and translocation during reproductive growth in soybean differing in leaf morphology. *J. Agric. Assoc. China* 135: 25–33, 1986.

Hsu, P. and P.D. Walton. The inheritance of morphological and agronomic characters in spring wheat. *Euphytica.* 19: 54–60, 1970.

Imamaliev, A.I., S.A. Rakhmankulov and A. Azizkhodzhaev. (The effect of hybridization on the structure and function of the photosynthetic apparatus in cotton). *Fiziol. Rast.* 22: 923–28, 1975.

Innes, P. and R.D. Blackwell. Some effects of leaf posture on the yield and water economy of winter wheat. *J. agric. Sci., U.K.* 101: 367–76, 1983.

Iordanov, I.G. and L.K. Pok. (Study of photosynthetic rate in maize in relation to heterosis). *Fiziol. Rast.* 13: 29–33, 1987.

Irvine, J.E. Photosynthesis in sugarcane varieties under field conditions. *Crop Sci.* 7: 297–300, 1967.

Irvine, J.E. Relation of photosynthetic rate and leaf canopy characters to sugarcane yield. *Crop Sci.* 15: 671–78, 1975.

Ito, Y., T. Omura and J. Kawakami. Measurement of photosynthesis of rice seedlings with an oxygen electrode. *In:* Ann. Rep. 1980. Division of Genetics, National Institute of Agricultural Sciences, Japan, Yatabe, Japan. pp. 12–13, 1981.

Izhar, S. and D.H. Wallace. Studies of the physiological basis of yield differences. III. Genetic variation in photosynthetic efficiency of *Phaseolus vulgaris* L. *Crop Sci.*, 7: 457–60, 1967.

Jiao, D., Q. Dai, Y. Nie, J. Liu and B. Guo. (Screening for rice varieties of high photosynthesis adapted to a wide range of light intensity). *J. Agric.Sci.*, China 1(3): 10–16, 1985.

Johnson, T.J., J.W. Pendleton, D.B. Peters and D.R. Hicks. Influence of supplemental light on apparent photosynthesis, yield, and yield components of soybeans (*Glycine max* L.). *Crop Sci.* 9: 577–81, 1969.

Jones, J.E. The present state of the art and science of cotton breeding for leaf morphological types. *In:* J.M. Brown (ed), Beltwide Cotton Production Res. Conf., Las Vegas, N.V. 3–7 Jan, 1980; National Cotton Council, Memphis, T.N. pp. 93–99, 1982.

Joshi, A.K., S.V. Chanda, P.N. Krishnan, P.P. Vaishnav and Y.D. Singh. Chlorophyll content, Hill reaction and PEPCase activity in relation to hybrid vigour in pearl millet during early seedling growth. *J.Agron. Crop Sci.* 157: 1–12, 1986.

Kaiser, H.W. Canopy limitations of crop photosynthesis and yield potential. *In*: Proc. 2nd South African Maize Breeding Symp., March 17–19, 1978. Sect. 3. Physiology and morphology in maize breeding. Tech. Commu., Dept. Agric. and Tech. Sources, South Africa No. 142: pp. 57–81; pp. 73–75, 1979.

Kaplan, S.L. and H.R. Kollar. Leaf area and CO_2 exchange rate as determinants of vegetative growth in soyabean plants. *Crop Sci.* 17: 35–38, 1977.

Karami, E. and J.B. Weaver, Jr. Dry matter production, yield, photosynthesis, chlorophyll relation to leaf shape and colour. *J. agric. Sci., U.K.* 94: 281–86, 1980.

Kawamitsu, Y. and W. Agata. (Varietal differences in photosynthetic rate, transpiration rate and leaf conductance for leaves of rice plants). *Jap. J. Crop Sci.* 56: 563–70, 1987.

Kerby, T.A. Fixation of ^{14}C in cotton canopies as influenced by leaf type. *Diss. Abstr. Int. B.* 37(11): 5476B–5477B, 1977.

Kerby, T.A., D.R. Buxton and K. Matsuda. Carbon source-sink relationships within narrow row cotton canopies. *Crop Sci.* 20: 208–13, 1980.

Kermanskaya, O.I. and R.A. Urazaliev. (Photochemical activity of isolated chloroplasts in winter wheat hybrids and varieties). Kazakstan Avyl Saruasylyk Gylymynyn. Habarsysy. 2: 22–25, 1979.

Kermanskaya, O.I., R.A. Urazaliev and V.P. Bedenko. (Photosynthetic characters as a test for high yield in heterotic wheat hybrids). *Selskokhozyaistvennya Biologiya.* 14: 593–96, 1979.

Khanna, R. Ph.D. Thesis, Indian Agricultural Research Institute, New Delhi. 1974.

Khodzhaev, A.S. and D.I. Igamberdieva. (Characteristics of the photosynthetic apparatus in cotton species of different ploidy). *Clzbekiston Biologiya Zurnali* 4: 26–29, 1986.

Khramova, G.A. (Features of the organization of the photosynthetic apparatus in rice varieties differing in yield). *In*: Geogr. aspekty realiz. Prod. progr. SSSR: tr. shk.-semin. mol. uchenykh biol. fak. MGU i In-ta pochvoved. i fotosinteza AN SSSR, Puchchino, 10–14 noyab, 1985. Dep. 6782V, 44–48, 1986.

Khramova, G.A., M.M. Yakubova and T.E. Kreudeleva. (Primary photosynthetic reactions in cotton mutants differing in yield). Ahboroti Akademijai Fanhoi RSS Tocikiston, Subai Fanhoi Biologi. 4: 35–41, 1979.

Kleese, R.A. Photophosphorylation in barley. *Crop Sci.* 6: 524–27, 1966.

Krasichkova, G.V., L.M. Asoeva, Yu.E. Giller and B.S. Sanginov. (Photosynthetic system of *Gossypium barbadense* at the early stages of development). Doklady Vsesoyuznoi Ordena Lenina i Ordena Trudovogo Krasnogo Znameni Akademii Selskokhozyaistvennykh Nauk Imeni V.I. Lenina, 12: 9–11, 1988.

Krieg, D.R., R.C. Ackerson, F.J.M. Sung and B. Leeton. Genotypic differences in photosynthetic activity of sorghum in response to water stress (Abstr), *In*: *Agron. Abstr.* Madison, U.S.A., Am. Soc. Agron. p. 79, 1978.

Kulshrestha, V.P. and S. Tsunoda. Role of 'Norin 10' dwarfing genes in photosynthetic and respiratory activity of wheat leaves. *Theo. appl. Genet.* 60 (2): 81–84, 1981.

Kumakov, V.A. (Plant photosynthetic activity from the breeding point of view). *Fiziol. fotosinteza.* Moscow, USSR. (1982): 283–93, 1982.

Kumakov, V.A., A.P. Igoshin and B.V. Berezin. (Evaluation of the role of individual organs in grain filling in wheat and its breeding aspects). *Fiziol. i Biokhim. Kulturnykh Rast.* 15: 163–69, 1983.

Kuo, C.G., F.H. Hsu, J.S. Tsay and H.G. Park. Variation in specific leaf weight and RuDPCase activity of mungbean. *Can. J.Pl. Sci.* 60: 1059–62, 1980.

Lambert, R.J. and A.R. Johnson, Leaf angle, tassel morphology and the performance of maize hybrids. *Crop Sci.* 18: 499–502, 1978.

Landivar, J. A., D.N. Baker and J.N. Jenkins. Application of GOSSYM to genetic feasibility studies. I. Analysis of fruit abscission and yield in okra leaf cottons. 23: 497–504, 1983a.

Landivar, J.A., D.N. Baker and J.N. Jenkins. Application of GOSSYM to genetic feasibility studies.II. Analysis of increasing photosynthesis, specific leaf weight and longevity of leaves in cotton. *Crop Sci.* 23: 504–10, 1983b.

Latche, J.C., G. Viala, J. Calmes and G.Cavalie. (Comparative study of the photorespiratory metabolism of different soybean varieties). Annales de l' Amelioration des Plantes. 28: 77–87, 1978.

Lavergne, D., E.Bismuth and M.L. Champigney. Physiological studies on two cultivars of *Pennisetum*. *P. americanum* 23DB, a cultivated species and *P. mollissimum,* a wild species. I. Photosynthetic carbon metabolism. *Z. Pflanzenphysiol.* 91: 291–303, 1979.

Lea, P.J., A.C. Kendall, A.J. Keys and J.C. Turner. The isolation of photorespiratory mutant of barley unable to convert glycine to serine. *Pl. Physiol.* 75 (1, suppl.): 155, 1984.

Liang, G.H., A.D. Dayton, C.C. Chu and A.J. Casady. Heritability of stomatal density and distribution on leaves of grain sorghum. *Crop Sci.* 15: 567–70, 1975.

Liu, Z.Y. and Z.Q. Liu, (Editors). (*Genetics and breeding for photosynthesis*). China, Guizhou People's Press. p. 352, 1984.

Lloyd, N.D.H. and D.T. Canvin. Photosynthesis and photo-respiration in sunflower selections. *Can. J. Bot.* 55: 3006–12, 1977.

Lugg, D.G. and T.R. Sinclair. A survey of soybean cultivars for variability in specific leaf weight. *Crop Sci.* 19: 887–92, 1979.

Lutkov, A.A. (Variation in maximum potential photosynthesis and special arrangement of the leaves in inbred lines of sugar beet). *In*: 3-ii s¹ezd Vses. ob-va genetikov i selektsionerov im N.I. Vavilova, Leningrad, USSR. p 315, 1977.

Mahon, J.D. Selection of field peas for attached leaf photosynthesis in the field.(Abstr). *Pl. Physiol.* 63(5, suppl): 39, 1979.

Mahon, J.D. and S.L.A. Hobbs. Selection of peas for photosynthetic CO_2 exchange rate under field conditions. *Crop Sci.* 21: 616–21, 1981.

Mahon, J.D., S.B. Lowe and L.A. Hunt. Variation in the rate of photosynthetic CO_2 uptake in cassava cultivars and related species of *Manihot. Photosynthetica.* 11: 131–38, 1977.

Markowski, A. and S. Grzesiak. (Physiological indicators of yield in various forms of maize breeding material. I. Photosynthetic heterosis effects). *Acta Agraria et Silvestria, Agraria.* 23: 33–44, 1984.

Martin, F.A. Ph.D. Thesis, Cornell Univ., Ithaca, N.Y. 1970.

Matsuda, T. (Studies on the breeding of high yield variety in air-cured tobacco. IV. Inheritance of apparent photosynthetic rate, rate of photorespiration and rate of respiration). *Utsunomiya Tab. Shikenjo Hokoku.* 16: 9–18, 1978.

Mauney, J.R., K.E. Fry and G. Guinn. Relationship of photosynthetic rate to growth and fruiting of cotton, soybean, sorghum and sunflower. *Crop Sci.* 18: 259–63, 1978.

Mc Donald, D.J., J.W. Stansel and E.C. Gilmore. Breeding for high photosynthetic rate in rice. *In*: Breeding researches in Asia and Oceania. Proc. 2nd General Congress of the Soc. Adv. Breed. Res. in Asia and Oceania. Session XV. Breeding for higher physiological efficiency. S. Ramanujam and R.D. Iyer (eds), New Delhi. pp. 1067–73, 1974.

Mc Hall, N.A., E.A. Havir and I. Zelitch. A mutant of *Nicotiana sylvestris* deficient in serine glyoxylate amino-transferase activity. *Theo. appl. Genet.* 76: 71–75, 1988.

Medrano, H. and E. Primo-Millo. Selection of *Nicotiana tabacum* haploids of high photosynthetic efficiency. *Pl. Physiol.* 79: 505–08, 1985.

Mehrotra, O.N., H.K. Saxena, K.B. Lal, R. Singh and S.S. Yadav. Varietal differences in photosynthesis and productivity of chickpea. *Indian J. agric. Sci.* 51: 306–12, 1981.

Melchers, G., M.D. Sacristan and A.A. Holder. Carisberg Res. Commu. 43: 203–18, 1978.

Meredith, W.R. Jr. Registration of eight sub-okra cotton germ-plasm lines. *Crop Sci.* 1035-36, 1988.

Meredith, W.R. Jr. and R.Wells. Normal versus okra leaf yield interactions in cotton. I. Performance of near isogenic lines from bulk populations. *Crop Sci.* 26: 219–22, 1986.

Miflin, B.J. and R.H. Hageman. Activity of chloroplasts isolated from maize inbreds and their F_1 hybrids. *Crop. Sci.* 6: 185–87, 1966.

Migus, W.N. Flag leaf photosynthetic rate and morphological characteristics in wheat (Abstr.). *Diss. Abstr. Int. B.* 43: 3432 B, 1983.

Miskin, K.E., D.C. Rasmusson, and D.N. Moss. Inheritance and physiological effects of stomatal frequency in barley. *Crop Sci.* 12: 780–83, 1972.

Mogileva, G.A., M.I. Zelenskii and O.V. Sakharova. (Features of the photosynthetic apparatus in *Triticum monococcum*). Byulleten Vsesoyuznogo Ordena Lenina i Ordena Druzhby Narodov Instituta Rastenievodstva Imeni N.J. Vavilova 87: 59–64, 1979.

Moli, A. (Rate of photosynthesis and yield in potato clones). *Potato Res.* 26: 191–202, 1983.

Morishima, H., H.I. Oka and T.T. Chang. Analysis of genetic variation in plant type of rice. I. Estimation of indices showing genetic plant types and their correlations with yielding capacity in a segregating population. *Jap. J. Breed.* 17: 73–84, 1967.

Muramoto, H., J.D. Hesketh and M. El-Sharkway. Relationships among rate of leaf area development, photosynthetic rate and dry matter production among American cultivated cottons and other species. *Crop Sci.* 5: 163–66, 1965.

Murata, Y. (Studies on photosynthesis in rice and its significance in crop production). *Bull. Nat. Inst. Agr. Sci.* Ser D 9: 1–169, 1961.

Murayama, S., K. Miyazato and A. Nose. Studies on matter production of F_1 hybrid in rice. I. Heterosis in the single leaf photosynthetic rate. *Jap. J. Crop Sci.* 56: 198–203, 1987.

Muresan, T., N. Hurdue, M. Terbea, O. Cosmin and T. Sarca. (The interdependence of the degree of heterosis, photosynthetic yield and some physiological indices in maize. III. The dynamics of leaf pigment content-chlorophyll, carotenoids). Analele Institutului de Cercetari pentru Cereale si Plante Tehnice, C. 40: 165–76, 1975.

Murthy, K.K. and M. Singh. Photosynthesis, chlorophyll content and ribulose diphosphate carboxylase activity in relation to yield in wheat genotypes. *J. agric. Sci.*, U.K. 93: 7–11, 1979.

Murty, K.S., S.K. Nayak, G. Sahu, G. Rama-Krishnayya, K.V. Janardhan and R.S.V. Rai. Efficiency of ^{14}C photosynthesis and translocation in local and high yielding rice varieties. *Plant Biochem. J.* 3: 63–71, 1976.

Murty, K.S. and R.K. Pattanaik. Net assimilation rate of traditional rice varieties at vegetative growth stage. *Oryza.* 23: 45–49, 1986.

Muszynski, S. and G.H. Schmid. Photorespiration in cultivars of peas and soybeans. *J. Pl. Physiol.* 124: 187–90, 1986.

Nagamine, T., T. Takita and J. Kawakami. Multichannel measurement of photosynthetic capacity of rice leaves by oxygen electrodes. Bull. Nat. Inst. Agrobiological Resources, Japan. 3: 115–25, 1987.

Nagy, A.H., A. Bokany, R. Bacs, N.G. Doman and A. Faludi-Daniel. Carboxylating enzymes in leaves of two lines of *Sorghum vulgare* cv. *frumentaceum* and their first generation hybrid. *Photosynthetica* 6: 7–12, 1972.

Nasyrov, Yu.S. Genetic control of photosynthesis and improving crop productivity. *Annu. Rev. Pl. Physiol.* 29: 219–37, 1978.

Nasyrov, Yu.S. (Physiological and genetical basis for increasing yield in crop plants). *Selskokhozyaistvennaya Biologiya.* 14: 762–66, 1979.

Nasyrov, Yu.S. (Genetical control of photosynthetic carbon metabolic pathways). *In*: Energ. metabolich. puti i ikh regulyatsiya v fotosinteze. Tez. dokl. Pushchino, USSR. pp. 40–41, 1981.

Nasyrov, Yu.S., Kh.A. Abdullaev and K.A. Asrorov, (Genetics of photosynthesis and methods for the future increase of yield in cotton). Ahboroti Akad. Fanhoi RSS Tocikiston, Subai Fanhoi Biologi, No. 4: 3–10, 1983.

Nichiporovich, A.A. Aims of research on the photosynthesis of plants as a factor in productivity. *In:* Photosynthesis of Productive Systems. A.A. Nichporovich (ed), Israel Program for Scientific Translations, Jerusalem. pp. 3–36, 1967.

Nosberger, V.J. Thesis "Separatabdruck Ans. Schweizerische Landwirtschaftliche Forschung". *Vol. IX.* No. 3/4 S. pp. 235–56, 1970.

Ojima, M. Improvement of photosynthetic capacity in soybean variety. *JARQ.* 8: 6–12, 1974.

Ojima, M. and R. Kawashima. Studies on the seed production of soybean. VIII. The ability of photosynthesis in F_3 lines having different photosynthesis in their F_2 generation. *Crop Sci. Soc. Japan Proc.* 39: 440–45, 1970.

Oliver, D.J. Inhibition of photorespiration and increase of net photosynthesis in isolated maize bundle sheath cells treated with glutamate or aspartate. Pl. Physiol. 62: 690–92, 1978.

Oliver, D.J. and I. Zelitch. Metabolic regulation of glycolate synthesis, photorespiration, and net photosynthesis in tobacco by α-glutamate. *Pl. Physiol.* 59: 688–94, 1977a.

Oliver, D.J. and I. Zelitch. Increasing photosynthesis by inhibiting photorespiration with glyoxylate. *Science.* 196 (4297): 1450–51, 1977b.

Orlyuk, A. P. (Inheritance and variation in the indices of photosynthetic activity in winter wheat hybrids under irrigation). *Selskokhozyaist-vennya Biologiya.* 9: 654–60, 1974.

Ovchinnikova, M.F. and A.Yakovlev. Sel. Sem. 2: 77–79, 1978.

Pallas, J.E. Jr. and Y.B. Samish. Photosynthetic response of peanut. *Crop Sci.* 14: 478–82, 1974.

Pearce, R.B. R.H. Brown and R.E. Blaser. Photosynthesis in plant communities as influenced by leaf angle. *Crop Sci.* 7: 321–24, 1967.

Peet. M.M. Physiological responses of *Phaseolus vulgaris* L. cultivars to growth environment (Abstr.) *Diss. Abstr.Int.B.* 36(11): 5415 B, 1976.

Peet, M.M., A. Bravo, D.H. Wallace and J.L. Ozbun. Photosynthesis, stomatal resistance, and enzyme activities in relation to yield of field grown dry bean varieties. *Crop Sci.* 17: 287–93, 1977.

Pehu, E.P. Analysis of anther derived plants of *Solanum phureja* : variation in ploidy, photosynthetic efficiency and structure of the nuclear genome. (Abstr.). *Diss. Abstr. Int. B.* 47: 1851 B, 1986.

Pehu, E., R.E. Veilleux and J.C. Servaites. Variation in RuBP carboxylase activity among anther-derived plants of *Solanum phureja*. *Physiol. Pl.* 72: 389–94, 1988.

Pendleton, J.W., G.E. Smith, S.R. Winter and T.J. Johnson. Field investigations of the relationships of leaf angle in corn to grain yield and apparent photosynthesis. *Agron. J.* 60: 422–24, 1968.

Pozsar, B.I. (Reduced photorespiration of the photoperiodically neutral barley variety Bankati Rorai and protein synthesis). *Agrobotanika.* 15: 189–94, 1974.

Rakhmankulov, S.R. and S.M. Gaziyants. (Rate of photosynthetic activity in interspecific cotton hybrids in relation to yield). *Selskhokhozyaistvennya Biologiya.* 15: 374–78, 1980.

Rakhmankulov, S.R., E.G. Zaprudes and A. Azizkhodzhaev. (Activity of photochemical reactions in the chloroplasts and the content of nucleic acids in different forms of cotton). *In:* Genet. i selektsiya rast. Taskent, Uzbek, SSR; Fan. pp. 58–63, 1975.

Rao, A.N. and V.S. Ram Das. Leaf photosynthetic characters and crop growth rate in six cultivars of groundnut (*Arachis hypogea* L.). *Photosynthetica.* 15: 97–103, 1981.

Reger, B.J., S.B. Ku, B.C. Prickril and J.E. Pallas, Jr. Genotypic variation of RuDPCase in peanuts (Abstr.). *Pl. Physiol.* 67(4, suppl): 35, 1981.

Roark, B. and J.E. Quisenberry. Environmental and genetic components of stomatal behaviour in two genotypes of upland cotton. *Pl. Physiol.* 59: 354, 1977.

Safarov, S.A. and G.A. Akhmedov. (Features of photosynthesis and photorespiration in winter wheat varieties differing in yield). *In*: Probl. fotoenerg. rast. i povysh. urozhainosti. Tez. dokl. Vses. knof., 3-5 apr, ca 84. L'vov, Ukrainian SSR. pp 84–85, 1984.

Saka, H. (Variation in the activities of several photo-synthetic enzymes at different growth stages in several genotypes and species of the genus *Oryza*). *Bull. Natl. Inst. Agric. Sci.*, Japan D No. 36: 247–79, 1985.

Santos, S. Dos: Germ-plasm evaluation and inheritance studies of stomata density and other plant characteristics in soybean (*Glycine max* (L.) Merrill) (Abstr.). *Diss. Abstr. Int. B.* 38 (12): 5750 B, 1978.

Santos, S. Dos, S. Nogueria, M. Albino and C. de Miranda. (Photosynthetic efficiency of two varieties of soybean (*Glycine max* (L.) Merr.) with different biological cycles.) *In*: Abstracts, 29th Ann. Meeting, Brazilian Society for Scientific Progress (Papers on pulses). *Ciencia e Cultura* 29 (7, suppl): p. 904 and p.14, 1977.

Shannag, E.M.S. Ecophysiological studies of selected wild and cultivated wheat species. *Diss. Abstr. Int. B.* 48: 1879 B, 1988.

Shmeleva, V.L., V.A. Kumakov, G.V. Klimova and B.N. Ivanov. (Photochemical activity of chloroplasts in varieties of spring wheat differing in yield). *Fiziol. Rast.* 30: 30–34, 1983.

Schmid, G.H., K.P. Bader, R. Gerster, C. Triantaphylides and M. Andre. Dependence of photorespiration and photosynthetic unit sizes on two interdependent nuclear gene factors in tobacco. *Z. Naturforschung, C.* 36: 662–67, 1981.

Sinclair, T.R. Leaf CER from post-flowering to senescence of field grown soybean cultivars. *Crop Sci.* 20: 196–200, 1980.

Sinha, S.K. and R. Khanna. The physiological, biochemical and genetic basis of heterosis. *Adv.Agron.* 27: 123–70, 1975.

Singh, M.K. and S.Tsunoda. Photosynthetic and transpirational response of a cultivated and a wild species of *Triticum* to soil moisture and air humidity. *Photosynthetica.* 12: 280–83, 1978.

Sivak, M.N., P.J. Lea, R.D. Blackwell, A.J.S. Murray, N.P. Hall, A.C. Kendall, J.C. Turner and R.M. Wallsgrove. Some effects of oxygen on photosynthesis by photorespiratory mutants of barley (*Hordeum vulgare* L.). I. Response to changes in oxygen concentration. *J. exp. Bot.* 39: 655–66, 1988.

Slack, C.R. and M.D. Hatch. *Biochem. J.* 103: 660–65, 1967.

Snoad, B. Annu. Rep. John. Innes Inst., Norfolk 63: pp. 29–33, 1972.

Snoad, B. A preliminary assessment of 'leafless' peas. *Euphytica.* 23: 257–65, 1974.

Snoad, B. and D.R. Davies. Breeding peas without leaves. *Span.* 15: 87–89, 1972.

Snoad, B. and G.P. Gent. Annual Rep., John Innes Inst., Norfolk. 67: 35–36, 1976.

Somerville, C.R. Genetic modification of photorespiration. *What's New in Plant Physiology.* 13: 29–32, 1982.

Srivastava, D.K. and M. Singh. Photosynthesis, ribulose diphosphate carboxylase activity and chlorophyll content in relation to yield in rice (*Oryza sativa* L.). *J. Nuclear Agric. Biol.* 8 (2): 41–143, 1979.

Stanev, V., M. Angelov, Ts. Tsonev and Zh. Danailov. Photosynthetic rate and chlorophyll content in leaves of tomato hybrids with heterosis for yield). *Fiziol. Rast.* 10: 32–39, 1984.

Stasik, O.O. (Photorespiration rate of the flag leaf in relation to yield in spring wheat plants). *Fiziol. i Biokhim. Kulturnykh Rastenii.* 19: 321–28, 1987

Stasik, O.O.; B.I. Gulyaev and I.I. Rozhko. (Rate of CO_2 gas exchange during development of flag leaf in wheat varieties differing in yield). *Fiziol. i Biokhim. Kulturnykh Rast.* 18: 347–55, 1986.

Su, S., R. Wang, H. Jiang, G. Wang and S. Wang. (Screening for rice germ-plasms with high photosynthetic rate collected from Tai Lake area). *J. agric. Sci.*, China 1(3): 17–24, 1985.

Subramanyan, K. and R.K. Pandey. Photosynthetic characteristics in blackgram (*Vigna mungo* (L.) Hepper). *J. Nuclear Agric. Biol.* 10: 41–43, 1981.

Sullivan, C.Y., M.D. Clegg and J.M. Bennett. *Agron. Abstr.* Madison, U.S.A., Am. Soc. Agron., 1976.

Suponina, S.L., G.V. Filatov and G.P. Kotova. (A study of the physiological and genetic aspects of photosynthetic activity in plants). *In*: Fiziol.-genet. osnovy povysh. produktiv. zern. Kultur. Moscow, USSR, Kolos. pp. 112–20, 1975.

Swain, P., R.K. Pattanaik, S.K. Nayak and K.S. Murty. Relationship between photosynthetic rate, growth components and yield in elite rice varieties. *J. Nuclear Agric. Biol.* 15: 18–22, 1986.

Tagmazyan, I.A. Photosynthetic activity in heterotic tomato hybrids and parental varieties. Byulleten.-Vsesoyuznogo Ordena Lenina i Ordena Druzhby Narodov Institute Rastenievodstva Imeni N.I. Vavilova. No. 99: 58–62, 1980.

Tan, G.Y. and C.M. Dunn. Relationship of stomatal length and frequency and pollen grain diameter to ploidy level in *Bromus inermis* Leyss. *Crop Sci.* 13: 332–34, 1973.

Timko, M.P. and A.C. Vasconcelos. Euploidy in *Ricinus*. Euploidy effects on photosynthetic activity and content of chlorophyll proteins. *Pl. Physiol.* 67: 1084–89, 1981.

Treharne, K.J. Biochemical limitations to photosynthetic rates. *In*: *Crop Processes in Controlled Environments*. A.R. Rees, K.E. Cockskull, D.W. Hand and R.G. Hurd (eds). Academic Press, London. 391: pp. 285–303, 1972.

Tsunoda, S. A developmental analysis of yielding ability in varieties of field crops. II. The assimilation system of plants as affected by the form, direction and arrangement of single leaves. Jap. J. Breed. 9: 237–44, 1989.

Tungland, L, L.B. Chapko, J.V. Wiersma and D.C. Rasmusson. Effect of erect leaf angle on grain yield in barley. *Crop Sci.* 27: 37–40, 1987.

Usmanov, P.D., Kh.A. Abdullaev, Yu.I. Pinkhasov and G.R. Bikasiyan. (The genetics, structure and function of the plastids in *Arabidopsis* and cotton plants with varigated leaves). *Genetica*. 11: 22–29, 1975.

Vasev, V.A. (Photosynthetic productivity of two single interline maize hybrids and their parental lines). *Selskokhozyaistvennya Biologiya*. 12: 934–37, 1977.

Vlasenko, V.S. (Aspects of assimilate attracting respiration in the ears of winter wheat during the grain formation and filling). Nauchno-teckhnicheskii Byulleten Vsesoyuznogo Selektsionno-geneticheskogo Instituta. No. 1: 49–52, 1982.

Vodogreeva, L.G. (Photosynthetic rate and earliness in *Gossypium barbadense*) Sb. nauch. rabot. Turkin. NII. selektsii i semenovodstva tonkovoloknist. khlopchatnika 14: 93–99, 1976.

Vodyanova, O.S. Development of leaf apparatus in induced polyploid forms of onion. *In*: Selektsiya i semenovod kartofelya i ovoshche-bakhchev, Kultur. Alma-Ata, Kazakh, SSR. pp. 66–73, 1979.

Volodarskii, N.I., E.E. Bystrykh and E.K.Nikolaeva. (Primary photosynthetic reactions in high yielding winter wheat varieties). Doklady Vsesoyuznoi Ordena Lenina i Ordena Trudovogo Krasnogo Znameni Akademii Selskokhozyaistvennykh Nauk Imeni V.I. Lenina. No. 2: 12–14, 1981.

Volodarskii, N.I., E.E. Bystrykh and E.K. Nikolaeva. (Photochemical activity of the chloroplasts in high yielding winter wheat varieties). *Selskokhozyaistvennya Biologiya*. 15: 366–73, 1980.

Walker, D.A. and M.N. Sivak. Improving photosynthesis by genetic means. *Span*. 29(2): 47–49, 1986.

Walton, P.D. The genetics of stomatal length and frequency in clones of *Bromus inermis* and the relationships between these traits and yield. *Can J. Pl. Sci.* 54: 749–54, 1974.

Warner, D.A. and G.E. Edwards. C_4 photosynthesis and leaf anatomy in diploid and autotetraploid *Pennisetum americanum* (pearl millet). *Plant Science,* Irish Republic. 55: 85–92, 1988.

Watanabe, N. (Photosynthetic oxygen evolution in leaf discs). *Res. Bull. Fac. Agric.*, Gifu Univ. 52: 41–45, 1987.

Wells, R., W.R. Meredith, Jr. and J.R. Williford. Canopy photosynthesis and its relationship to plant productivity in near isogenic cotton lines differing in leaf morphology. *Pl. Physiol.* 82: 635–40, 1986.

Wells, R., W.R. Meredith, Jr. and J.R. Williford. Heterosis in upland cotton. II. Relationship of leaf area to plant photosynthesis, 1988.

Wettstein, D. von. Genetic engineering in the adaptation of plants to evolving human needs. *Experientia.* 39: 687–713, 1983.

Wiebold, J.W. Heritability of net photosynthesis and related leaf characters in soybeans. M.S. Thesis. Iowa State Univ., Ames, Iowa, 1975.

Wiebold, J.W., R. Shibles and D.E. Green. Selection for apparent photosynthesis and related leaf traits in early generations of soybean. *Crop Sci.* 21: 969–73, 1981.

Wildner, G.F. and J. Henkel. *Biochem. Biophys. Res. Commu.* 69: 268–75, 1976.

Wilson, D. Variation in photorespiration in *Lolium. J. exp. Bot.* 23: 517–25, 1972.

Wilson, D. Response to selection for dark respiration rate of mature leaves in *Lolium prenne* and its effects on growth of young plants and simulated swards. *Ann. Bot.* 49: 303–12, 1982.

Wilson, D. and J.P. Cooper. Assimilation of *Lolium* in relation to leaf mesophyll. *Nature* 214: 989–91, 1967.

Winter, S.R. and A.J. Ohlrogge. Leaf angle, leaf area, and corn (*Zea mays* L.) yield. *Agron. J.* 65: 395–97, 1973.

Wittwer, S.H. Carbon dioxide fertilization of crop plants. *In: Crop Physiology*, U.S. Gupta (ed), Oxford & IBH Pub. Co., New Delhi, 310–33, 1978.

Wu, E.F. (Study on photosynthetic traits in maize germplasm). Scientia Agricultura Sinica 5: 25–32, 1982.

Yang, S.J., L.B. Zhang and G.M. Wang. (The theory and method of ideal plant morphology in rice breeding). *Scientia Agricultura Sinica.* 3: 6–13, 1984.

Yoshida, S. Physiological aspects of grain yield. Annu. Rev. *Pl. Physiol.* 23: 437–64, 1972.

Yoshida, T. (On stomata number in barley. I. The correlation between stoma number and photosynthesis). *Jap. J. Breed.* 26: 130–36, 1976.

Yoshida, T. Effect of stomatal frequency on photosynthesis and its use for breeding in barley. *Bull. Kyushu Agric. Exp. Stn.* 20: 129–93, 1978.

Yoshida, T. Relationship between stomatal frequency and photosynthesis in barley. *JARQ.* 13: 101–05, 1979.

Zaigraev, S.A. (Photosynthetic activity of spring wheat varieties in Transbaikalia). *In:* Fiziol. i produktion. rast. v Zabaikale Ulan-Ude, USSR. pp. 25–34, 1978.

Zelenskii, M.I., G.A. Mogileva and I.P. Shitova. Varietal diversity of spring wheats in photochemical activity of the chloroplasts. Byulleten Vsesoyuznogo Ordena Lenina i Ordena Druzhby Narodov Instituta Rastenievodstva Imeni N.I. Vavilova. No. 87: 36–40, 1979.

Zelenskii, M.I., G.A. Mogileva, I. Shitova and F. Fattakhova. Hill reaction of chloroplasts from some species, varieties and cultivars of wheat. *Photosynthetica.* 12: 428–35, 1978.

Zelenskii, M.I., G.A. Mogileva, I.P. Shitova and T.N. Tikhomirova. (Genotypic diversity and stability of some photosynthetic characters in spring wheat). *Selskokhozyaistvennaya Biologiya,* 17: 482–87, 1982.

Zelitch, I. *"Photosynthesis, Photorespiration and Plant Productivity."* Academic Press, N.Y. p. 347, 1971.

Zelitch, I. Plant productivity and the control of photorespiration. *Proc. Natl. Acad. Sci.* U.S.A. 70: 579–84, 1973.

Zelitch, I. Improving the efficiency of photosynthesis. *Science.* 188: 626–33, 1975.

Zelitch, I. The biochemistry of photorespiration. In: *Commentaries in Plant Science.* H. Smith (ed), Oxford, U.K., Pergamon Press. pp. 51–61, 1976.

Zelitch, I. Basic research in biomass production: Scientific opportunities and organizational challenges. In: *Linking Research to Crop Production*. R.C. Staples and R.J. Kuhr (eds), Plenum Press, New York and London. p. 235 and pp. 101–14, 1980.

Zelitch, I. and P.R. Day. The effect on net photosynthesis of pedigree selection for low and high rates of photorespiration in tobacco. *Pl. Physiol.* 52: 33–37, 1973.

Zhao, T.Y., P. Shi and J.A. Guo. (Relationship between heterosis of photosynthetic characters and yield components in sorghum). *Shanxi Nongye Kexue* 9: 18–21, 1983.

Zhebin, D.F. (Heterosis and components of photosynthetic assimilation in hybrids of sweet corn). Sb. nauch. tr. Belorus. NII Zemledeliya, 14: 122–27, 1970.

CHAPTER 2

PHOTOSENSITIVITY

Introduction
Variability in photosensitivity
Selection for photosensitivity
Genetic studies on photosensitivity
   Inheritance behaviour
Breeding achievements
   Conventional breeding
   Mutation breeding
   Composites
Grafting experiments.

INTRODUCTION

From the time of the great experiments of Garner and Allard (1920) on Maryland Mammoth variety of tobacco, we know about the short day (SD) and long day (LD) requirement of crop plants. Then in 1940, essentiality of the continued dark period was proved (Hammer, 1940) and thus the plants should be called as long night, short night, and night neutral, but as a convention the terms SD, LD and ND (day neutral) are still being used. On the same analogy, the term 'photoinsensitivity' is also not appropriate. In this write up, the conventional terminologies will still be used.

Garner and Allard (1923) predicted that a detailed study of the reaction of cereal crops to day length would lead to interesting valuable results. Probably the most significant confirmation of their prediction has been the broad adaptability of day length — insensitive wheat varieties of *Triticum aestivum* and *T. durum*, notably those developed at CIMMYT in Mexico. The part of the genotype × environment interaction that was a function of day length – response was virtually eliminated as a breeding problem (Krull et al., 1968). Insensitive plants flower under both long or short – day conditions.

Most spring wheats developed in the northern latitudes have a LD requirement and, consequently, have limited adaptation to other parts of the wheat-growing world as commercial varieties and parental stocks. An added advantage of insensitivity is that seed stocks of new varieties can be increased very rapidly. Quick (1971) estimated that 1 bu of durum wheat could be multiplied to 4m bu in four croppings grown over a 16-month period through plantings in SD conditions of winter in Arizona and California.

Species which contain only photosensitive types have a small distribution area and are geographically and ecologically isolated. The response to photoperiod of many crops has been extensively investigated. In wheat, rice, oats and maize cultivars wide adaptations have been developed by selecting photoperiod insensitivity (Krull et al., 1968; Lebsock et al., 1973; Martinic, 1978; Chang and Vergara, 1972; Burrows, 1973). Wheat varieties with a low sensitivity to photoperiod are believed to have a short life cycle while those with high sensitivity are believed to have a long life cycle. However, Martinic (1978) mentioned the possibility of breeding spring or winter wheat varieties in which there is no clear relationship between photoperiod and the life cycle. Also, in *Avena*, Burrows (1973) bred less photosensitive varieties by incorporating genes from *A. byzantina* into *A. sativa*. In northern latitudes, the strains flower too early to give high yields and have therefore been selected for lateness. Photothermic insensitivity could prove an asset where for example, high temperatures soon after sowing cause some high-yielding wheat varieties such as Kalyan Sona and Sonalika to flower too quickly.

The productivity of the U.S.A. soybean cultivars (mostly SDP) grown under short day lengths (less than 14 hours) often has been limited because flowering and subsequent reproductive growth are initiated before sufficient vegetative growth has occurred. Inadequate vegetative growth results in plants with reduced yield and stature which are not suitable for machine harvesting. A day neutral genotype on the contrary, produces adequate vegetative growth under all photoperiod conditions.

In *Phaseolus vulgaris*, it has been suggested that early flowering genes be combined with genes for SD response to produce a bean variety for the northern U.S.A. (Coyne, 1970). Lawn and Byth (1973) planted 18 cultivars of soybean on different dates and observed that genetically controlled lateness was associated with increasing sensitivity to photoperiod at all stages of development. The earliest maturing cultivars were day neutral but responded significantly to photoperiod during flowering.

High-yielding photoinsensitive strawberry has been bred by backcrossing hybrids of *Fragaria virginiana* var. *glauca*, which shows dominance for day neutrality, with California SD varieties (Bringhurst, 1976). Large fruitedness and commercially acceptable quality were resorted to within three BC generations. Day neutral lines produce fruit throughout the year under California conditions, whereas SD lines begin fruiting in February and give unreliable production in late summer and autumn. In these crops mentioned above and several other crops, the genetics of photoperiod response is known and can be used in adapting a cultivar for a specific photoperiod (Kohel and Richmond, 1962; Li, 1970; Coyne, 1970; Marx, 1969; Murfet, 1973).

VARIABILITY IN PHOTOSENSITIVITY

Out of 30 upland rice varieties tested (Alluri and Vergara, 1975) for photoperiod response, 17 were insensitive, five weakly sensitive and eight strongly so. Further, when 600 strains of rice belonging to 25 different species were classified into photosensitive and insensitive groups (Katayama, 1977), the majority were photosensitive. Of the cultivated species, *Oryza sativa* had more photoinsensitive strains than *O. glaberima*. As *O. officinalis* and the groups Minuta and Latifolia which carry the C genome of this species, had a high frequency of insensitive strains, the C genome was thought to be responsible for the suppression of photosensitivity (Katayama, 1977).

Trofimovskaya and Ivanova (1976) reported that most of the early barley varieties studied were LDP's; under short photoperiod their heading stage was delayed or, in some cases, was absent. However, some varieties had a weak photoperiodic reaction, for example, Chinese forms K5195 and Tinirgazevskii 85 and the Bankeata Korai were photoperiodically neutral. Presowing vernalization of the seeds accelerated plant development under SD in some varieties, but this including the day neutral, Tinirgazevskii 85, did not react to vernalization.

A large number of cowpea collections were screened for photosensitivity (Bruster, 1970). The results showed that *Vigna sinensis* was SD, *V. catgang* L.D. and *V. resquipedalis* DN. Lu and Yen (1974) studied the interaction between photo- and thermal-sensitivity of four soybean cultivars; PI 181698 was strongly photo- and thermal-sensitive, Sangoku and Chung-hing 2 were intermediate in sensitivity and Nungyuan 1 was weakly sensitive. Akinola and Whiteman (1975) identified SD, DN, nearly DN and intermediate photoperiodic forms of pigeonpea varieties. Permadi (1983) reported that flowering response differed among 26 accessions of winged bean under SD conditions, five flowered early, five were intermediate and 14 late. Suzuki et al. (1975) noted that lucerne varieties of the erect type from low latitudes were photo-insensitive while the prostrate varieties from high latitudes were highly sensitive.

Mendoza (1975) studied the photoperiodic behaviour of 18 potato clones from four groups (Andegena, Phureja and *Tuberosum diploid* and *T. tetraploid*) under controlled environments. A wide variability for critical day length was observed among clones, the higher values corresponding to *T. tetraploid* and the lower values to the Phureja. Tuber weight decreased as day length increased for the groups. Andegena, Phureja and *T. diploid* showed a SD reaction for tuber yield. *T. tetraploid* did not show any significant change indicating DN behaviour. Studies of Churata-Masca (1975) on the photoperiodic behaviour of two lines and four varieties of okra showed increased flowering with reduction in day length. Triangulo Mineiro, line A and Chifre de Veado flowered in less than 10 to 14 hours, 10 to 16 hours and 14 to 18 hours day lengths respectively. However, Clemson Spineless flowered in all the day lengths studied. Clemson Spineless and IAC$_1$ did not suffer reductions in flowering under LD conditions.

SELECTION FOR PHOTOINSENSITIVITY

Sarkar (1973) selected a high yielding photoinsensitive rice variety KUAg 1 from the indigenous stock in West Bengal, India. When seeds of some triticales from Mexico and the U.S.A. were sown in a green house in Hungary, some forms flowered under SD (Kiss, 1975). The day length insensitivity from six strains selected were incorporated into hexaploid triticales. Photoinsensitive early maturing types which matured 30 to 40 days earlier than the LD forms in the green house were selected (Kiss et al., 1978).

Although sorghums are mainly SDP's, some DN and LD biotypes have been found among broom corn and other forms (Yastrebov, 1977). The DN forms were successfully cultivated in more northernly areas of USSR. Yastrebov developed a technique for rapid selection of DN forms from the varietal populations. Further, Spencer (1979) developed a technique in which: (i) photoperiod sensitive and photoperiod insensitive maize lines are intercrossed under SD conditions, and (ii) the segregating F_2, from the selfed F_1, is selected (5 per cent selection intensity) for earliness under LD conditions, selfed and the progeny backcrossed to the photoperiod sensitive parent under SD conditions. After three cycles of selection and backcrossing, the photoperiod-sensitive inbred 11 Mex 44 became insensitive.

Cowpea cultivars IVu 1305 and IVu 1503 and soybean Grant and Clark 63 are relatively insensitive to both photoperiod and night temperature (Huxley and Summerfield, 1974). Another two soybean varieties were photoperiod insensitive but night temperature sensitive. P3-1, a F_3 selection from Pusa Phalguni × Cream Pea is insensitive to day length and matures synchronously (Singh et al., 1974). Further, Mital et al. (1980) evaluated 210 lines of *Vigna unguicula* and reported that Rituraj, Red Seeded, Sel. 24/8-2, EC42712, EC4312 and EC30040 are early, photoinsensitive and have good pod length.

Ariyanayagam (1977) studied 101 pigeonpea varieties for their photosensitivity under SD and LD conditions. None was found insensitive to photoperiod but there was considerable variation and the material could be classified into four groups. Jitendra Mohan (1974) identified sunnhemp plants which were uniformly early flowering with high seed yield. The day neutral type was designated as T6.

GENETIC STUDIES ON PHOTOINSENSITIVITY

Wheat and other cereals generally have been recognized as LDP (Cooper, 1923; Garner and Allard, 1920) although some cultivars are day neutral. Times of spikelet initiation, heading, anthesis and ripening have been used as indicators of photoperiod response in wheat. Photoinsensitivity in winter wheat is partially dominant in F_1 (Tesemma, 1974) and is controlled by two major genes with dominant epistasis and some modifying genes (Welsh et al., 1973). Photoperiodic response studies of the Chinese Spring/Hope — set of chromosome substitution lines (Halloran and Boydell, 1967) showed the presence of a gene; or genes of major effect on photoperiodic sensitivity on chromosome 4B and genes of lesser effect on chromosomes 1A, 3B, 6B and 7D of the cultivar Hope.

The F_1 and F_2 progenies from crosses of the photoperiod insensitive wheat variety Sonora 64 with 21 different monosomic lines of the sensitive winter variety Cheyeme were grown with the parental cultivars under a 10 hours photoperiod (Pirasteh and Welsh, 1975). The vernalized winter parent was still in the vegetative state after 152 days while the spring parent headed 54 days after planting. The F_1 population headed 66 days after planting. Disomic F_2 population segregation indicated that three gene pairs control heading with one pair exhibiting dominant epistasis over the other two. The heading response of the F_2 progenies of the line monosomic for chromosome 2D showed that this chromosome is the carrier of the Ppd1 gene, which has a dominant epistatic effect. Two of the four chromosomes 2B, 4B, 3D and 6D are possible carriers of the other two genes.

For producing photoinsensitive plants, chromosome 2D of Chinese Spring is most potent, followed by 2A, 2M of *Aegilops comosa*, 2Cu of *Ae. umbellulata* and $2R^m$ of *Secale montanum* (Law et al., 1978). Law and coworkers also observed dominant epistasis towards insensitivity.

Studies with photosensitive wheat varieties, Larcer and Warrior, and the insensitive Sonora 64, indicate that two genes designated Ppd1 and Ppd2 control photoperiodic response with dominant epistasis for insensitivity (Kein et al., 1973). However, Halloran (1976) showed that in spring wheat varieties photoperiodic sensitivity was dominant over insensitivity. Analysis of F_1 and F_2 data from crosses of Novosibirsk 67 and its initial form,

Novosibirsk 7, with monosomic lines of Chinese Spring located the genes for photoperiodic reaction, Ppd1, Ppd2 and Ppd3 on chromosomes 2D, 2B and 2A, respectively (Lbova, 1980). Evidence was found that Novosibirsk 67 and Novosibirsk 7 carry the dominant allele Ppd1, inhibiting sensitivity to photoperiod, and recessive alleles Ppd2 and Ppd3, conferring sensitivity to SD, while Chinese Spring carries Ppd1, Ppd2 and Ppd3. The last two genes control the length of the phase from stem extension to heading (mainly Ppd3) and from emergence to stem extension (mainly Ppd2), while Ppd1 controls the length of the onset of heading. Lbova and Chernyi (1981) further elucidated the effect of *vrm* genes for spring/winter habit (response to vernalization) on straw length, flag leaf length and width, ear length, number of spikelets per ear and number of grains per ear. Dominant alleles of the Ppd and *vrn* genes were associated with a short and narrow flag leaf, a short ear with reduced number of spikelets, increased grain number per ear and short straw. Further, Scarth et al. (1985) report that the primary influence of the Ppd genes is on ear growth. In the plants carrying the insensitive alleles, Ppd1 and Ppd2, the relative growth rate (RGR) of the floral apex was faster than that of plants with the sensitive alleles, Ppd1 and Ppd2; when grown under SD. There were no differences in the rate of spikelet initiation, but the spikelets of the Ppd lines grew and developed more slowly. The Ppd2 material segregated for another gene located on chromosome 2B was affecting duration of the life cycle. This gene also affected the RGR of the ear. They suggested that the major effect of Ppd1 and Ppd2 and the second genetic factor on chromosome 2B is on floral growth rate. Differences in apex morphology, stem growth and ear emergence are thought to be due to the differences in floral apex growth and size.

Marwan et al. (1981) made four crosses between three photosensitive and three photoinsensitive wheat varieties. Data from the parents, F_1, BC, and F_2 indicate that the heading date was controlled by minor genes together with one to two dominant major genes. Transgressive segregation towards lateness was observed in two crosses. Broad sense and narrow sense heritability estimates were in the range of 86.1 to 88.9 per cent and 24.7 to 78 per cent, respectively. The degree of dominance for photoinsensitivity over sensitivity varied among crosses. Further, 60 photosensitive and 60 insensitive nearly isogenic wheat lines derived from crosses between sensitive var. Era and insensitive var. ECM403 and ECM409 plus their controls were grown (Elsayed, 1981). The 20 highest yielding lines (10 sensitive and 10 insensitive) and 10 pairs of isogenic lines plus their controls were evaluated for yield. Both high yielding insensitive and isogenic insensitive lines were significantly earlier than the sensitive lines. The segregation data on the heading date suggest that Inia 66 and Ciano 67 have different photoinsensitivity factors, and that possibly two major genes plus modifiers are involved. Further to this Busch et al. (1984) reported that the insensitive lines headed earlier than the sensitive lines. In general, the sensitive lines tended to be higher yielding in the northerly sites (Minnesota). At the more southerly site, and under delayed sowing, the insensitive lines were higher yielding than their respective sensitive lines.

In rice, photoinsensitivity is controlled by a single recessive gene in some crosses but by two genes in others (Oka, 1977; Tripathi and Rao, 1984). In IR30, the major gene responsible for insensitivity was designated Ef1 but in var. BR4, the two major genes responsible were designated Ef1 and Ef2 (Masiruddin and Dewan, 1978). Further, the crosses involving *indica* and *japonica* varieties indicated that weak photoinsensitivity was recessive to strong photosensitivity (Li, 1981). All the F_1 progenies of crosses between strongly photosensitive varieties from low altitudes were strongly photosensitive. Such varieties showed a higher conservative inheritance over strongly photosensitive japonica varieties from high altitudes.

Crosses between the *indica* variety Peta and the *japonica* varieties Kac-hsiung 68 and Chia-nung 242 indicated that the length of the vegetative period under 10 days was controlled by two pairs of Ef alleles with cumulative but unequal effects (Lin, 1972). A short vegetative period was dominant. Several genes appeared to control sensitivity to photoperiod. Weak sensitivity was associated with a short vegetative period. The monogenic control of the basic vegetative phase and photosensitive phase in rice was reported by Yokoo and Kikuchi (1982). The effects of alleles of Lm on the two phases of vegetative growth, namely the basic vegetative phase and the photoperiod-sensitive phase were analyzed by growing two isogenic lines with Lm^e and Lm^u under natural and controlled day lengths. Under natural day length (LD) conditions, $Lm^e Lm^e$ and $Lm^u Lm^u$ lines headed 73.5 and 118.4 days after sowing, respectively. After SD treatment for seven weeks, Lm^u headed in 56 days and $Lm^e Lm^e$ headed in 60.5 days. The results also indicate that Lm^e and Lm^u have some effect on the basic vegetative phase.

In barley, photoinsensitivity is controlled by a single recessive factor (Young, 1977). But in sorghum, the range of photoperiodic reactions is very wide indicating a quantitative polygenic control (Inst. Res. Tropical Agric. and Food Crops, 1972). The short strawed lines from the cross Toko Bessenou × Combine Kafir 60 are less sensitive to photoperiod than those from Guineensia Togo × Hazera 610. In this SD genus, photoinsensitivity is under genetic control which results in an excess synthesis of auxin (Quinby, 1977). Quinby reported that var. SM100 is earlier than 100M because of a higher auxin content in SM100. Further SM 80 is later than SM100 because SM80 contains too much auxin to flower early, and HKM^3 which flower in 80 days when planted on the first June at Plainview is later than HKM^5 which flowers in 65 days because HKM^3 contains too much auxin to flower early.

In soybean, photosensitivity (SD-requirement) appears to be dominant over insensitivity (Polson, 1972): Insensitivity is controlled by a single recessive gene (Anon, 1976). Under SD, early flowering is dominant over late flowering and is controlled by a single gene (Anon, 1977). Selected LD cultivars from maturity group 00 were crossed to adapted SD cultivars from maturity group IV (Younes, 1982). In the segregating populations several insensitive plants were observed. The insensitive lines were evaluated in field plantings from early May to late July with SD and LD cultivars as controls. The insensitive lines flowered, set seed and matured normally on large vigorous plants in approximately the same period of time regardless of planting date or day length. Segregation data from crosses between insensitive and SD cultivars revealed that this trait is probably under polygenic control.

In *Phaseolus* beans, days to first flowering are affected by day length in photosensitive but by temperature in photoinsensitive cultivars (Enriquez, 1975). Crosses between the two types of lines indicated that photosensitivity is controlled by alleles at one or two loci. Redkloud which is photoinsensitive flowers and matures earlier than Redkote which is slightly sensitive to photoperiod (Massaya et al., 1977). These differences are due to two genes which become more active as temperature and day length increase, giving a range of sensitivities to photoperiod. Plants sensitive to photoperiod contained less ABA in axillary branches than did the insensitive plants. However, the observations of Coyne (1978) on the P, F_1 and F_2 of the cross GN1140 × PI207262, grown under a 14 hour photoperiod indicated that delayed flowering in PI207262 under a long photoperiod is determined by a single recessive gene.

Tiwari and Ramanujam (1976) studied the inheritance of days to flowering and photo-sensitivity between a late and photosensitive green gram variety, and five photoinsensitive varieties, three of which were early and two medium in flowering. The F_1's of all the five crosses flowered as early as the early or medium parent in summer and *kharif* (July to September) seasons indicating dominance of earliness and photosensitivity. Permadi (1983) intercrossed the winged bean accessions, Mn^{2-5}, Mn 2-8 and Mn 3-5 and produced two F_1, two F_2 and two BC generations. There seemed to be a day neutral response for flower initiation, and F_2 segregation data suggested that this trait is controlled by one or more major genes.

Hybrid seeds produced from crosses between the photoinsensitive onion var. Violet de Galmi and photosensitive varieties or lines were sown. The photoinsensitivity of Violet de Galmi was dominant over photosensitivity (Anais and Schweisguth, 1978). However, in jute (*Corchorus*) this character is recessive (Islam et al., 1975). An almost sterile hybrid was obtained between SD *C. olitorius* variety Tossa and photoinsensitive wild species *C. depressus*. Approximately 80 per cent of the F_3 were SDP and 20 per cent day neutral, suggesting that day neutrality is recessive. In allotetraploid roselle hemp, the situation is different (Adamson and Bryan, 1981). Analysis of F_1 and F_2 plants from crosses of the photosensitive line 360M35 with six different photosensitive lines indicated that two different genetic systems are operating. In the food type roselle (a typical line 205 from J6992 and line 144 from J69147), photosensitivity was conditioned by two dominant genes both of which are required for expression (duplicate recessive epistasis). In the fibre type (line 150 from A59-56 and line 165 from THS2) and wild types (lines 141 and 158 from A64-567), two dominant genes were involved and both produced photosensitivity (dupli-cate dominant epistasis).

Inheritance Behaviour

The inheritance of photoperiodic response was studied in crosses involving four spring wheats and three winter wheats (Klaimi and Quaiset, 1973). The parental cultivars were classified into a photosensitive group and a relatively photoinsensitive group, based on their heading response when vernalized and grown under different day length regimes. The F_1 data indicated that photoinsensitivity was not dominant over photosensitivity and that the dominance relationship with respect to photoperiodic response depends on the alleles present in the parents. The heading patterns after vernalization and growth under SD of F_1, F_2, F_3 and BC generations of a four parent diallel cross could be satisfactorily explained on the basis of two major loci with three alleles at each locus. Genes with minor effects also influenced the photoperiodic response in a quantitative manner. Diallel cross analysis of the number of days to heading indicated significant additive and dominant genetic variances, a high average degree of dominance for earliness (photoinsensitivity) and a preponderance of recessive alleles in the parents acting in the direction of lateness (photosensitivity). Estimation of the genetic components of variation contained in the generation means of individual crosses showed that epistasis was also an important factor in the genetic control of photoperiodic response. Halloran (1976) also crossed varieties of spring wheat in a diallel fashion and the hybrids vernalized. The data indicated almost complete dominance in the inheritance of high photoperiodic sensitivity.

Single recessive genes condition photosensitivity in barley varieties (Ramage and Suneson; 1958; Takahashi and Yasuda, 1971; Dorruling and Gustafson, 1969). However, each gene appears to cause debilitating effects that limit their use. Other reports indicate that several genes may condition inheritance of photoperiod response in barley. Fisher (1974) reported that the loci on at least four chromosomes determine flowering response under SD in two cultivars of barley. Tew (1977, 1978) also concluded that the inheritance of photoperiod response was quantitative in crosses of sensitive and insensitive barleys. F_1 and F_2 data indicated that photoinsensitivity was partially dominant to sensitivity. Data from advanced generations were indicative of epistasis (Tew, 1978). Additive effects, dominance effects and additive × additive epistatic effects influence photoperiod response (in SD environment) in the F_1, F_2, F_3 and BC generations from crosses of the photoinsensitive barley var. Manker with the photoinsensitive C18044, Steptoe, C15064 and C17452 (Barham and Rasmusson, 1979). Although major genes were not detected, the high heritability values and the selection results indicated that breeders should have little difficulty when selecting for photoperiod response in these and similar barley populations (Barham and Rasmusson, 1981).

Chang et al. (1969) suggested that cumulative action of one or two dominant genes of varying effect (Ef1-3) control a short basic vegetative phase in rice. Either one or two dominant genes (Se 1, 2) control photosensitivity. Se genes are epistatic to Ef genes. These results were from crosses between strongly sensitive and insensitive parents. Similar results were obtained for the basic vegetative phase using weakly sensitive and insensitive parents. The weak photoperiod response was attributable to genes showing partial dominance (Anon, 1977). Li (1970) using strongly sensitive parents, found additivity with some dominance in favour of a short basic vegetative phase. Other results suggesting control by a limited number of genes and dominance for earliness were reported by Chang et al. (1969) and Li and Chang (1970).

Ganashan and Whittington (1976) studied the inheritance of five characters under the day lengths of 10 and 14 hours in a diallel cross of six varieties of rice. Early flowering was dominant to late flowering in both environments, but the varieties flowering early in one environment were late flowering in the other. Analysis of F_1 and F_2 data from the cross Heenati 310 × IR8 indicated digenic control of early flowering in SD with complementary interaction. The crosses Heenati 310 × IR8 and Taivan × MI273 (m) were high yielding.

Kiss and Videkine Kusz (1972) in F_1 hybrids between day-neutral, medium tall spring triticales and LD dwarf and semidwarf winter forms, day length requirement was generally intermediate but closer to the LD-type, in B463 A272 × GTA205-21, the LD-type was dominant. In F_1's between Hungarian and other triticales, the day-neutral character showed intermediate inheritance, and some neutral F_2 plants were obtained (Kiss, 1973).

Relatively photoinsensitive maize lines have been identified (CIAT, 1974) and sensitivity is indicated to be inherited simply. Insensitivity to photoperiod appears to be controlled by a dominant gene but in crosses between varieties with very different responses, more than one gene seems to be involved (Sousa Rosa and Trujillo Figueroa, 1972). A maternal effect was observed in crosses Klein Inpacto × Ciano 67 and Klein Inpacto × Bb-Inia 66, although not consistently.

Photoinsensitivity in soybean is dominant over sensitivity(Goswami and Kumari, 1972). Varma (1971) reported photoinsensitivity to be a dominant character in a cross between *mung* bean strain T_1 and ST7. Ramanujam et al. (1974) reported flowering in F_1 to be significantly earlier than the mid-parent in 12 out of 25 crosses among 10 parent *mung* bean strains. Swindell and Poehlman (1978) identified a dominant or partially dominant gene for sensitivity to photoperiod in *mung* bean strain PI 180311 and lebeled Ps. The gene was expressed when strain PI 180311, or crosses involving PI 180311 were grown in 16 or 14 hours photoperiods but was not expressed when the populations were grown in 12 hours photoperiod. Dominance × dominance epistatic effects were indicated as governing days to flower in the absence of the obvious effect of the Ps gene.

Lewis and Richmond (1960) investigated the flowering response of cotton (*Gossypium barbadense*). In the progeny of a cross between SD 'Lengupa' and DN 'Pima-S-1', they found that a single gene controlled flowering, and the flowering was recessive to non-flowering. In crosses between SD 'Mariegalante' (*G. hirsutum*) and DN 'Pima S-1 (*G. barbadense*) and DN 'Upland' (*G. hirsutum*), flowering response in both sets of hybrid progeny was inherited as a complex trait (Kohel et al., 1965). Kohel et al. (1974) crossed the *G. hirsutum* DN lines Texas Marker 1 and M11 and SD lines Texas 220 and Texas 371 in all possible combinations with the DN *G. barbadense* lines Pima S-1 and 3-79, and the Sd Lengupa. The F_1, F_2, BCP1 and BCP2 and a test cross to Texas Marker 1 were grown for each combination. The results indicated that these species have non-homologous systems controlling flowering response. In *G. barbadense,* flowering response is recessive to non-flowering and under single gene control, whereas in *G. hirsutum* flowering response is under multigenic control which is partially dominant or additive to non-flowering. In interspecific cross combinations, the segregation of flowering response followed a multi-genic pattern. In the wild subspecies *G. hirsutum* subsp. *mexicanum*, photoperiodic reaction appears to be controlled by three polymeric loci, all the three loci p_1 p_1 p_2 p_2 p_3 p_3 being recessive (Mukhamedkhanov, 1977). Reciprocal F_1 and F_2 hybrids between the wild form and the variety S4727 were grown in a natural day, a 10 hour day and a 24-hours day (Sadykov et al., 1976). The F_1 hybrids fruited under natural day length and were uniform in morphological characteristics. The day neutrality of S4727 was dominant over the SD of subsp. *mexicanum*. The early variety S4727 was dominant in the F_1 in a 10-hour day as regards height of the first fruiting branch and length of the growth period. The photoperi-odic reaction of the SD wild parent dominated in the F_1 hybrids in a 24-hour day.

Mendoza (1975) studied the photoperiodic behaviour of 18 potato clones from four groups (Andigena, Phureja and Tuberosum diploid and T. tetraploid) under controlled environments. A wide variability for critical day length was observed among clones, the higher values corresponding to T. tetraploid and the lower values to the Phureja. Tuber weight decreased as day length increased for the groups. Andigena, Phureja and T. diploid showed a short day reaction for tuber yield. T. tetraploid did not show any significant change, indicating DN behaviour. Three DN tuberosum clones, one DN and four SD Andigena clones were used as parents in an inheritance study. F_1 generation means from SD × DN crosses showed complete dominance of SD reaction (late tuber initiation) over DN reaction (early tuber initiation). The data indicate a variable degree of parental heterozygosty for the loci controlling tuber initiation. It appears that several major and minor genes control tuber initiation. Broad sense heritability estimates of 90 per cent at 11 hours and 55 per cent at 15 hours day length were obtained.

The F_2 of crosses between the DN sesame variety Aceitera and the SD varieties Glauca and Oro showed continuous variation for flowering data (Kotecha et al., 1975). Transgres-sive segregation was observed for both early and late flowering. A minimum of three loci may be involved in the inheritance of photoperiodic response of the sesame varieties mentioned.

BREEDING ACHIEVEMENTS

Conventional Breeding

Lebsock et al. (1973) selected 52 day length (DL)-sensitive and 52 DL insensitive F_3 lines of wheat from four crosses between northern varieties and a DL-insensitive selection from Israel. Ten pairs of F_6 lines, isogenic except for DL-response were developed from three of the crosses. Insensitive lines tended to be significantly earlier and shorter than the sensitive lines and their per day grain yield was very high (Table 2.1) and the quality characters were comparable (Table 2.2).

Table 2.1: Mean values for days to head and grain yield of day length-sensitive and day length-insensitive F_3 *durum* wheat bulks (After Lebsock et al., 1973)

Cross and day length response	Days to head	Grain yield (kg/ha)
Z/B/LK/3/Leeds		
Sensitive	133	2775
Insensitive	73	2742
Z/B/LK/3/61-120		
Sensitive	136	2802
Insensitive	73	2708
Z/B/LK/3/61-2		
Sensitive	124	2675
Insensitive	67	2708
Z/B/Wis/3/61-2		
Sensitive	126	2816
Insensitive	65	2816
All crosses		
Sensitive	129	2769
Insensitive	69	2755
Parental means		
Sensitive	129	2668
Leeds	139	2876
60-120	120	2876
61-2		
Insensitive		
Z/B/LK	74	2916
Z/B/Wis	60	2964

Jenkins and Kirby (1975) selected barley lines with low photoperiod-sensitivity from a cross between the east European variety Kruglik 21 (1303) and the Japanese winter barely Shimabara, which are complementary in their responses to vernalization, photoperiod and frost. Further, it was shown that an advanced generation photoperiod insensitive line developed from CM67 Mona had a wider adaptability than most other barleys (Anon, 1976a.)

Table 2.2: Mean values for quality characteristics of DL-sensitive and DL-insensitive near isogenic pairs of *durum* wheat (After Lebsock et al., 1973)

Quality characteristics	Sensitive	Insensitive
Kernel appearance, score	2.2	2.0
Kernel weight, g/1000	39.1	38.4
Wheat protein content, %	15.1	15.2
Semolina yield, %	51.3	51.0
Spaghetti colour, score	9.0	8.3
Semolina colour, score	7.0	4.7
Absorption, %	33.7	33.6

In sorghum, F_1 hybrids from crosses of SD forms with each other and with DN- or LD-forms showed a SD photoperiod reaction (SD character dominant) and yielded more than the parental varieties as a result of a combination of heterosis for yield and heterosis for rate of development (Yastrebov and Tsybulko, 1971). However, Vidyabhusanam (1977) reports that synchronization of flowering has proved a problem in the sorghum hybrids CSH5 (2077A × CS3541) and CSH6 (2219A × CS3541). The male line CS3541 is relatively insensitive to photoperiod and temperature whilst 2077A and 2219A are both sensitive.

The F_1 hybrids between crosses of 'Tift 23A' (cytoplasmic male sterile line of *Pennisetum americanum* and five populations of 'Maiwa' (*P. maiwa*) were studied for various characteristics including restoration of male fertility (Bhardwaj and Webster, 1971). Tift 23A is DN, dwarf, and early maturing, whereas Maiwa is photosensitive, tall, and late maturing. Results indicate that genes for non-restoration of cytoplasmic male sterility are present in the Nigerian Maiwa populations. It also seems possible to develop high yielding, DN or photosensitive strains from the segregating generation of Tift 23A × Maiwa crosses.

Out of the 101 varieties of pigeonpea (*Cajanus cajan*) collected from different countries, Ariyanayagam (1977) could not observe any photoinsensitive variety. Varieties which flowered in less than 90 days with January planting and which had less than 21 days difference in flowering between SD and LD were crossed with large seeded but photosensitive material from the West Indies and early profusely flowering types were selected. Greater variability for quantitative characters occurred among segregates derived from a modified diallel cross than in generations subjected to pedigree selection. One apparently homozygous variety selected from a F_3 pedigree generation was insensitive to photoperiod. It is suggested that it is suitable for cultivation at any time of the year. Further, Ariyanayagam and Spence (1978) looked for the insensitivity gene from *Atylosia*. *A. platycarpa* and *A. scarabacoides* reach 50 per cent flowering in 38 and 53 days respectively under SD of 12 hours and 45 minutes. Under 14, 16 and 19 hours day lengths, *A. platycarpa* exhibited earliness and photoperiod-insensitivity. *C. cajan* 'Code 11', the shortest time to reach 50 per cent flowering is 114 days in the field. Earliness in progeny of the cross *A. platycarpa* × Codel, under a 14-hour photoperiod was similar to that of *A. platycarpa*.

Later, Saxena et al. (1980) identified two early flowering male sterile plants in F_2 of the cross late flowering M53A × photoinsensitive QPL1. Under a 16-hours photoperiod, they flowered at the same time as the photoinsensitive control.

Stephens (1976) reported the development of a DN form of *Gossypium barbadense* through introgression between primitive SD-sensitive forms of *G. barbadense* and *G. hirsutum*. Intersubspecific groundnut hybrids were developed and their photoperiod response determined (Wynne and Emery, 1974). Three *Arachis hypogea* lines from three geographic areas of South America representing Valencia, Virginia and Spanish types were used in a diallel cross. Six F_1's and their three parents were evaluated for photoperiodic response. When the parents and F_1's were grown under a S.D. (nine hours) photoperiod, they were small but produced more pods than when grown under a LD treatment produced by nine hours light plus a three-hour interruption of the dark period. The Valentia and Spanish lines flowered earlier and produced more flowers and pods than the Virginia line. All the F_1's exceeded the midparental value for pod yield under SD whilst only two exceeded the midparental value for pod yield under LD. Reciprocal crosses involving the Spanish line showed significant differences for five to seven quantitative characters measured. The genetic and environmental controls of flowering in *Phaseolus vulgaris* were studied (Massaya, 1978). A DN and a SD cultivar, the F_1 and F_2 of a cross between them, and selected S_3 and S_4 progenies were tested under SD and LD in controlled environments with relatively high temperatures, and under LD in the field. The early flowering and early maturing DN character was dominant in the growth chamber and recessive in the field.

Mutation Breeding

Induced mutations in some crops have proved successful in developing DN mutants. Upadhya et al. (1972) reported that in potatoes, N-nitroso-N-methyl urea at 200 ppm gave the highest frequency of DN mutants (7.86 per cent) of the four chemical mutagens used. Diethyl sulphate induced mutants only in the hybrid B 1882. Further, Upadhya et al. (1976) reported that sodium metabisulphite induced DN mutants in the true seeds of potato var. Kufri Jyoti but the highest frequency was observed when seeds were irradiated with 40 KR of γ rays and then treated with 0.5M hydrazinium dichloride solution.

Following treatment of seeds of cotton var. MCU5, a high yielding but late flowering *G. hirsutum* variety, with 25 and 30 KR γ rays, two true breeding mutants were isolated in the M_2 which were relatively insensitive to day length, taking 60 days to flower and 160 days to mature irrespective of site or season (Raut and Jain, 1976). They appear suitable for cultivation in Orissa (India) in the wet season and as an off-season winter crop. Further, following selfing of an early flowering jute mutant isolated in the M_1 after γ-irradiation of seed from *Corchorus capsularis* D154, two plants which flowered within 60 days were selected in the M_3 and M_4 (Hossain and Sen, 1978). When days to flowering were compared between these early mutants in the M_5 and three photoinsensitive cultivars, using five sowing dates and three fertilizer concentrations, the mutants proved to be insensitive to photoperiod.

Composites

Alexander and Spencer (1982) described that in maize photoperiod insensitive, Composite 1, was developed from unequal amounts of nine sensitive and 14 insensitive lines. Photoperiod insensitive Composite II was made up of 13 insensitive or intermediate lines, and photoperiod insensitive Composite III was made up of six insensitive tropical populations crossed as females with Composite I.

GRAFTING EXPERIMENTS

In grafting experiments with six photoinsensitive and three photosensitive soybean varieties serving as either donors or receptors of the floral stimulus, flowering times of receptor stocks did not significantly differ from those of the donor varieties when the donors were photosensitive (Niwa, 1976). Photosensitive varieties flowered later, when used as receptor scions than did photoinsensitive varieties. Varietal differences in the flowering time of receptor scions were more marked when these were on later stock varieties.

Chailakhyan et al. (1976, 1977) performed grafting experiments with SD, LD and DN tobacco varieties. They (1976) worked with three varieties namely, DN *N. tabacum* 'Trebizond', SD '*N. tabacum*' 'Mamont' and LD *N. sylvestris* 'Sylvestris'. Compared with ungrafted plants, flowering was accelerated in DN var. when it was grafted onto LD var. under a LD, but when DN var. was grafted onto SD var. under a SD, flowering was at the same rate as when DN var. was grafted onto itself. When SD var. was grafted onto DN var., it flowered under a LD but when grafted onto itself, it failed to flower. When LD var. was grafted onto DN var., it flowered in SD, whereas ungrafted plants of LD failed to flower. Further, Chailakhyan et al. (1977) reported that when the SD form *N. tabacum* var. Maryland Mammoth was grafted on flowering plants of the LD. *N. sylvestris*, the plants of he SD forms flowered in LD. When the LD form was grafted on flowering plants of the ;D form, the plants of the LD form flowered in a SD.

REFERENCES

Adamson, W.C. and J.E.O. Bryan. Inheritance of photosensitivity in roselle, *Hibiscus sabdarilffa*. *J. Heredity*. 72: 443–44, 1981.

Akinola, J.O. and P.C. Whiteman. Agronomic studies on pigeonpea (*Cajanus cajan* (L.) Mill sp.) I. Field response to sowing time. *Austr. J. Agric. Res.* 26: 43–56, 1975.

Alexander, D.E. and J. Spencer. Registration of South African photoperiod insensitive maize composites I, II and III (Reg. No. GP90 to GP92). *Crop Sci.* 22: 158, 1982.

Alluri, K. and B.S. Vergara. Importance of photoperiod response and growth duration in upland rice. *Philippine Agriculturist*, 59: 147–58, 1975.

Anais, G. and B. Schweisguth. (Behaviour in the tropics of F₁ onion hybrids *Allium cepa* L. between photosensitive varieties and an indifferent variety). *Ann. de l'Amelioration des Plantes* 23: 223–29, 1978.

Anon, Barley. In: *Cereal production in times of change*. Crop Conference, Cambridge, Dec. 1976. National Inst. of Agric. Bot., Cambridge, U.K. 15–17, 1976a.

Anon, Soybean. In: Asian Vegetable Research and Development Centre, Taiwan Progress Report for 1976. p.77, 1976b.

Anon, Soybean. In: Asian Veg. Res. and Dev. Centre, Taiwan. Progress report for 1977. p. 90, 1977.

Ariyanayagam, R.P. Breeding for year round production in pigeonpeas. *Nonvellas Agronomiques des Antilles et de la Guyane*. 3: 608–14, 1977.

Ariyanayagam, R.P and J.A. Spence. A possible gene source for early, day-length neutral pigeon peas. *Cajanus cajan* (L.) Mill sp. *Euphytica*. 27: 505–09, 1978.

Barham, R.W. and D.C. Rasmusson. Inheritance of photoperiod response in barley, *H. vulgare*. In: *Agron. Sbstr*. Am. Soc. Agron., Madison, Wis. p. 55, 1979.

Barham, R.W. and D.C. Rasmusson. Inheritance of photoperiod response in barley. *Crop Sci*. 21: 454–56, 1981.

Bhardwaj, B.D. and O.J. Webster. Studies of F₁ hybrids between photoperiod sensitive and nonsensitive *Pennisetum* millets. *Crop Sci*. 11: 289–91, 1971.

Bringhurst, R.S. Day neutral versus short-day strawberry breeding advantages and exploitation potential. *Fruit varieties J*. 30: 25, 1976.

Bruster, D.P. (Biological characteristics of cowpea and breeding results with it in Moldavia). *In:* Biol. genet. selektsiya zernobob, Kultur v Moldavii. pp. 43–79, 1970.

Burrows, V.D. Day length insensitive oats. *Canada Agriculture* 18: 7–9, 1973.

Busch, R.H., F.A. ElSayed and R.E. Heiner. Effect of day length insensitivity on agronomic traits and grain protein in hard red spring wheat. *Crop Sci.* 24: 1106–09, 1984.

Chailakhyan, M.Kh, Kh. K. Khazhakyan and L.B. Aganiyan. (Flowering of plants after the reciprocal grafting of photoperiodically neutral and sensitive species). Doklady Akad. Nauk, SSSR. 230: 1002–05, 1976.

Chailakhyan, M. Kh, L.I. Yanina and Kh. K. Khazhakyan. (Plant flowering after reciprocal grafts of short day and long day species). *Doklady Akad. Nauk, SSSR.* 237: 1248–51, 1977.

Chang, T.T., C.C. Li and B.S. Vergara. Common analysis of duration from seeding to heading in rice by basic vegetative phase and the photoperiod sensitive phase. *Euphytica.* 18: 79–91, 1969.

Chang, T.T. and B.S. Vergara. Ecological and genetic information of adaptability and yielding ability in tropical rice varieties. *In:* Rice Breeding, IRRI, Los Banos, Philippines. pp. 431–53, 1972.

Churata,-Masca, M.G.C. (Effects of photoperiod on flowering in lines and varieties of okra (*Abelmoschus esculentus.* L. Moench). Abst. Ciencia e cultrua. 27: 1336, 1975.

CIAT. Annual Report for 1974. Maize. Cali Colombia, p. 192, 1974.

Cooper, H.P. The inheritance of the Spring and winter growing habit in crosses between typical spring and typical winter wheats, and the responses of wheat plants to artificial light. *J. Am. Soc. Agron.* 15: 15–25, 1923.

Coyne, D.P. Genetics of flowering in dry beans (*Phaseolus vulgaris* L.). *Am. Soc. hort.Sci.* 103: 606–08, 1978.

Coyne, D.P. Genetic control of a photoperiod-temperature response for time of flowering in beans (*Phaseolus vulgaris* L.). *Crop Sci.* 10: 246–48, 1970.

Dorruling, T. and A. Gustafson. Phytotron cultivation of early barley mutant. *Theor. Appl. Genet.* 39: 51–61, 1969.

ElSayed, F.A. Day length insensitivity in spring wheat (*Triticum aestivum* L.) effect of agronomic traits and its inheritance. *Dis. Abstr. Int. B.* 42: 2157B, 1981.

Enriquez, G.A. Effect of temperature and day length on time of flowering in beans (*Phaseolus vulgaris* L.). *Dis. Abstr. Int. B.* 36: 1528B–1529B, 1975.

Fisher, V. (Investigations on inheritance of the spring-winter habit in barley with regard to flowering time and winter hardiness.). Z. Pflanzenzüchtung. 71: 69–84, 1974.

Gamashan, P. and J. Whittington. Genetic analysis of the response to day length in rice. *Euphytica.* 25: 107–15, 1976.

Garner, W.W. and H.A. Allard. *J. Agric. Res.* 18: 553, 1920.

Garner, W.W. and H.A. Allard. Further studies in photoperiodism, the response of the plant to relative length of day and night. *J. Agric. Res.* 23: 871–1920, 1923.

Goswami, L.C. and S. Kumari. Studies on the breeding behaviour of cowpea (*Vigna sinensis* Endl.). *J. Assam Sci. Soc.* 15: 20–28, 1972.

Halloran, G.M. Genetic control of photoperiodic sensitivity and maturity in spring wheat within narrow limits of adaptation. *Euphytica.* 25: 489–98, 1976.

Halloran, G.M. and C.W. Boydell. Wheat chromosomes with genes for vernalization response. *Can. J. Genet. Cyt.* 9: 632–39, 1967.

Hammer, K.C. Interrelation of light and darkness in photoperiodic induction. *Bot. Gaz.* 101: 658–87, 1940.

Hossain, M. and S. Sen. Early and photoinsensitive mutants in jute. *Indian J. Genet. Pl. Breed.* 38: 179–81, 1978.

Huxley, P.A. and R.J. Summerfield. Effects of night temperature and photoperiod on the reproductive ontogeny of cultivars of cowpea and soybean selected for wet tropics. *Plant Sci. Lett.* 3: 11–17, 1974.

Inst. of Res. in Tropical Agric. and Food Crops. Annu. Rept. 1972. Sorghum. Agron. Tropicale. 29: 377–54, 1972.

Islam, A.S., M. Haque and M.B. Dewan. An attempt to produce a photoneutral strain of jute through interspecific hybridization. *Jap. J. Breed.* 25: 349–54, 1975.

Jenkins, G. and E.J.M. Kirby. Selection for developmental responses in winter barley. *In:* Third Int. Barley Genetics Symp. July 7–12, 1975.

Jitendra Mohan, K.V. Identification of a day-neutral type of sunnhemp (*Crotalaria juncea* L.) and its potentialities. *Jute Bull.* 36: 229–30, 1974.

Katayama, T.C. Studies on the photoperiodism in the genus *Oryza. JARQ.* 11: 14–17, 1977.

Kein, D.L., J.R. Welsh and R.L. McConnell. Inheritance of photoperiodic heading response in winter and spring cultivars of bread wheat. *Can. J. Pl. Sci.* 53: 247–50, 1973.

Kiss, J. (Experiments to produce day-neutral triticale). *Zoldsegtermesztesi Kutato Intezet Bulletinje.* 8: 81–84, 1973.

Kiss, J. Tests to produce day length insensitive triticales. *In:* Triticale, studies and breeding. Materials of an international symposium, Leningrad, 3–7 July, 1973. V.F. Dorofeev and U.K. Kurkiev (eds). pp. 224–26, 1975.

Kiss, A., J.M. Kiss and G. Sallay. (Improvement of some unfavourable traits: tillering type, earliness and seed quality, in short hexaploidization in plant breeding). Proc. 8th Congr. of Eucarpia. III. Alloploidy, Madrid, Spain, 1978.

Kiss, J. and K. Videkine Kusz. (Study of long day winter and day neutral spring triticales). *Zoldsegtermesztesi Kutato Intezet Bulletinje.* 7: 69–74, 1972.

Klaimi, Y.Y. and C.O. Quaiset. Genetics of heading time in wheat (*Triticum aestivum* L.). I. The inheritance of photoperiodic response. *Genetics.* 74: 139–56, 1973.

Kohel, R.J., C.F. Lewis and T.R. Richmond. The genetics of flowering response in cotton. V. Fruiting behaviour of *Gossypium hirsutum* and *Gossypium barbadense* in interspecific hybrids. *Genetics.* 51: 601–04, 1965.

Kohel, R.J. and T.R. Richmond. The genetics of flowering response in cotton. IV. Quantitative analysis of photoperiodism of Texas 86, *Gossypium hirsutum* race *latifolium*, in a cross with an inbreed line of cultivated American upland cotton. Genetics. 47: 1535–42, 1962.

Kohel, R.J., T.R. Richmond and C.F. Lewis. Genetics of flowering response in cotton. IV. Flowering behaviour of *Gossypium hirsutum* L. and *G. barbadense* L. hybrids. *Crop Sci.* 14: 696–99, 1974.

Kotecha, A.K., D.M. Yermanos and F.M. Shropshire. Flowering in cultivars of sesame (*Sesamum indicum*) differing in photoperiodic sensitivity. *Econ. Bot.* 29: 185–219, 1975.

Krull, C.F., A. Cabrera, N.E. Borlaug and I. Narvaez. Results of the second international spring wheat yield nursery. *In: CYMMIT. Res. Bull.* No. 11, 165–66, 1968.

Law, C.N., J. Sutka and A.J. Worland. A genetic study of day-length response in wheat. *Heredity.* 41: 185–91, 1978.

Lawn, R.J. and D.E. Byth. Response of soybeans to planting date in south-eastern Queensland. I. Influence of photoperiod and temperature on phasic developmental patterns. *Aust. J. Agric. Res.* 24: 67–80, 1973.

Lbova, M.I. (Monosomic analysis of some characters in the radiation-induced spring wheat variety Novosibirsk 67 and its initial form. III. Location of genes controlling sensitivity to photoperiod). *Genetika USSR.* 16: 1068–76, 1980.

Lbova, M.L. and I.V. Chernyi. (Effect of genes controlling sensitivity to photoperiod and habit in wheat on the expression of some yield components). *Genetika USSR.* 17: 150–59, 1981.

Lebsock, K.L., L.R. Joppa and D.E. Walsh. Effect of day length response on agronomic and quality performance of durum wheat. *Crop Sci.* 13: 670–74, 1973.

Lewis, C.F. and T.R. Richmond. The genetics of flowering response in cotton. IV. Inheritance of flowering response in a *Gossypium barbadense* cross. *Genetics.* 45: 79–85, 1960.

Li, C.C. Inheritance of the optimum photoperiod and critical photoperiod in tropical rices. *Bot. Bull. Acad. Sinica.* 11: 1–13, 1970.

Li, C.C. and T.T. Chang. Diallel analysis of agronomic traits in rice (*Oryza sativa L.*). *Bot. Bull. Acad. sin. Taipei.* 11: 61–78, 1970.

Li, Z.Y. (Preliminary report of the study of dominance and recessivity in relation to photosensitivity in F₁ progenies of crosses of local yunnan rice varieties). *Heredistas China.* 3 (2): 24–26, 1981.

Lin, F.H. (The genetics of light sensitivity in crosses between Hsien and Keng rice varieties). *Taiwan Agric. Quant.* 8: 168–75, 1972.

Lu, Y.C. and H. Yen. (Studies on the interaction between sensitivity to light and temperature in soybean and their effects on yield components). *J. Agric. Forestry.* 23: 5–34, 1974.

Martinic, Z. Photoperiodism in relation to the life cycle of common wheat varieties. *Genetics.* 74 (2,II): 173, 1978.

Marwan, M.A., H.E. galal, A.A. Gomma and A.H. Abdel Latif. Inheritance of heading period as a photoperiodic response in some spring bread wheat varieties. (*Triticum aestivum L.*) *Res. Bull. Fac. Agric.,* Ain Shams Univ. No. 1617: 23, 1981.

Marx, G.A. Some photodependent responses in *Pisum.* I. Physiological behaviour. *Crop Sci.* 9: 273–76, 1969.

Massaya, P. Genetic and environmental control of flowering in *Phaseolus vulgaris L. Diss. Abstr. Int. B.* 39: 1625–26B, 1978.

Massaya, P., D.H. Wallace and P.L. Ludford. Photoperiod temperature and hormonal control of maturity in beans. *Agron. Abstr.,* Madison. Wis, U.S.A., p. 89, 1977.

Mendoza, H.A. Genetics of photoperiodic behaviour in tuber bearing *Solanum. Diss. Abstr. Int. B.* 35: 4941B, 1975.

Mital, S.P., B.S. Dabas and T.A. Thomas. Evaluation of the germ-plasm of vegetable cowpea for selecting desirable stocks. *Indian J. agric. Sci.* 50: 323–26, 1980.

Mukhamedkhanov, U. (Inheritance of photoperiodic reaction in cotton). *In:* Materially 9-i Kont. molod. uchonykh Uzbekistan po.S. Kh. Taskent, Uzbek. pp. 71–76, 1977.

Murfet, I.C. Flowering in *Pisum:* Hr, a gene for high response to photoperiod. *Heredity.* 31: 157-64, 1973.

Nasiruddin, Md. and S.B.A. Dewan. The inheritance of photoperiod sensitivity in floating rice. *SABRAO Journal.* 10 (2): 103–08, 1978.

Niwa, M. Varietal differences in the sensitivity of growing points of soybean to the floral stimulus as revealed by grafting experiments. *Jap. J. Breed.* 26: 213–19, 1976.

Oka, H.I. A breeding experiment in *Oryza glaberrima* p. 97. *In:* National Inst. of Genetics, Japan Annual Report. No. 27: 112, 1977.

Permadi, A.H. 1983. Variation in photoperiodic response and flower structure of winged bean (*Psophocarpus tetragonolobus L.*) *Diss. Abstr. Int. B.* 44: 657B.

Pirasteh, B. and J.R. Welsh. Monosomic analysis of photoperiod response in wheat. *Crop Sci.* 15: 503–05, 1975.

Polson, D.E. Day-neutrality in soybeans. In: *Agron. Abstr.* Madison, Wis, U.S.A. p. 27, 1972.

Quick, J.S. Improving durum wheat. Crop Prod. Rep. Qual. Council, pp. 18–22, 1971.

Quinby, J.R. Pioneer Hi-Bred International, Insensitivity to photoperiod in sorghum. *Sorghum Newsletter.* 20: 112, 1977.

Ramage, R.T. and C.A. Suneson. A gene marker for the g chromosome of barley. *Agron. J.* 50: 114, 1958.

Ramanujam, M., A.S. Tiwari and R.B. Mchra. Genetic divergence and hybrid performance in *mung* beans. *Theor. appl. Genet.* 45: 211–14, 1974.

Raut, R.N. and H.K. Jain. A day length insensitive mutant of cotton. *Mutation Breed. Newsl.* 8: 2–3, 1976.

Sadykov, S.S., V.A. Avtonomov and M.M. Kiktev. (Effect of light on inheritance of photoperiodic reaction in cotton hybrids). In: Genet i selektsiya Khlopchatnika, Tashkent, Uzbek USSR. pp. 86–91, 1976.

Sarkar, K.P. Selection of a high yielding photoinsensitive variety of paddy KUAg₁ from indigenous to stock in stock in West Bengal. *Sci. and Cult.* 39: 190–91, 1973.

Saxena, K.B., E.S. Wallis and D.E. Byth. (Development of photoinsensitive early flowering male sterile line of pigeonpea (*Cajanus cajan* L. Millsp.). *Trop. Grain Legume Bull.* 20: 22–23, 1980.

Scarth, R., E.J.M. Kirby and C.N. Law. Effects of the photoperiod gene Ppd1 and Ppd2 on growth and development of the short apex in wheat. *Ann. Bot.* 55: 351–50, 1985.

Singh, H.B., S.P. Mital, B.S. Dabas and T.A. Thomas. Breeding for photoinsensitivity and earliness in grain cowpea. *In*: Breeding research in Asia and Oceania. Proc. 2nd General Congr. Soc. Adv. Breed. Res. in Asia and Oceania. Session XII. Improvement of grain legumes S. Ramanujam and R.D. Iyer (eds), pp. 77–86, 1974.

Sousa Rosa, O.De and R. Trujillo Figueroa. (Inheritance of date of heading in wheat. II. Inheritance). *Agrociencia B.* 8: 119–33, 1972.

Spencer, J. Progress in photoperiod research. *In:* Proc. 3rd South African Maize Breeding Symp. held at Potchafstroom, March 21–23, 1978. pp. 104–08, 1979.

Stephens, S.G. Some observations on photoperiodism and the development of annual forms of domesticated cottons. *Econ. Bot.* 30: 409–18, 1976.

Suzuki, S., S. Inami and Y. Sakurai. (Relationship of day length and temperature to plant height and habit in varietal groups of lucerne). *J. Jap. Soc. Grassl. Sci.* 21: 245–51, 1975.

Swindell, R.E. and M. Poehlman. Inheritance of photoperiod response in *mung* bean (*Vigna radiata* (L.) Wilczek). *Euphytica.* 27: 325–33, 1978.

Takahashi, R. and S. Yasuda. Genetics of earliness and growth habit in barley. Barley Genetics II. pp. 388–408, 1971.

Tesemma, T. Analysis of yield and related traits in a diallel cross of winter wheat with reference to long and short photoperiod treatments. *Diss. Abstr. Int. B.* 34 (10): 4791B, 1974.

Tew, T.L. Inheritance of photoperiod response in barley. Ph.D. Thesis, Univ. of Minnesota, St. Paul, 1977.

Tew, T.L. Inheritance of photoperiod response in barley. *Diss. Abstr. Int.* B. 38: 4582B, 1978.

Tiwari, A.S. and S. Ramanujam. Genetics of flowering response in *mung* bean. *Ind. J. Genet. Pl. Breed.* 36: 418–19, 1976.

Tripathi, R.S. and M.J.B. Rao. Inheritance of photoperiod sensitivity and apiculus pigmentation in rice. *Ind. J. Agric. Sci.* 54: 157–64, 1984.

Trofimovskaya, A. Ya. and O.A. Ivanova. (Photoperiod reaction of early barley varieties in relation to breeding problems). *Vestmik Selskokhozyaistvennvi Nanki.* 7: 39–44, 1976.

Upadhya, M.D., R. Chandra and M.J. Abraham. Mutagenic treatments towards increasing the frequency of day-neutral mutations and standardization of procedures for tissue culture, in potato. pp. 151–70. *In:* Improvement of vegetatively propagated plants and tree crops through induced mutations. Proc. 2nd research coordination meeting held at Wageningen, Netherlands, 17–21 May, 1976. Int. Atomic Energy Agency, Vienna, Austria. p. 186, 1976.

Upadhya, M.D., T.R. Dayal, B. Dev, V.P. Chaudhri, R.T. Sharda and R. Chandra. Chemical mutagenesis for day neutral mutations in potato. *In*: Polyploidy and Induced Mutations in Plant Breeding. IAEA-PL-503/17-503/32. *Int. Atomic Energy Agency,* Vienna, Austria. pp. 379–83, 1972.

Varma, S.N.P. Inheritance of photosensitivity in *mung* beans (*Phaseolus aureus* Roxb.). *Mysore J. agric. Sci.* 5: 477–80, 1971.

Vidyabhushanam, R.D. Seed production with hybrid sorghums. *Indian Fmg.* 17: 15–17, 1977.

Welsh, J.R., D.L. Kein, B. Ciastek and R.D. Richards. Genetic control of photoperiod response in wheat. *In:* E.R. Sears and L.M.S. Sears (eds.) Proc. of the 4th Int. Wheat Genet. Symp., Columbia, M.O., Univ. of Missouri, Columbia. pp. 879–84, 1973.

Wynne, J.C. and D.A. Emery. Response of interspecific peanut hybrids to photoperiod. *Crop Sci.* 14: 878–80, 1974.

Yastrebov, F.S. Selection for photoperiod in sorghum breeding. *In:* IX Meeting of Eucarpia maize and sorghum section. USSR, Krasnodar, August 7–13, 1977, pp. 110–111, 1977.

Yastrebov, F.S. and V.S. Tsybulko. (Heterosis for yield and the rate of development in sorghum hybrids in relation to photoperiodic conditions.) Selektsiya i semenovodistvo. Resp. mezhved. temat. nanch sp. 19: 74–81, 1971.

Yokoo, M. and F. Kikuchi. (Monogenic control of the basic vegetative phase and photoperiod-sensitive phase in rice.) *Jap. J. Breed.* 32: 1–8, 1982.

Younes, M.H. The characterization of a day length — neutral trait in soybean (*Glycine max* (L.) Merrill). *Diss. Abstr. Int. B.* 42: 4271B, 1982.

Young, C.T. (8-parent diallel cross analysis on heading date of barley). *J. Kor. Soc. Crop Sci.* 22: 71–79, 1977.

CHAPTER 3

DETERMINATE HABIT

INTRODUCTION

In many crop species we have either natural or induced determinate or semideterminate genotypes in addition to indeterminate ones. The determinate and semideterminate genotypes are characterised with shorter stem, fewer and thicker internodes at maturity, and have a larger reproductive structure as against their indeterminate counterparts. Thus theoretically these genotypes should have a higher yielding ability, which is generally true, but not under all situations. We will discuss some of these aspects with a view to increasing yield by introducing and inducing such genotypes under commercial cultivation. Among other crops, we will concentrate more on soybean which is most worked.

In indeterminate cultivars, substantial stem elongation and leaf production continue long after the onset of flowering; during pod and seed set, their top vegetative growth increases about three-fold (Hanway and Weber, 1971), and nearly two units of LAI are formed (Weber et al., 1966), suggesting a strong competition for assimilates between reproductive and vegetative 'sinks'. Thus, some attenuation of vegetative growth was thought desirable in indeterminate soybean cultivars to reduce competition of reproductive structures, and consequently increase seed yields (Greer and Anderson, 1965; Weber, 1968; Shibles and Green, 1969; Shibles et al., 1975) as a greater proportion of photosynthate will be partitioned to reproductive structures.

Sjodin (1971) bred determinate *Phaseolus vulgaris* var. Primus, which has distinct advantages over the indeterminate types, for example, a shorter plant less prone to lodging, a shorter life cycle and an early pod set which results in aphid avoidance, that is, there is an absence of a growing apex during the period of high potential aphid infestation and a harvest index which is as high or higher than that of the indeterminate plant. The effect of the genetic prevention of the indeterminate stem growth could be to reduce the number of competing sinks within the plant and result in an advanced pod development. In fact, Egli et al. (1985) observed that determinate soybean cultivars produced fewer nodes on the main stem between initial flowering and the onset of seed filling, but produced more branches, thus giving a higher total node production than indeterminate cultivars. Beaver et al. (1985) also compared soybean cultivars Elf (determinate), Williams (indeterminate) and a determinate (dt$_1$) isoline of Clark. Although Williams produced more main stem nodes, the other genotypes set just as many pods on the main stems. They also produced a greater number of pods and a higher seed weight from branches than did Williams.

Determinate and semideterminate growth habits have some other advantages, for example, decreased lodging in highly productive environments where plant growth is great enough to contribute to lodging problems. Leaves of determinate *Vicia faba* plants retain their chlorophyll level for a longer period of time than the indeterminate ones (Austin et al., 1981). Leaves of indeterminate plants abscised after 66 days while those of determinate plants did not abscise until some 40 days later. The determinate genotypes had a longer LAD and a higher rate of photosynthesis. The determinate varieties of *Phaseolus lunatus* possess high drought resistance while the indeterminate ones have poor resistance (Tan Boun Suy, 1978).

While many workers have observed that determinancy results in significant yield increases (Shibles and Green, 1969; Singh and Whitson, 1976; Wilcox and Sediyama, 1981; Thseng, 1982), others found no difference (Green et al., 1977; Hartung et al., 1981; Foley et al., 1986) or even reductions in yield (Green et al., 1977; Wilcox, 1980) when compared with their indeterminate counterparts.

GROWTH AND DEVELOPMENT

Wilson and Cole (1968) made a detailed study on the agronomic potential of determinate soybeans. Three pairs of lines, near-isogenic for determinate and indeterminate stem termination, were evaluated in three row spacings with three plant populations. Yields of the determinate types were less on the average than the indeterminate types, but approached the yields of the indeterminate types at a narrow row spacing and high plant population. Further, Hicks et al. (1969) made similar studies on the agronomic performance of determinate, semideterminate and indeterminate isolines of Clark and Harosoy in four different planting patterns. Semideterminate types were significantly shorter (9 per cent) and higher yielding (4.5 per cent) than indeterminate types, with some what less lodging occurring in the former. Although the determinate types were significantly shorter (5 per cent) than the semideterminates and exhibited essentially no lodging, they generally yielded less (3 per cent). Shibles and Green (1969) also observed a significant yield advantage (6 per cent) of a semideterminate isoline of Harosoy compared with indeterminate Harosoy. They postulated that stem determinancy might be beneficial in minimizing the overlap of vegetative and reproductive growth, and thus provide more efficient partitioning of photosynthates.

Egli and Leggett (1973) compared the dry matter accumulation in various plant components of a determinate (D66-5566) soybean strain and an indeterminate (Kent) cultivar, both of similar maturity and yielding ability. At initial flowering D66-5566 had reached 84 per cent of its maximum height, compared with 64 per cent for Kent. At this point the stem dry weight of D66-5566 was 67 per cent of the maximum compared with 30 per cent for kent. Although stem elongation of the determinate type was nearly complete when flowering began, it continued to increase in dry weight. At initial flowering, Kent had produced 58 per cent of its total vegetative material, compared to 78 per cent for D66-5566. Both types produced very little vegetative growth after a measurable amount of dry weight had accumulated in the pods and seeds. Kent produced more vegetative material during flowering and pod set and the potential for competition for photosynthate between reproductive and vegetative growth was greater. The rate of accumulation of dry matter in the pods and seeds was greater for D66-5566 than Kent. D66-5566 produced higher grain yield (2614 kg/ha) than Kent (2446 kg/ha) and also the effective filling period, grain yield, and rate of increase of dry weight in reproductive material was higher in the former (33.1 days) as against the latter (29.3 days).

Bernard (1972) identified the semideterminate characteristic in soybean and showed that it was controlled by a single dominant gene, Dt_2. Bernard developed isolines of the cultivars Harosoy and Clark that differed primarily in the Dt_2 gene. The semideterminate isoline of Harosoy was 15 per cent shorter than the indeterminate Harosoy, and lodged appreciably less than the latter (Table 3.1). The semideterminate isoline of Clark was also stunted (about 13 cm less) but lodged more than the indeterminate Clark. While the determinate isoline of Harosoy (dt_1) was very dwarfed with least lodging, it was also a poor yielder. The dwarf determinate isoline of Clark, however, gave only slightly less yield (about 2.5 per cent less). Similar results were also observed by Shannon et al. (1971) when the performance of the Harosoy was compared in hill plots.

Table 3.1. Mean agronomic performance of Harosoy and Clark soybean isolines tested in six environments (Adapted from Bernard, 1972).

Genotype	Height (cm)	Lodging score	Yield(kg/acre)
Harosoy	99	2.0	29.7
Harosoy dt_1	39	1.0	21.3
Harosoy Dt_2	84	1.7	29.0
Clark	105	1.7	33.0
Clark dt_1	58	1.2	32.2
Clark Dt_2	92	2.0	32.9

Green et al. (1977) studied a random selection of 60 F_2 derived soybean lines selected in the F_3 (30 indeterminates and 30 semideterminates) in the F_4 and F_5 generations of each of the two classes. They found that, in one cross, indeterminates averaged 6.5 per cent greater seed yield than did semideterminates; the yield superiority of indeterminates coincided with later maturity, longer reproductive period, and larger seed size. In the other cross, the two stem types matured at the same time, had reproductive periods of similar lengths, similar seed size, and gave equal seed yield. Also Wilcox (1980) grew 40 each of the indeterminate and semideterminate stem types at 0.5 and 1 m row spacings and noted that indeterminates averaged 5 per cent higher in seed yield.

Flowering time in near isogenic soybean lines is two to three weeks earlier in determinate types and a few days earlier in semideterminate types, compared with the indeterminate ones (Bernard, 1972; Rode, 1979). Bernard observed that semideterminate types were 12 to 15 per cent shorter and somewhat earlier in maturity than indeterminate types, but were similar in yield. The determinate isoline of cv. Clark was similar in yield to its indeterminate counterpart, but was 45 per cent shorter in height which effectively accounted for the absence of lodging. But another determinate isoline of cv. Harosoy was significantly lower yielding probably because 61 per cent height reduction was too great.

Sowing Time

Seed yield of indeterminate soybean cultivars declined steadily after the early May planting date, whereas seed yield of the determinate cultivars did not decrease until planting date was delayed past early June. They are early in maturity (Thseng, 1982) and lodge less (Wilcox and Sediyama, 1981). Beaver and Johnson (1981) also reported that sowing of determinate soybeans can be delayed by a month with no loss in yield. Wilcox and Frankenberger (1987) further compared the performance of five determinate and five indeterminate soybean isolines. The indeterminate strains became shorter and had fewer nodes with sowing dates after May 24, while determinate isolines either increased or remained the same for these characters. In general, determinate isolines yielded more when sown in late May or early June, while indeterminate isolines gave maximum yields when sown from early to late May.

Planting Density

Determinate cultivars in general, occupy less space and can thus be planted more densely. Slight reduction in per plant yield is richly compensated by increasing plant density. The determinate soybean cultivars perform well at 400,000 to 500,000 seeds/ha (Beaver and Johnson, 1981). Since determinate cultivars are more synchronous in maturity, they can be harvested earlier and uniformly matured grains obtained.

A significant interaction of cultivar and row spacing has also been recorded for bush beans (Mack and Hatch, 1968) and equidistant close spacing has generally proved more advantageous for bush than for vine type beans (Crothers and Westermann, 1976). At higher plant densities, determinate varieties are subjected to less competitive stress than the indeterminate ones and thus exhibit greater seed yield potential (Westermann and Crothers,1977).

There is a theoretical possibility of increasing grain yield by sowing tall determinate and dwarf determinate cultivars either in blends or in alternate rows, by allowing better sun harvesting. But out of several combinations, Sumarne and Fehr (1980) observed encouraging results only in 1 : 1 blend of Chippewa 64 with Amsoy 71.

Synchrony in Maturity

Polyanskaya (1985) explored on the synchrony of cucumber fruiting for once-over mechanical harvesting. The semideterminate varieties F_1 Signal 235, Parad and Kustovoi 98 gave the best total and marketable yields at the higher planting densities and were most suitable for mechanical harvesting. In peas (*Pisum sativum*) a hybrid between cv. Svoboda and one of its determinate mutants with small stipules and accacia-like leaves, had a higher degree of synchronous ripening and produced uniform seeds, indicating their suitability for once-over mechanical harvesting (Popova and Polunin, 1986).

Earliness

A direct correlation between the 'index of determinate habit' and earliness has been observed in tomato varieties (Paponov and Mezentseva, 1980). Earliness of determinate tomatoes was confirmed by Olabi (1985). Garnako (1983) compared the characteristics of 52 determinate tomato varieties and hybrids with those of 50 indeterminate ones. Determinate genotypes were earlier in maturity and higher yielding in the first month of fruiting, though had overall lower productivity and market quality. With soybeans, Chang et al. (1982) concluded that with the semideterminate stem termination type it is possible to have lines that are earlier maturing and shorter with less lodging under conditions of luxuriant growth without sacrificing yield. Foley et al. (1986) also observed more than three days earliness. However, in potatoes, the determinate varieties have longer growth period (Kleinkopf et al., 1981) probably because of the competition by a larger number of developing fruits as in other crops, is not there. The increase in number and weight of underground tubers results from the savings of metabolites form the reduction in continued growth of the indeterminate pseudoshoots.

Water Requirement

With regard to irrigation, the indeterminate soybean cultivar gave the highest yield with one irrigation at the third or fourth developmental stage, whereas the determinate cultivar needed irrigation at all the six developmental stages, but gave higher yield with constant seed number and size (Kadhem et al., 1985). The determinate soybean line gave 24.9 per cent higher grain yield and used 5.6 per cent more water over the indeterminate isoline and thus showed a greater water use efficiency by 18.1 per cent (Singh and Whitson, 1976).

Grain Filling

Park (1979) noted that seed development was slower in indeterminate than in determinate soybean varieties. But Metz et al. (1985) observed that the time from the beginning of flowering to the beginning of rapid seed filling of determinate soybean line averaged 12.6 and 7.3 days less than those of indeterminate lines, while the seed filling period and days from the beginning of rapid seed filling to physiological maturity respectively, averaged 11.7 and 6.7 days longer (desirable character) in determinate lines than for indeterminate lines. The realized heritability estimates of seed filling period ranged from 0.40 to 1.02, except for indeterminates where considerably less genetic variance was observed. Reproductive period and seed filling period of selected F_4 populations were lengthened by F_3 selection up to 6.4 and 6.1 days, respectively, relative to population means, without significantly changing physiological maturity. Metz et al. (1985) suggested that selections from two of the three crosses should be useful in incorporation of longer seed filling period and reproductive period into adapted germ-plasm to enhance yield potential without delaying the date of physiological maturity.

Grain Yield

The climbing bean (*Phaseolus vulgaris*) cultivars gave 6 to 26 per cent higher grain yield than the bush cultivars (Francis et al., 1978a, b; Robinson, 1975). The indeterminate cultivars of cowpea (*Vigna unguiculata*) also yielded more than the determinate, Bush Purple Hull (Fernandez and Miller, 1985). However, Ojomo (1976) in his yield trials with 34 cowpea lines at three sites recorded higher grain yields from the bunched determinate lines over the prostrate indeterminate ones.

The observations of Foley et al. (1986) on 21 determinate and 21 indeterminate soybean lines indicate that the determinate types are potentially useful for improving yield by affecting various yield components and lodging resistance (Tables 3.2 and 3.3).

Table 3.2: Means of degree of stem termination (DST), days to beginning bloom (BB), node of first flower (NOFF), height at beginning bloom (HTBB), flowering period (FP), and growth during flowering (GF) (After Foley et al., 1986).

		Dst (Nodes)	BB (days after planting)	NOFF (node)	HTBB (cm)	FP (days)	GF (cm)
Cross 1	Determinate	3.4	59.0	6.4	49.1	13.8	27.4
	Indeterminate	7.3	60.0	6.4	51.2	32.7	54.3
Cross 2	Determinate	7.1	49.7	3.0	34.0	19.9	50.4
	Indeterminate	8.5	50.8	3.0	34.6	36.8	66.3
Cross 3	Determinate	7.8	49.9	3.1	36.3	19.5	56.0
	Indeterminate	7.7	50.9	3.0	36.9	37.0	64.3
Overall	Determinate	6.1	52.9	4.2	39.8	17.7	44.6
	Indeterminate	7.8	53.9	4.2	40.9	35.5	61.6
	S E	0.3	0.4	—	0.7	0.3	1.2

Table 3.3: Means of pod height (PHT), maturity (MAT), height (HT), lodging (LDG), seed yield (YIELD), and seed quality (QUAL) (After Foley et al., 1986).

		PHT (cm)	MAT (days)	HT (cm)	LDG (score)[*]	YIELD (kg/ha)	QUAL (score)[**]
Cross 1	Determinate	7.9	37.9	70.0	1.5	2750	1.8
	Indeterminate	11.3	38.9	103.4	2.4	2720	2.1
Cross 2	Determinate	8.1	21.1	72.6	2.1	2440	2.0
	Indeterminate	11.6	27.3	97.6	2.3	2690	1.9
Cross 3	Determinate	7.7	19.1	76.0	2.0	2410	2.3
	Indeterminate	10.2	22.2	95.4	2.4	2520	2.3
Overall	Determinate	7.9	26.0	72.9	1.9	2530	2.1
	Indeterminate	11.0	29.5	98.6	2.4	2640	2.1
	S E	0.6	0.7	8.2	0.4	110	0.03

[*] 1 = erect, 5 = prostrate

[**] 1=very good, 5 = very poor

Harvest Index

A mean harvest index (HI) of 0.46 for indeterminate and 0.49 for determinate soybean varieties at a planting density of 50 plants/m^2 has been recorded (Baker et al., 1983). It was further confirmed by Dayde and Ecochard (1985) and Beaver et al. (1985) that the determinate soybean varieties had higher HI. Also with 34 cowpea lines, Ojomo (1976) recorded higher HI for the determinate types. Further, Fernandez and Miller (1985) recorded three times higher HI for the determinate cowpeas than the indeterminate ones.

GENETIC STUDIES ON AND BREEDING FOR DETERMINANCY

Woodworth (1932, 1933) studied the F_2 of a cross between indeterminate 'Manchu' and determinate 'Ebony' varieties of soybean and classified 0.75 of the plants as indeterminate and 0.25 as determinate based on main stem appearance. Woodworth proposed a gene pair Dtdt to explain this, and most genetic studies since then have agreed with his results (Ting, 1946; Nagata, 1960; van Schaik and Probst, 1958; Haque, 1964, 1965; Caviness and Prongsirivathana, 1968). Kawahara (1963) reported three 'viny' to one 'normal' F_2 ratios in 10 different crosses. However, Nagata (1960) mentioned an indeterminate type that usually segregated, but results were variable because of genetic modifiers.

Inheritance for the common determinate stem type (dt_1) versus indeterminate stem (Dt_1) in soybeans was later confirmed by Bernard (1972) as monogenic. The heterozygote Dt_1dt_1 showed a distinct indeterminate or semideterminate phenotype in the genetic backgrounds studied. A stem type resembling this Dt_1dt_1 heterozygote was found in a few true breeding varieties and was shown to be controlled by a single dominant gene designated Dt_2. Crosses between the two types gave F_2 ratios of one indeterminate : 11 semideterminate : four determinate, the expected ratio for independent segregation with dt_1 epistatic to Dt_2dt_2. The primary effect of both dt_1 and Dt_2 is to hasten the termination of apical stem growth, which decreases both plant height and the number of nodes; but dt_1 has a much greater effect.

Bernard (1972) described two genes, dt_1 (determinate) and Dt_2 (semideterminate), controlling stem termination with dt_1 epistatic to Dt_2 or dt_2. Bernard developed isolines of the soybean cultivars, Harosoy and Clark, which differed only in stem termination. Studies involving the determinate, semi-determinate and indeterminate isolines of Clark and Harosoy (Hartung et al., 1980; Hartung, 1981; Shannon et al., 1971) have shown that both the dt_1 and Dt_2 decreased plant height, number of seeds on the main stem, and maturity. Lodging was reduced by dt_1 and sometimes by Dt_2. Yields of the semideterminate were better than or at least equal to indeterminate isolines, while the determinates yielded equal to or less than the indeterminates. Chang et al. (1982) also tested near-isogenic pairs of semideterminate and indeterminate soybean lines derived from three crosses. The semideterminates were shorter, earlier in maturity, more lodging resistant, and equal in yield to their indeterminate counterparts.

A negative partial dominance occurred in progenies of indeterminate × semideterminate soybean forms, and a positive overdominant heterosis was seen in the F_1s of indeterminate and determinate parents (Chen et al., 1982). Where an indeterminate or semideterminate form was used as the pollen parent and crossed with a geographically distant determinate female parent, the F_1s showed high heterosis and matured later than either of the parents. Combinations with a determinate form as a parent, especially as the female parent, showed comparatively great hybrid vigour for yield components.

Miranda (1976) reports that in the F_1 of the cross between Masterpiece, a determinate cultivated variety of French bean (*Phaseolus vulgaris*) used as female parent, and Durange 47, an indeterminate wild type, all the plants were of indeterminate habit, and in the F_2 the segregation ratio was three indeterminate : one determinate. But in the reciprocal cross, 3 : 1 segregation did not occur, rather the indeterminate type predominated. This was so because when the wild type was used as female parent, selective fertilization occurred in the F_1 between gametes bearing the wild type. Further, Coyne and Steadman (1977) crossed PI 165426 (indeterminate) and the Pinto breeding line P73-121-7 (determinate) of French bean. Data from the F_1 and F_2 indicate that the determinate habit is controlled by a single recessive gene. However, Davis (1979) recorded that single dominant genes control indeterminate versus determinate habit. Later, progenies of four crosses made among two early determinate and two late indeterminate varieties were investigated (Freire Filho, 1980) and the dominance of the indeterminate habit was established.

In crosses of *vicia faba* amongst Kristall (indeterminate), FP3930/80 (determinate, with *ti-1*) and STP 470/79 (a semi-determinate); semideterminate habit was controlled by a single gene dominant to both indeterminate and determinate growth habits (Frauen and Brime, 1983). Further, Brime (1983) established from a study of crosses between two determinate and five semideterminate mutants of the indeterminate variety Kristall, that determinate habit is controlled by two non-allelic genes, *ti-1* and *tp* (topless). The semideterminate mutants possess different alleles at the *ti* locus, designated tis_1, tis_2, tis_3 and *Tis*. *Tis* is also dominant over ti_1 and *Ti* for indeterminate habit. When homozygous, *tp* is epistatic to *Tis*, it is producing a determinate form. The determinate and semideterminate forms were shorter and earlier and had more shoots than the indeterminate forms. F_1 performance for virtually all characters was superior to that of the better parent. The determinate mutants TP 3930 and TP 215 had a negative effect on yield in the F_2, while STP 470 had a positive effect.

In crosses between *ti* (determinate) and normal male partners with extreme values of certain yield components such as 1000-seed weight, Steuckardt and Dietrich (1986) observed that these extreme values were transmitted to *ti* progenies. Further, Habetinek et al. (1988) crossed three sources of the gene *ti* for determinate habit with seven varieties of *Vicia faba* having indeterminate habits. Of the 79 F_3 strains with *ti*, the most promising were derived from the crosses Wieselburg small seeded × TL–G, Chlumecky × TL–P, and White Compact × TL–P.

Waldia and Singh (1987a) concluded that hybridization between determinate and indeterminate lines should be attempted to improve pigeonpeas, as the determinate lines have higher genotypic variation coefficients than indeterminate lines, and hence are genetically more variable. A study of the F_1–F_3 from two crosses between determinate and indeterminate lines showed that indeterminate habit was dominant over determinate habit. With two genes being involved in the expression of this character (Waldia and Singh, 1987b), the gene for indeterminate habit, designated *Id*, had an epistatic (inhibitory) effect on gene for determinate habit, *D*. The genotypes of the parental lines are given as *DDidid* (determinate) and *ddIdId*, (indeterminate).

In peas, a study of the F_1 , F_2 and BC derived from a diallel cross of six lines, four with determinate and two with indeterminate habit; Krarup (1974) suggested that complementary genes may be involved.

In crosses of determinate with indeterminate cotton varieties, the indeterminate type of sympodium was dominant. Segregation in the F_2 from the above crosses was in the ratio of 3 : 1, determinate habit being recessive (Sagdullaev, 1971; Gulamov et al., 1973; Uzakov and Khalugitov, 1974; Chernomaz, 1976; Uzakov and Kim, 1981). Further, the *Gossypium hirsutum* varieties S8257 and 2, with determinate sympodial fruiting branches, were crossed with *G. barbadense* S6030, which has very short sympodial fruiting branches of the zero type (Uzakov and Kim, 1981). The F_1 hybrids had determinate branches and the three classes segregated in the F_2 (zero type, determinate, and indeterminate). BCs of S8257 × S6030 to S6030 resulted in 20.7 per cent zero type plants, 57 per cent determinate,

22.2 per cent indeterminate and 6.1 per cent deformed sterile plants, BCs of 2 × S6030 to S6030 gave a similar segregation namely 17.7 per cent zero, 50.8 per cent determinate, 24.3 per cent indeterminate and 7.2 per cent deformed and sterile. BCS to the determinate parents gave only determinate progeny. Classification of determinate and indeterminate types together as sympodial gave 3 : 1 ratio of sympodial : zero type.

Uzakov et al. (1980) crossed *G. hirsutum* varieties with the determinate type of fruiting branch (S8255, S8230 and 2) and Taskent 1 with indeterminate branching type. F_2 determinate hybrids had an emergence-flowering period of six to seven days shorter than hybrids of the indeterminate type. This difference was somewhat less in the F_3. The hybrids inherited lower attachment of the first fruiting node.

By crossing gynoecious hybrid, Poineer, with the dwarf determinate cucumber Minnesota 82.59, the number of fruits per dwarf hybrid plant increased significantly (Prend and John, 1976). In the same year, Tkachenko (1976) reported that hybridization of the determinate Prolong 121–1, Sort–154 and Karlik Pyzhenkova with the indeterminate Shehedryl 118 indicated that determinate habit is controlled by two independent dominant genes. A positive correlation was found between determinate habit and female sex; determinate forms contained more plants of the female type than did indeterminate forms. Further to this, crosses were made involving Dwarf Marketmore and Dwarf Tablegreen, on the one hand and the indeterminate varieties Marketer, Marketmore and Tablegreen 65, and three determinate breeding lines, on the other (Miller and George, 1979). F_1, F_2 and BC_1 data revealed that the dwarf varieties possessed : (i) a recessive determinate gene allelic with *de*, and (ii) a recessive gene for delayed growth, *de*, which while not a modifier, reduced the length of the hypocotyl and the first few internodes and the growth rate, and appeared to be weakly linked with *de*. The F_2 data from crosses between Dwarf Tablegreen and two of the determinate breeding lines suggested the presence of modifiers of determinate habit, other than *In-de*, which was present in the determinate breeding lines. Such modifiers, it is suggested, make the breeding of cucumbers with a range of determinate sizes feasible.

Three tomato varieties with determinate habit and three with indeterminate habit were studied by Conti (1974). The hybrids homozygous for indeterminate habit were as early as the homozygous varieties; the hybrids homozygous for determinate habit were only a little earlier than the varieties; the heterozygous hybrids, phenotypically indeterminate, were intermediate between the parents.

Some Promising Mutants

In some crop species determinate plants have been induced by irradiation and their genetics studied, with a view to developing desirable varieties. Ashri (1981) found four determinate sesame plants originating from a single mutation, in the M_2 of the Israeli variety 45 irradiated with 50 Krad gamma rays. The determinate habit was true breeding, monogenic and recessive. In the same year, Starzycki and Goral (1981) found a broad bean mutant, in the M_2, derived by gamma irradiation of the seeds of Wierboon with 5 Krad which bore an inflorescence above the topmost leaf. They crossed it with two normal lines of the original variety. F_2 segregation showed that the character was monogenic and recessive.

Following gamma-ray treatment of black gram, *Vigna mungo*, accession B10, an erect synchronous and determinate mutant B 10 (23) was selected from the M_2 generation (Shaik and Majid, 1982). It has upright pods compared to the downward or horizontally borne pods of the original variety. Its increased number of pods/plant results in higher seed yield/

plant and harvest index. Also, from the gamma-irradiated seeds of cluster bean, *Cyamopsis tetragonoloba*, B5-54, an early flowering determinate mutant was found (Singh et al., 1981). Crosses with normal plants indicated that two genes control determinate habit and that at least two dominant alleles are required for expression of the character.

REFERENCES

Ashri, A. Increased genetic variability for sesame improvement by hybridization and induced mutations. FAO Plant Production and Protection Paper. 29: 141–45, 1981.

Austin, R.B., C.L. Morgan and M.A. Ford, A field study of the carbon economy of normal and 'topless' field beans, *Vicia faba. World Crops* 4: 60–79, 1981.

Baker, D.A., G.P. Chapman, M. Standish and M. Bally. Assimilate partitioning in a determinate variety of field bean. *In*: D.G. Jones and D.R. Davis (eds), *Temperate Legumes : Physiology, Genetics and Nodulation*, Pitman, London, pp. 191–99, 1983.

Beaver, J.S., R.L. Cooper and R.J. Martin, 1985. Dry matter accumulation and seed yield of determinate and indeterminate soybean. *Agron. J.* 77: 675–79, 1985.

Beaver, J.S. and R.R. Johnson. Response of determinate and indeterminate soybeans to varying cultural practices in the Northern U.S.A. *Agron. J.* 73: 833–38, 1981.

Bernard, R.L. Two genes affecting stem termination in soybeans. *Crop Sci.* 12: 235–39, 1972.

Brime, M.C. (Inheritance of determinate/indeterminate habit and its effect on yield in broad bean (*Vicia faba* L.). Thesis, Gottengen Univ., German Federal Republic. 51, 1983.

Caviness, C.E. and C. Prongsirivathana, Inheritance and association of plant height and its components in a soybean cross, *Crop Sci.* 8: 221–24, 1968.

Chang, J.F., D.E. Green and R. Shibles. Yield and agronomic performance of semideterminate and indeterminate soybean stem types. *Crop Sci.* 22: 97–101, 1982.

Chen, Y., X.Y. Weng and B.R. Wang. (The inheritance of growth period and yield traits in F_1 soybean hybrids with different podding habits). *Chinese Oil Crops*. 1: 42–46, 1982.

Chernomaz, M.B. (Inheritance of types and subtypes of sympodial branches in cotton hybrids.) *Nauch. tr. Taskent s-kh. in-t.* 67: 24–28, 1976.

Centi, S. (Research on the heterosis and components of phenotypic variance in long-fruited tomato hybrids). *Rivista di Agronomica* 8: 383–91, 1974.

Coyne, D.P. and J.R. Steadman. Inheritance and association of some traits in a *Phaseolus vulgaris* L. cross, *J. Heredity* 68: 60–62, 1977.

Crothers, S.E. and D.T. Westermann. Plant population effects on the seed yield of *Phaseolus vulgaris* L. *Agron. J.* 68: 958–60, 1976.

Davis, J.H.C. (Basic concepts on genetics in French beans). *CIAT.* p. 11, 1979.

Dayde, J. and R. Ecochard. (Dry matter production in soybean. I. Comparison of determinate and indeterminate types). *Agronomie* 5 : 127–33, 1985.

Egli, D.B., R.D. Guffy and J.E. Leggett. Partitioning of assimilate between vegetative and reproductive growth in soybean. *Agron. J.* 77: 917–22, 1985.

Egli, D.B. and J.E. Leggett. Dry matter accumulation patterns in determinate and indeterminate soybeans. *Crop Sci.* 13: 220–22, 1973.

Fernandez, G.C.J. and J.C. Miller, Jr. Yield component analysis in five cowpea cultivars, *J. Am. Soc. hort. Sci.* 110: 553–59, 1985.

Foley, T.c., J.H. Orf and J.W. Lambert. Performance of related determinate and indeterminate soybean lines. *Crop Sci.* 26: 5–8, 1986.

Francis, C.A., M. Prager and D.R. Laing. Genotype × environment interactions in climbing bean cultivars in monoculture and associated with maize. *Crop Sci.* 18: 242–46, 1978a.

Francis, C.A., M. Prager, D.R. Laing and C.A. Flor. Genotype × environment interactions in bush bean cultivars in monoculture and associated with maize. *Crop Sci.* 18: 237–41, 1978b.

Frauen, M. and M. Brime. The inheritance of semideterminate growth in *Vicia faba* L. *Z. fur Pflanzenzuchtung.* 91: 261–63, 1983.

Freire Filho, F.R. (Inheritance of number of days to flowering and the growth habit in French bean *Phaseolus vulgaris* L.) Thesis, Univ. Federal de Vicosa, Brazil, p. 38, 1980.

Garanko, I.B. (Influence of habit on the economically valuable characteristics of tomatoes grown in plastic green houses). Trudy po Prikladnoi Botanike, Genetike i Selektsii, 81: 67–71, 1983.

Green, D.E., P.E. Burlamaqui and R. Shibles. Performance of randomly selected soybean lines with semideterminate and indeterminate growth habits. *Crop Sci.* 17: 335–39, 1977.

Greer, H.A.L. and I.C. Anderson. Response of soybeans to tri-iodobenzoic acid under field conditions. *Crop Sci.* 5: 229–32, 1965.

Gulamov, V.K., S. Narimov, M. Atazhnov and A. Azimova. (Some results of studying the genetics of the cotton species, *Gossypium hirsutum* L. using radiation induced mutants). *In:* Voprosy geneteki, selektsii i semenovodstva Khlopchatnika i lyutserny. 1. Taskent, Uzbek SSR; Izdatelstvo Mskh. pp. 151–67 1973.

Habetinek, J., M. Ruzickova and J. Soucek. (Transfer of the gene *ti* into the genome of bread bean (*Faba vulgaris*) with determinate habit.) *Rostlinna Vyroba* 48: 237–50, 1988.

Hanway, J.J. and C.R. Weber. Dry matter accumulation in soybean (*Glycine max* (L.) Merril) plants as influenced by N, P, and K fertilization. *Agron. J.* 63: 263–66, 1971.

Haque, M.F. Linkage between resistance to *Phytophthora* root rot and plant and seed characters in soybean. *Indian J. Genet. Pl. Breed.* 24: 99–105, 1964.

Haque, M.F. Linkage in soybean. *Indian J. Genet. Pl. Breed.* 25: 102–04, 1965.

Hartung, R.C. Modification of soybean plant architecture by genes for stem growth habit and maturity. *Crop Sci.* 21: 51–56, 1981.

Hartung, R.C., J.E. Specht and J.H. Williams. Agronomic performance of selected soybean morphological variants in irrigation culture with two row spacings. *Crop Sci.* 20: 604–9, 1980.

Hartung, R.C., J.E. Specht and J.H. Williams. Modification of soybean plant architecture by genes for stem growth habit and maturity. *Crop Sci.* 21: 51–56, 1981.

Hicks, D.R., J.W. Pendleton, R.L. Bernard and T.J. Johnson. Response of soybean plant types to planting patterns. *Agron. J.* 61: 290–93, 1969.

Kadhem, F.A., J.E. Specht and J.H. Williams. Soybean irrigation serially timed during stages R_1 to R_6. I. Agronomic responses. II. Yield component responses. *Agron. J.* 77: 291–98; 299–304, 1985.

Kawahara, E. Studies on the gene analysis of soybeans. Tohoku Agric. Exp. Sta. Acad. Rep. 26: 79–147, 1963.

Kleinkopf, G.E., D.T. Westermann and R.B. Dwelle. Dry matter production and nitrogen utilization by six potato cultivars. *Agron. J.* 73: 799–802, 1981.

Krarup, H.A. (Complementary genes may control determinate versus indeterminate habit in pea). *Agro. Sur.* 2: 28–29, 1974.

Mack, H.J. and D.L. Hatch. Effects of plant arrangement and population density on yield of bush snap beans. *Proc. Amer. Soc. hort. Sci.* 92: 418–25, 1968.

Metz, G.L., D.E. Green and R.M. Shibles. Reproductive duration and date of maturity in populations of three wide soybean crosses. *Crop Sci.* 25: 171–76, 1985.

Miller, G.A. and W.L. George, Jr. Inheritance of dwarf and determinate growth habits in cucumber. *J. Am. Soc. hort. Sci.* 104: 114–17, 1979.

Miranda, C.S. Direct and reciprocal hybridization in French bean (*Phaseolus vulgaris* L.). *Agrociencia* 25: 65–72, 1976.

Nagata, T. Agronomic studies of the gene ecological differentiation of soybeans in Japan and the world. Sci. Rep. Hyogo Univ. Agr. 4 (Series Agr.): 96–122, 1960.

Ojomo, O.A. The current outlook on cowpea improvement at University of Ife. *In:* Proc. IITA collaborators meeting on grain legume improvement, Plant improvement. R.A. Luse and K.O. Rachie (eds), Ibadan, Nigeria. pp. 14–18, 1976.

Olabi, M. (Tomato forcing : Plant density and the role of the cultivar). *Kerteszet es Szoleszet*. 34 (12),: 4, 1985.

Paponov, A.N. and A.I. Mezentseva. (Use of the 'Index of determinate habit, in evaluating tomato varieties). *Tr. Perm. s-sk. in-t.* 136: 72–73, 1980.

Park, K.Y. Seed development and germination of soybeans at various filling stages. *In*: World Soybean Res. Conf. II. Colorado, U.S.A. p. 71, 1979.

Polyanskaya, O.F. (Breeding cucumber for suitability for once-over mechanical harvesting). *Selektsiya i Semenovodstvo, USSR*. 5: 19–20. 1985.

Popova, I.a. and Ya. Ya. Polunin. (Using a determinate mutant to produce garden pea varieties). *In*: Khimicheskii mutagenez v sozdanii sortov s novymi svoistvami, Moscow, USSR (1986): 150–53, 1986.

Prend, J. and C.A. John. Improvement of pickling cucumber with the determinate (*de*) gene. *HortScience* 11: 427–28, 1976.

Robinson, R.G. Pulse or grain legume crops for Minnesota. *Minn. Agric. Exp. Sta. Bull.* p.19 and p. 513, 1975.

Rode, M.W. Growth and development of the soybean plant as affected by dt_1 and Dt_2. Proc. 9th Soybean Seed Res. Conf. Chicago Ill., pp. 1–11, 1979.

Sagdullaev, F. (The inheritance of the type of branching of fine-fibred cotton and its correlation with certain economically valuable characters). *In*: Genet. issled, Khlopchatnika. Taskent, Uzbek, SSR. pp. 96–110, 1971.

Saikh, M.A.Q. and M.A. Majid. Altered plant architecture in a blackgram mutant. *Mutation Breed, Newsl.* 19: 11, 1982.

Shannon, J.g., J.R. Wilcox and A.H. Probst. Response of soybean genotypes to spacing in hill plots. *Crop Sci.* 11: 38–40, 1971.

Shibles, R.M. and D.E. Green. Morphological and physiological considerations in breeding for narrow rows. *In*: W.R. Fehr (ed), Proc. Soybean Breed. Conf, Iowa State Univ., Ames., pp. 1–12, 1969.

Shibles, R.M., I.C. Anderson and A.H. Gibson. Soybean *In: L.T.* Evans, *Crop Physiology : Some Case Histories*. Camb. Univ. Press, London. pp. 151–89, 1975.

Singh, B.P. and E.N. Whitson. Evapotranspiration and water use efficiency by soybean lines differing in growth habit. *Agron. J.* 68: 834–35, 1976.

Singh, V.P., R.K. Yadav and R.K. Chowdhury. Note on determinant mutant of clusterbean. *Indian J. agric. Sci.* 51: 682–83, 1981.

Sjodin, J. Induced morphological variation in *Vicia faba* L. *Hereditas*. 67: 155–80, 1971.

Starzycki, S. and M. Goral. (A broad bean mutant with terminal mutant). Hodowla Roslin, Aklimatyzacja i Nasiennictwo. 25 (3/4) : 77–85, 1981.

Steuckardt, R. and M. Dietrich. Differences with respect to yield structure in *ti*-forms of *Vicia faba* after crossing and backcrossing with normal types. *Biologisches Zentralblatt*. 105: 137–45, 1986.

Sumarno, and W.R. Fehr. Integenotypic competition between determinate and indeterminate soybean cultivars in blends and alternate rows. *Crop Sci.* 20: 251–54, 1980.

Tan Boun Suy. (Contribution to the study of drought resistance in two species of the genus *Phaseolus* and in their hybrids). *Bull. Rech. Agron. det Gemblour.* 13: 73–81, 1978.

Thseng, F.S. The significance of indeterminate growth habit in soybean breeding. *In*: Proc. Symp. Plant Breeding, Taiwan. S.C. Hsiech and D.J. Liu (eds). pp. 139–55, 1982.

Ting, C.L. Genetic studies on the wild and cultivated soybeans. *J. Am. Soc. Agron.* 38: 381–93, 1946.

Tkachenko, N.N. (Producing cucumber varieties for mechanical harvesting). Trudy po Prikladnoi Botanike, Genetike i Selektsii, 56 (2): 20–28, 1976.

Uzakov, Yu. and Kh. Khalugitov. (The inheritance of branch type in remote hybridization). *Khlopkovodstvo*. 12: 19, 1974.

Uzakov, Yu., Kh. Khalugitov and R. Kim. (Breeding cotton with determinate branching type). Tr. VNII Selektsii i semenovid. Khlopchatnika. 18: 87–94, 1980.

Uzakov, Yu. and R. Kim (Inheritance of type of fruiting branch in interspecific hybrids). *Khlopkovodstvo.* 6: 30–31, 1981.

Van Schalk and A.H. Probst. The inheritance of inflorescence type, peduncle length, flowers per nodes and per cent flower shedding in soybeans. *Agron. J.* 50: 98–102, 1958.

Waldia, R.S. and V.P. Singh. Growth habit vis-a-vis mature plant characteristics in pigeonpea. *Int. Pigeonpea Newsl.* 6: 28–29, 1987a.

Waldia, R.S. and V.P. Singh. Inheritance of stem termination in pigeonpea (*Cajanus cajan* (L.) Millsp). *Euphytica.* 36: 525–27, 1987b.

Weber, C.R. Physiological concepts for high soybean yields. *Field Crop Abstr.* 21: 313–17, 1968.

Weber, C.R., R.M. Shibles and D.E. Byth. Effect of plant population and row spacing on soybean development and production. *Agron. J.* 58: 99–102, 1966.

Westerman, D.T. and S.E. Crothers. Plant population effects on the seed yield components of beans. *Crop Sci.* 17: 493–96, 1977.

Wilcox, J.R. Comparative performance of semideterminate and indeterminate soybean lines. *Crop Sci.* 20: 277–80, 1980.

Wilcox, J.R. and E.M. Frankenberger. Indeterminate and determinate soybean responses to planting data. *Agron. J.* 79: 1074–78, 1987.

Wilcox, J.R. and T. Sediyama. Interrelationships among height, lodging and yield in determinate and indeterminate soybeans. *Euphytica.* 30: 323–26, 1981.

Wilson, R.R. and R.H. Cole. The potential of dwarf soybean varieties. *Univ. Delaware Agr. Exp. Sta. Bull.* 367: 17, 1968.

Woodworth, C.M. Genetics and breeding in the improvement of the soybean. *Illinois Agric. Exp. Sta. Bull.* 304: 297–404, 1932.

Woodworth, C.M. Genetics of the soybean. *J. Am. Soc. Agron.* 25: 36–51, 1933.

CHAPTER 4

DWARF STATURE

INTRODUCTION

Recent researches have shown that all the leaves produced by most plants are not necessary, and in fact, reduction in leaf area (even up to 50 per cent) either mechanically or genetically (leafless peas and okra and superokra type of cotton plants) has resulted in increased yield on unit land area (Harvey and Goodwin, 1978; Hedley and Ambrose, 1981; Karami et al., 1980; Meredith, 1984; Meredith and Wells, 1986, 1987). Vegetative growth produced by most cereals, legumes and sweet potato, etc., is in excess, and its reduction by introducing dwarfing genes (dwarf plants of the green revolution) or resorting to growth retardants (CCC, TIBA, B_9, etc.) has almost always resulted in increased grain yield and harvest index (HI) (Gupta, 1978).

The popularity of Norin 10 dwarfing gene was greatly enhanced by its incorporation into the Mexican wheat programme by Dr N.E. Borlaug in 1954. Pitic 62 was the first of many semidwarf varieties released by CIMMYT. The genetic system underlying Norin 10 dwarfism is understood to consist of two partially recessive independent genes which act additively together. Reduced height gene (Rht_1) is carried on chromosome 4A (Gale and Marshall, 1976), and Rht_2 maps at a homologus position on the short arm of chromosome 4D (Gale et al., 1975). The Rht_1 allele is carried at the same locus as Rht_3 the stronger, partially dominant gene (Zeven, 1969) carried by Tom Pounce (Tom Thumb) wheat from USSR.

The higher yielding ability of Norin 10 dwarf wheats has been explained : Bingham

(1972) suggests that the reduction in stem growth may reduce competition with ears while Fisher (1973) argues that the stronger apical dominance in the spike and individual spikelets allows a greater number of spikelets and florets to develop. Later, Brooking and Kirby (1981) confirmed that the greater proportion of the total growth is apportioned to the ear and less to the stem in dwarf lines as against the tall lines. Ear weight at anthesis, number of fertile florets, number of grains per spikelet, harvest index and grain yield are all greater in dwarf than in the tall lines. The semidwarf wheats have higher N content and grain yield than the tall or double dwarf wheats (Kim and Paulsen, 1986). However, the double dwarf lines have significantly a higher kernel number per spike and higher HI (0.58) than the other lines (0.54), but the total grain yield is not the highest because of the lower biomass. The semidwarf varieties give higher grain yield than the tall ones, especially at N and P applications of ≥ 100 kg/ha (Mellado, 1987). The semidwarf varieties are more efficient in N and P utilization, the semidwarfs yielded 26.2 kg/kg N+P while the tall variety yielded 14.6 kg/kg of N+P.

How all these happen is far from clear. Gupta (1978) attempted to explain this in his article "Production potential of dwarf genomes" on the basis of literature then available. By now we all know beyond doubt that dwarfening has generally resulted in increased grain yield. In this book we will review the literature pertaining to crop improvement through dwarfening.

SELECTION

Johnson et al. (1986) report that selection for dwarfness in maize resulted in a linear reduction in plant height from 282 to 179 cm (2.4 per cent per cycle). Optimum plant density for yield increased from 48,000 to 64,000 plants/ha, an increase of 2.1 per cent per cycle. As height was reduced, yield when determined at an optimum plant density for yield for each generation, increased by 4.4 per cent per cycle. Later cycles showed reduced total lodging, reduced barrenness, earlier flowering and fewer leaves per plant. Yield improvement was associated with a linear increase in HI from 0.30 to 0.45 from original to the 15th cycle. Detailed data on reduction in plant height, cob position, growth duration, grain filling period, grain yield and HI as a result of 15 cycles of selection may be seen in Table 4.1.

Table 4.1:   Effects of 15 cycles of recurrent selection on reduction of plant height, cob position, growth duration, grain filling period, grain yield and harvest index of maize (After Johnson et al., 1986).

Cycle	Plant height (cm)			Duration (days)			Grain	HI
	Total Plant	Below ear	Above ear	Emergence to silking	Emergence to maturity	Effective filling period	Yield (t/ha)	
0	282	191	91	73.4	124.6	35.1	4.14	0.30
6	219	132	87	67.1	121.7	35.4	5.44	0.40
9	211	117	94	66.7	122.1	30.9	5.45	0.40
12	202	113	89	65.7	118.6	32.9	5.94	0.41
15	179	99	80	61.6	116.1	33.0	6.31	0.45

Mishra and Chaubey (1987) subjected a composite mixture of 13 diverse genotypes of sunflower to two cycles of modified mass selection. As a result, plant height was reduced by 37 per cent compared with that of the unselected open pollinated population; achene yield *per se* was not adversely affected, but capitulum size, hull content, stem girth and maturation period were significantly reduced. Since achene yield showed a reduction in heritability estimate, cessation of selection for reduced height after cycle two is regarded as essential, to avoid yield reduction in subsequent cycles. However, modified mass selection is considered to be a useful means of reducing height if confined to two selection cycles.

With regard to winter rye, Skorik (1986) reports that selection pressure in the course of 10 years reduced plant height by 49.8 per cent in comparison with the initial population.

GENETIC STUDIES

Wheats of different heights have been classified as 1-gene, 2-gene, 3-gene or 4-gene dwarfs or a single, double, triple or quadruple dwarfs. Although semidwarfism is thought to be a relatively simple inherited character, one to three major factors along with some modifiers have been reported to control this trait (Jha and Swaminathan, 1969; Reddi and Heyne, 1970; Anand and Aulak, 1971). However, some workers using monosomic and nullisomic analysis and substitution lines (Sears, 1954; Allan and, Vogel, 1963) concluded that at least 11 to 16 of the 21 chromosomes of wheat are involved in height.

Four independently segregating loci account for most of the differences in height of spring wheat cvs. Ramona 50, Olesen, D 6301 and D 6899 (Fick and Qualset, 1973). Two major recessive genes control dwarfness in Olesen and the Norin 10 derivative in D6301. Olesen also carries a third dwarfing gene which is partially dominant in its effects over genes for tallness. Dwarfing in D 6899, a derivative of Tom Thumb, is controlled partially by a single gene with many additive effects. Dwarfness in Tom Thumb is controlled by a pair of major genes located on chromosome 4A and/or more minor genes located on chromosome 2D (Jha and Singh, 1977). The location of the four hybrid dwarfing genes D_1, D_2, D_3 and D_4 was confirmed by Worland and Law (1980) using a monosomic series of Cappelle Desprez and a number of intervarietal chromosome substitution lines.

Wang et al. (1982) indicated that dwarfness is controlled by two partially dominant genes located on chromosome 2A and 4D, and some modifiers. However, Stelmakh and Bondar (1980) indicated that in the semidwarf wheat variety Inia 66, length of the internodes 2 to 4 is controlled by genes on chromosomes 3B and 4A.

In Norin 10 derived semidwarf wheats both the genes controlling height (Rht) and GA-insensitivity (Gai) are located on chromosome 4D (Gale et al., 1975). Rht_1 is located on chromosome 4A (Gale and Marshall, 1976) and Rht_2 is located on chromosome 4D (Gale and Marshall, 1976). By means of monosomic analysis Morris et al. (1972) located Rht_3 on chromosome 4A. Tom Thumb dwarfism is thought to be genetically independent from Norin 10 dwarfism (Piech, 1968; Chaudhary, 1973). However, segregation of GA-insensitivity in crosses between Rht_1 and Rht_3 varieties, show that these two genes are alternative alleles at the same locus of chromosome 4A (Gale and Marshall, 1976). Rht_3 confers a more severe GA-insensitivity than the Norin 10 genes. There are many other genes contributing to dwarfism in wheat which have been summarized in Table 4.2.

Table 4.2:   Major genes for dwarfism in wheat (After Gale and Youssefian, 1985)

Gene	Comments
Rht_1 and Rht_2	The Norin 10 genes from Daruma. World-wide use in bread wheat. Rht_1 used in durum and triticale varieties.
Rht_3	The Tom Thumb gene. No commercial use but may have potential for use in triticale, hybrid wheat or as a means of sprouting damage control.
Rht_4	Irradiance induced mutant in Burt with no effect on coleoptile length. No commercial use yet.
Rht_5	EMS induced mutant in Marted. No commercial prospect because of deleterious effect on yield.
Rht_6	In Burt and possibly other standard height varieties.
Rht_7	EMS induced mutant in Bersee. Deleterious effects on yield.
Rht_8 and Rht_9	The Akakomugi genes. Extensive use in Italian wheats and European and possibly CIMMYT breeding programmes. Rht_9, probably transferred to Italian durums.
Rht_{10}	The Ai-bian 1 gene. Very potent gene with similar characteristics to Rht_3.
Rht (Karcag 522M7K)	Irradiation induced mutant in Karcag 522.
Rht (C_pB 1232)	Irradiation induced mutant in Cappelli, released as variety Castelporziano. Used further in varietal production in durums.
Rht (Leeds M 131)	EMS induced mutant in Leeds, being considered for use in durum breeding.

The GA-insensitive character of Norin 10 and Tom Thumb was first noted by Allan et al. (1959) who found that these varieties differed from most other tall and dwarf genotypes in that applied GA did not result in increased stem elongation. Despite the physiological relationship between plant height and gibberellins, the relationship between the Norin 10 and Tom Thumb dwarfing genes and GA insensitivity has been disputed. Initial reports (Konzak et al., 1973; Hu, 1974; Jalalyar, 1974) indicate that the two characters are controlled by different, but linked genes. This gave rise to the assignment of a second series of gene symbols, Gai 1, Gai 2 and Gai 3 (McIntosh, 1979). Subsequent genetic and physiological evidence (Gale and Marshall 1975; 1976; Gale and Gregory, 1977) has indicated that the two phenotypes are simply different, but pleiotropic effects of the same genes. This controversy may have resulted from linkage, in some genotypes, of the GA-insensitive Norin 10 dwarfing genes with other genes affecting plant height.

The hybrid dwarfness genes D_1 and D_2 exist in multiple alleles with different degrees of phenotypic expression (strength) ranging from strong (D_1s) through moderate (D_1m) to weak (D_1w), and that D_2D_3 in combination with Th_o or Th_2 can condition hybrid dwarfness (Kazaryan, 1986). In producing vigorous hybrids of the dwarf 3 type, capable of giving good yields, use of varieties with weak alleles of D_1 (such as Vandilla) or with the genotype $D_2D_3Th_2$ (Stepnyachka 30, Bel'tskaya 32 and Odesskaya 12) is recommended.

Jia et al.(1988) identified a group of 50 dwarf and semidwarf wheat varieties as insensitive to GA_3 and a further group of 43 as sensitive. The dwarfing sources of most of the GA-insensitive group were Suweon 86, St 2422/44, Norin 10, Youbao and Huixian-hong or they were the result of natural or induced mutations. Among the GA-sensitive group, the dwarfing source of 27 varieties was Red Wheat via the cross (Rieti × Wilhelnim) × Red Wheat; the genes involved were Rht_8 or Rht_9. Among the others, Yuanzhu was a mutant induced by gamma radiation and colchicine treatment; it is 70 cm high, 200 grains/ ear and an average protein content of 20 per cent.

Additive gene action is predominant for height but non-additive gene action has also been observed (Sharma and Malik, 1975; Guseinova, 1987). Dominance of the tall parent has been noted in crosses between varieties showing little difference in height. In crosses where the initial forms differ markedly in this character, intermediate inheritance of height is generally observed (Rudenko, 1979). The intermediate inheritance of height in crosses of the tall × short type is seen as evidence of the considerable number of additive genes involved in the control of this character. The degree of dominance for tallness varies depending on the genotypes of the initial varieties ranging from intermediate through partial and complete dominance to overcome (Table 4.3).

Table 4.3:   Variability in inheritance of plant height in wheat varieties.

Nature of inheritance	Authority
Intermediate	Babadzhanyan et al., 1974. Glukhotseva, 1975. Varlnitsa et al., 1978. Mardzilovich, 1979. Rudenko, 1979. Shindin and Karacheva, 1979.
Partial dominance of tallness	Britto and Cassalett, 1975. Nasr et al., 1976. Nalepa and Pochaba, 1980. Sharma and Singh, 1976. Alderov, 1979a; b.
Dominance of tallness	Alderov, 1979a; b; c. Dechev, 1985. Rudenko, 1979.
Overdominance of tallness	Alderov, 1979b.
Partial dominance of dwarfness	Nalepa and Pochaba, 1980.
Overdominance of dwarfness	Alderov, 1979c.
Heterosis for short straw	Dechev, 1985.

In the winter wheat genotypes, single doses of Rht_1 or Rht_2 genes gave pleiotropic effects on yield ranging from −1 to 24 per cent of their rht controls and Rht_3 genotypes had 4 per cent lower yields compared to their Rht_3 controls (Gale and Youssefian, 1985). Grain weights of winter wheats with either Rht_1 or Rht_2 were 5 to 18 per cent lighter than their rht control, whereas grain weights of Rht_3 genotypes were 23 per cent lighter than their rht_3 controls. They indicated that winter wheat genotypes with either one or both of the Rht_1 and Rht_2 genes had 4 to 60 per cent more kernels per spike than their controls. Spike numbers of winter wheat with either Rht_1 or Rht_2 ranged from −2 to 9 per cent of their controls. Gale and Youssefian (1985) cited studies of winter wheat in which single doses of the Rht_1 and Rht_2 genes reduced plant height by 13 to 23 per cent compared to their normal counterparts; the Rht_3 gene reduced height by 46 per cent and the combined effects of Rht_1 and Rht_2 reduced height by 49 per cent.

Wheat near isogenic lines with single doses of Rht_1 and Rht_2 had mean grain yields of 11 per cent greater than lines with both genes together (Allan, 1983). Lines with single doses of Rht_1 and Rht_2 had mean grain yields of 16 per cent greater than their non-semidwarf counterparts. The combination of Rht_1 and Rht_2 produced neutral effects on grain yield in six populations, but in three populations with inherently tall plant stature, the Rht_1 plus Rht_2 lines had mean yields that were 16 to 31 per cent greater than those of their normal height sibs.

Dwarfism in the *japonica* semidwarf rice variety Dan-kan-baik-mang is controlled by a simple recessive gene while in crosses involving Taichung Native 1, inheritance of dwarfness is more complex (Bai, 1973). In varieties such as Aiyinfang and Aigui, inheritance of height is controlled by major gene(s) while in varieties Nongken 57 and Nihonbare it is the minor gene(s) (Gu and Zhu, 1981). Dwarfness behaves as a monogenic recessive trait functioning in association with modifier complexes of various strengths (Singh et al., 1979). Three major groups of dwarfs have been recognized. Vahiduzzaman and Ahmad (1980) report that tall and very tall stature are dominant over short, shortness being controlled by a major allele.

F_2 segregation data indicate that the variability for height in a cross between the *indica* rice, VLK 39 and the *japonica* rice VL 8 is due to digenic segregation with complementary gene effects, each parent having one of the dominant genes (Chauhan and Tandon, 1983). The shortest plants in the segregating population are much shorter than the short parent and are double recessive, making them a significant new source of dwarfing genes.

In a 6 × 6 diallel cross of cultivars and dwarf mutants, height was governed predominantly by additive effects with some dominance gene action (Kumar and Rangasamy, 1986). Tripathi and Misra (1986) also noted additive gene effects, and overdominance was evident as well. Dominance effects were not significant but significant unidirectional dominance was indicated, with asymmetrical distribution of positive and negative alleles in the parents. The direction of dominance was towards tallness. However, Kumar and Rangasamy (1984) with some other rice cultivars had earlier noted that non-additive gene action (dominance or epistasis) conditioned plant height, panicle length and yield/plant.

Most grain sorghum hybrids are 3-dwarfs, and have three recessive dwarfing genes; dw 1, dw 2 and dw 3. Shorter hybrids as 4-dwarfs are also of interest. One of the two male parents used by Schertz et al. (1974) was a double haploid quadruple dwarf (recessive for the four major height genes, dw1, dw2, dw3 and dw4) of T × 403 and the other (TM1) was a tall mutant line ($DW_3 Dw_3$) from the double haploid. One of the two female parents, AT × 3118 was a four-gene dwarf and the other was a tall mutant line Dw_3Dw_3 derived from the four-gene dwarf. The hybrids derived from these lines have the genotypes dw_3dw_3, Dw_3dw_3 and Dw_3Dw_3, and are recessive at the other major height loci, dw_1, dw_2 and dw_4. The Dw_3dw_3 and Dw_3Dw_3 tall hybrids exceed the dwarf dw_3dw_3 in height and total grain yield. The Dw_3Dw_3 hybrids do not significantly out-yield the Dw_3dw_3 hybrids. In pearl millet (*Pennisetum americanum*) Rao et al. (1986) discovered new sources of dwarfing genes to which they designated as d_3 (from IP 10401) and d_4 (from IP 40402).

Ray (1965; 1970) described a cotton cultivar with short, compact habit producing few vegetative branches but initiating fruiting branches at lower nodes on the main stem, and is early in maturity. Plant height is associated with additive genetic effects without any indication of dominance effect. The main stem internode length has a genetic system different from that for fruiting branch internode length. Diallel analysis indicated that the genetic variation for plant height and maturity traits among the semidwarf stocks is primarily additive, with a small though significant dominance component associated with all the traits except the number of main stem nodes. Heritability is relatively low for plant height and main stem internode length, intermediate for number of main stem nodes and node of the first fruiting branch, and high for fruiting branch internode length and mean maturity date (Quisenberry, 1977). In crosses involving tall, dwarf and semidwarf cotton varieties, intermediate inheritance of plant height has been observed in F_1 (Akmuradov, 1979).

In legumes, in addition to height, determinancy is another important criterion. High yielding indeterminate (Dt_1) soybean cultivars are crossed with determinate (dt_1) cultivars and then high yielding determinate cultivars are selected (Cooper, 1981). From 150 crosses each year, short statured, determinate progeny were selected in F_2, and subsequent generations were evaluated for yield. The determinate gene (dt_1) and other genes which complement dt_1, were most useful in producing high yielding short statured types (Cooper, 1981). These lines are more responsive to narrow rows and high seeding rates (high plant population) but are sensitive to moisture stress and thus less stable; further improvement for deep root system is needed.

In peas, height is influenced by both additive and non-additive gene action (Kumar, 1976). Narrow sense heritability was considerable. In the hybrids involving dwarf, medium and tall varieties of peas, tallness was dominant in F_1 (Khmelev and Rozvadovskii, 1984). Additive gene effects were predominant in the medium × medium, and medium × tall crosses, while dominance effects predominated in the dwarf × medium, and dwarf × tall crosses. There was a marked heterosis in the dwarf × medium cross.

In French bean (*Phaseolus vulgaris*) cultivars, additive gene action is dominant for height (Santos and Vencovsky, 1986). In black gram (*Vigna mungo*) a mutant with shorter internodes was induced (Dwivedi and Singh, 1985). In this mutant, dwarfism is controlled by a single recessive gene. In pegeon pea, however, tallness is dominant over dwarfness (Mohamed et al., 1975), height being controlled by a single pair of genes. Dwarfness is governed by duplicate, recessive genes designated t_1 and t_2 (Waldia and Singh, 1987).

BREEDING FOR DWARFNESS

Dwarf wheats have been bred by introducing the recessive dwarfing genes from Norin 10, dominant genes from Tom Pounce (Tom Thumb) and many short statured mutants like Krasnodar Dwarf 1 and KMB 1 (Vasilenko, 1977). Dwarf forms of durum wheat with resistance to lodging and unfavourable environmental conditions, solid straw and high grain yield have been obtained from crosses between *Triticum durum* with *T. dicoccum* (Khangildin et al., 1977). In the F_2 of *T. dicoccoides* var. *arabicum* × *T. carthlicum* var. *stramineum* 'Dika 9/14', some dwarf plants with compact ears segregated. In the F_{10}, four high yielding tetraploid forms with short straw and compact ear were obtained (Naskidashvili, 1979). The F_6 of a cross involving C_pB_{132} with short straw and black awns, and Adur, with long straw and white awns, yielded four lines including Probstdorfer Miradur, which gave yields greater than that of Adur, and were also shorter than Adur (Hansel, 1976). They possessed a greater 1000-grain weight and were more resistant to lodging than Adur.

Dwarfening in tall hexaploid triticale 64 has been obtained by crossing with wheat var. Tom Thumb which transmitted dwarfness in a partially dominant manner (Kiss, 1973). After several years of breeding work, semidwarf triticale strains were obtained which are as fertile as triticale 64.

In Niger, Chanterean and Etasse (1976) obtained dwarf populations, IRAT-P_{10}, P_{15} and P_{16} of pearl millet after crossing the tall local varieties, Souna, Hainei Kirei and Ex Bornu with the Indian line I 472 carrying the gene d_2 for reduced internode length. These were then crossed and recurrent selections were made. Plant height ranged from 35 cm (P_{16}) to 55 cm (P_{15}).

Non-conventional methods of improvement, for example, tissue culture and radiation have also been used for developing dwarf genomes (Ono and Takaiwa, 1983; Das and

Kumar, 1974). From the callus culture of a rice plant, induced in 1 per cent NaC1, 24 out of the 35 regenerated plants (D_1) were 20 cm shorter (D_2) as against the original source (Ono and Takaiwa, 1983). Dwarfness was comparatively stable in D_7 after selfing.

Mutation was induced in tall scented rice Basmati 370 by using 25 Krad gamma rays and the four promising dwarf mutants induced were compared with the parent variety in the field (Cheema and Awan, 1988). The mutants were 18 to 24 per cent shorter than Basmati 370 and had comparable grain quality. They also gave significantly higher grain yields, possessed higher tillering ability and lodging resistance.

Ling and Li (1986) developed short culm wheat mutants having shorter but similar number of internodes as the tall isolines. Genetic studies revealed that mutant 633-1 carries a recessive dwarfing gene, 650 and 1300 carry two such genes and 527 carries a partially dominant gene. These genes differ from the Norin 10 genes Rht_1 and Rht_2 in that they do not show a pleiotropic effect on coleoptile growth. The coleoptiles of mutants 527, 633-1 and 1300 are as long as those of their respective parents. It is considered that these mutants could be of value in developing short culm varieties with better emergence than those carrying Norin 10 genes.

Radiation-induced dwarf mutants have also been produced in grain legumes like field peas and cowpeas. Such induced dwarfs of pea, for example, "Duke of Albany" are monogenic recessive (Das and Kumar, 1974). Both additive and dominance components contribute to dwarfness. Since the additive gene effect is considerable, selection of dwarf progenies was effective. The M_2 mutant plants obtained by 40 Krad gamma ray irradiation of dry cowpea var. V_{16} seeds were 50 per cent shorter and increased the number of pods per plant with 1000-seed weight (a 40 per cent increase) and seed yield per plant, and also matured 10 to 12 days earlier than V_{16} (Chowdhary, 1983).

PHYSIOLOGICAL ASPECTS OF DWARFNESS

Photosynthesis, Respiration and Translocation

Flag leaf, considered to contribute maximum towards grain filling is only weakly related with grain weight in wheat regardless of plant stature (Alimov, 1980). However, photosynthetic rate of the whole plant recorded at different developmental stages was always highest in the dwarf wheat var. Oviachik 65 than in the tall Sevindzh or the semidwarf Sark varieties (Aliev and Kazibekova, 1977). Further, the photosynthetic superiority of semidwarf wheats carrying two dwarfing Norin genes has been shown over tall, 1-gene dwarf and 3-gene dwarf wheat varieties (Khan and Tsunoda, 1970; Lupton, 1972; Pearman et al., 1979; Kulshrestha and Tsunoda, 1981). A comparative study of eight cultivars classified by height as tall (T), semidwarf (D_1), dwarf (D_2) or very dwarf (D_3) revealed that cultivars carrying Norin 10 dwarfing genes (D_1, D_2 and D_3) have a significantly higher photosynthetic rate per unit leaf area than the tall (T) cultivars that lack these genes (Kulshrestha and Tsunoda, 1981). Among the groups carrying the Norin 10 genes, D_2 had the highest photosynthetic and respiration rates, followed by D_3 and D_1 (Tables 4.4 and 4.5).

Tanaka (1972) reports that tall rice varieties have lower growth efficiency (dry matter produced per unit substrate used) and a larger proportion of maintenance respiration (respiratory rate without plant growth) as against the dwarf varieties, which corroborates with the earlier reports.

Table 4.4: Mean observations on oxygen exchange in photosynthesis and respiration of different height groups of wheat varieties (After Kulshrestha and Tsunoda, 1981).

Height group	Cultivar	Oxygen exchange μmole/dm²/h.	
		Photosynthesis	Respiration
T	NPB 52	178.3	11.2
T	K 68	183.6	12.9
D_1	Kalyan Sona	216.2	21.3
D_1	Sonalika	211.7	17.2
D_2	Arjun	238.8	63.9
D_2	Shera	234.2	46.6
D_3	HD 2160	229.5	29.5
D_3	Moti	219.3	21.2

Table 4.5: Comparison of different height groups of wheat varieties for oxygen exchange in photosynthesis and respiration (After Kulshrestha and Tsunoda, 1981).

Height group	Group mean (O_2 exchange μmole/dm²/h)	
	Photosynthesis	Respiration
T	181.5	12.1
D_1	213.9	19.2
D_2	236.5	55.2
D_3	224.4	25.4

In dwarf wheat cultivars, contribution of an ear to total photosynthesis is greater and that of the straw smaller than in tall cultivars (Igoshin and Kumakov, 1982). A more complete and efficient assimilate transport into the ear from the upper and particularly the second top leaf takes place in dwarf cultivars (Wojcieska and Slusarczyk, 1975; Igoshin and Kumakov, 1982). The rate of ^{14}C transport from stem to grains has been noted to be about 50 per cent faster in a semidwarf wheat cultivar than in the tall cultivar and thus the dwarf cultivar showed a higher harvest index (Thorne, 1982).

Hormone Relations

Gibberellins: Gale and Marshall (1973) observed that dwarf wheat varieties did not show stem elongation with GA_3 treatment although the number of tillers increased. This insensitivity to GA (controlled by Gai) is associated with plant height in derivatives of Norin 10, which carry Rht_1 and Rht_2 for height reduction (Rao, 1980; Filev, 1984; Saxena and Singh, 1986). Monosomic analysis of GA sensitivity has indicated that Gai Rht_1 and Gai Rht_2 are located on chromosomes 4A and 4D, respectively (Rao, 1980; Worland and Petrovic, 1988). The Rht gene reduced height which resulted in increased spikelet fertility and in the total number of grains developing in the ear, but reduction in grain size nearly eliminated the advantage of increased grain number.

In an experiment, six *japonica* and eight *indica* rice cultivars and nine *japonica - indica* hybrids treated with GA_3 at the second leaf sheath showed that most *japonica* - cultivars had high GA response, whereas the *indica* cultivars except Dee-geo-woo-gen which possesses a semidwarfing gene have low GA response. Among the hybrids, six had high, while three had low GA response (Song et al., 1981). In the same year, Matsunaga et al. (1981) screened 122 rice varieties and lines for GA response. Generally the *japonica* varieties showed low, and *indica* and upland *japonica* varieties showed high response. Varieties with the semidwarfing gene of Dee-geo-woo-gen varied in response, suggesting that semidwarfism is independent of sensitivity to GA. However, Singh et al. (1981) observed that five dwarf rice strains showed a significant increase in height in response to exogenous GA and had much lower endogenous GA concentrations than the tall strains. But the other two dwarf strains did not respond to GA and had much higher endogenous GA concentrations. The former five dwarf strains exhibited a partial block in GA biosynthesis while the latter two exhibited a partial block in GA utilization (Singh and Singh, 1982).

Pisum sativum lines differing at the internode loci *Na/na* and *Le/le* (tall/dwarf) were assayed for GA-like substances (Potts and Reid, 1983). Lack of detectable levels of GA in shoots of the extremely dwarf *na* plants suggests that this mutation blocks an early step in GA synthesis. Extracts of shoots of *naLe* plants treated with GA_{20} (the major active GA in dwarf types) showed GA1-like activity compared with extracts from similarly treated *nale* plants. They suggested that *Le* allows conversion of less active GA (GA_{20}) into one more active in stimulating elongation (GA1-like). Allison-Creese et al. (1985) also confirmed that in intact plants *na* prevents expression of *Le* or *le*. Allele *na* blocks a step early in the GA biosynthetic pathway while *le* prevents conversion of GA_{20} to GA_1. When grafted onto leafy *Na* stocks, *lelenana* and *LeLenana* scions became photosynthetically dwarf and tall, respectively. Hence, *Na* stocks produce a graft-transmissible substance which promotes elongation of *na* scions and allows expression of the *Le* versus *le* difference.

With beans on the other hand, Morales and Greene (1972) showed that the dwarf-1 mutant has identical GA content as the normal height plants, but in addition the dwarf mutant contains a substance which specially inhibits the neutral fraction gibberellins (abscisic acid?). Park (1978) also reported that the super tall soybean plants from the F_6 contained a higher GA and lower ABA content than the parental varieties. Dwarf soybean genotypes, T209, T244, T256 and M64-503-Duddy respond to GA_3 but their growth as a percentage of initial shoot length was less than that of the normal isolines following GA_3 treatment (Birnberg et al., 1987). Dwarf genotype T210 responded to GA_3 treatment with a much smaller increase in height than the other dwarf genotypes.

Börner and Melz (1988) also noted that some of the dwarf rye lines showed significant growth elongation with GA_3 application, while others with gene ct_2 did not react. They concluded that the diploid rye contains both GA-sensitive and GA-insensitive dwarfs.

Abscisic Acid: The actively growing tissues (leaf bases and stems) of two dwarf rice mutants contained up to 50 per cent more ABA than the normal height plants (Tietz, 1979). No differences in ABA concentration were detected among the leaves of dwarf mutant and the tall cultivar. However, significant differences have been observed in ABA accumulation between the pearl millet cultivars Serere 39 (tall, high ABA) and B282 (dwarf, low ABA) and between six S_1 derivatives within each cultivar (Tietz, 1979). ABA accumulation in the F_1 of B282 × Serere 39 is intermediate between parental values which indicates additive genetic control. Selfed progeny of this cross were selected for low and high ABA. A six-fold range in values was found among 207 F_2 plants, indicating to almost nine-fold in the F_4. The tall F_3 plants contained significantly more ABA than the dwarf ones.

Auxin : Kirillova and Bocharnikova (1986) report that compared to tall mutant, the dwarf tomato mutant had lower levels of growth promoters, particularly IAA, in the stem apices and a higher rate of growth inhibitor accumulation. At the early stages of development, dwarfness was caused not only by limited formation of gibberellins, but also by the inactivation of auxins, notably IAA. Also in dwarf wheats, Saxena and Singh (1986) report that greater oxidation of IAA and higher levels of peroxidase could be responsible for cessation of elongation growth.

Ethylene: Aharoni et al. (1973) observed higher ethylene production from the dwarf varieties of *Phaseolus vulgaris* and *Triticum aestivum* as against their tall varieties. Maximum production and the greatest varietal differences occurred in young plants.

Enzyme Activity

Peroxidase activity measured in four pairs of nearly isogenic sorghum lines differing in one gene for height showed that the enzyme activity was higher in leaves and stems of shorter plants than in those of the corresponding tall isogenics. Roots showed the highest enzyme activity followed by leaves and stem (Liang et al., 1973). Similarly, the grains of dwarf wheat varieties also had higher peroxidase activity than the grains of taller ones. Saxena and Singh (1986) report that in wheat cytoplasmic IAA oxidase and peroxidase show significant inverse correlation with plant height at maturity, indicating that oxidation of IAA proceeds faster in the dwarf than in the tall cultivars. High levels of peroxidase in the shoots of dwarf cultivars could be responsible for cessation of elongation growth. Lower levels of α-amylase activity induced by Norin 10 gene could be another reason of height reduction in wheat (Tupitsyn, 1985).

In three dwarf and three tall (at least double) mutants of pea, peroxidase activity was negatively correlated with growth, the dwarf mutants having the highest and the tall ones the lowest activity (Ancheva and Vasileva, 1983). Similarly, Chary and Bhalla (1983) report that in pigeon pea peroxidase activity was higher in the dwarf mutants and lower in the tall mutants than in the parent cultivars which indicate an inverse relationship between peroxidase activity and plant height.

The RuDP carboxylase activity in the first leaf of wheat genotypes was highest in the semidwarfs as against the tall ones (Pyke and Leech, 1985) which could be responsible for the higher rates of photosynthesis and grain yield. However, the lower nitrate reductase activity in some of the dwarf wheat varieties may be responsible for their lower grain protein content (Edwards et al., 1973).

ADAPTATION AND STABILITY

Stress Tolerance

Allan et al. (1959) showed with wheat seedlings that the semidwarfs with relatively short coleoptiles are slow in emergence and produce inadequate stands. The short coleoptiles in some semidwarf wheats are due to fewer cells per longitudinal plane while in others, shorter coleoptile parenchyma cells are responsible (Allan et al., 1962). In winter wheats, coleoptile length is positively correlated with emergence and plant height (Sunderman, 1964). In rice, mesocotyl (the internode between the coleoptile node and the seed) elongates in the absence of light. Turner et al. (1982) tried to separate whether short mesocotyl or coleoptile creates emergence problem in rice. Turner et al. selected three semidwarf cultivars (Bellemont, M-101 and IR 24) and two tall cultivars (Labelle and Starbonnet) and straw hulled Red Rice which were planted at 1, 2.5, 4, 6 and 10 cm below the soil surface. Mesocotyl and coleoptile lengths measured on 14-day old seedlings incubated at 30 °C showed that semidwarf plant types had shorter mesocotyls at all planting depths and coleoptiles equal to or longer than the taller cultivars when planted 1, 2.5, 4, and 6 cm deep. For the 10 cm planting depth mesocotyl lengths were 7.9, 13.5 and 35.6 mm for the semidwarfs Bellemont, M-101 and IR 24, while the tall types, Labelle, Starbonnet and Red Rice, exhibited mesocotyl lengths of 83.5, 53.9 and 24.0 mm, respectively. Coleoptile lengths were 38.3, 47.1 and 63.2 mm for the same semidwarfs and 21.0, 53.9 and 62.7 mm for the tall plant types, respectively. Thus mesocotyl and coleoptile contribute differently towards the emergence problem of a specific variety.

Soil temperature also affects coleoptile length and thus the sowing depth. A study of Radford (1987) with eight wheat cultivars showed that the mean coleoptile length decreased from 10.8 cm at 15 °C to 3.1 cm at 35 °C which indicates that in warm soil shallow sowing is necessary for satisfactory establishment. Fluctuating temperatures (± 5 °C from the mean) further reduced coleoptile length. The temperature × cultivar interaction was significant, differences among cultivars in coleoptile length occurring more at low than at high temperatures. The mean coleoptile lengths of the eight cultivars varried from 8.7 to 14.5 cm at 15 °C, but only 2.7 to 3.6 cm at 35 °C. Two tall cultivars generally had longer coleoptiles than the six semidwarf. The tall wheats also had near maximum coleoptile length at a wide range of temperatures than the semidwarfs.

Pylnev et al. (1986), on the basis of the close correlation established between coleoptile length (a measure of field emergence) and the length of the second and the third internodes, proposed a method of selection based on the length of these internodes relative to total stem length : short lines are selected in which the second internode constitutes ≥ 10 per cent and the third internode ≥ 15 per cent of the total length.

High temperature affects fertility and grain filling adversely. Tests of the wheat varieties Bersee and Maris Huntsman showed that the fertility of near-isogenic lines carrying Rht_1, Rht_2 or Rht_3 is more sensitive to high temperatures between the flag leaf and ear emergence stages than that of their tall counterparts (Law et al., 1985). Further, Hoogendoorn and Gale (1988) studied near isogenic lines of the wheat varieties Maringa and Nainari 60, differing at Rht_1, Rht_2 and Rht_3 dwarfing loci and concluded that very short varieties may not be suited to environments where heat commonly shortens the grain filling period.

A study of Tsilke and Gerasimenko (1987) on two short, two medium and six tall wheat varieties differing in duration of the emergence to heading phase, and of 14 F_1 and F_2 hybrids involving them, growth rate varied according to genotype and growing conditions. Under severe drought, the tall Saratovskaya 29 had the highest growth rate. The F_1s generally had lower growth rates than the F_2s, a fact attributed to the effect of dominant genes inhibiting growth; the segregation of homozygotes with recessive genes partially offset this inhibitory effect in the F_2. But Dobachev and Germantsev (1987) have suggested that short stature and relative drought resistance could be combined in a single genotype. In fact, many dwarf genotypes have been improved for their rooting character and thus drought resistance (*see* Chapter on 'Root type' in this volume).

Aleshin et al. (1984) grew three rice varieties differing in height and salt tolerance – Spalchik (dwarf and tolerant), Solnechnyi (dwarf and moderately tolerant) and Krasnodar-skii 424 (tall and susceptible) at three levels of salinity. As the salinity level increased, accumulation of starch and sucrose in the stems at heading decreased in all varieties. However, the decrease was least in the tolerant dwarf (Spalchik) and greatest in the susceptible tall variety (Krasondarskii 424). Yields were depressed by 69.3, 82.4 and 90.9 per cent at the highest salinity level (0.5 per cent NaC1 by weight) in the dwarf tolerant, dwarf moderately tolerant and the tall sensitive varieties, respectively.

Rooting Pattern

Roots of most dwarf wheat varieties penetrate deeper than the taller varieties (Subbaiah et al., 1968; Lupton and Bingham, 1970). The semidwarf wheat varieties have more roots below 30 cm than the taller varieties suggesting a possible advantage in dry seasons (Welbank, 1974). At 40 to 60 cm depth, roots of semidwarf wheat lines are more extensive and absorb more phosphate (Lupton et al., 1974). The number of seminal roots and total root weight are positively correlated with 1000-grain weight and thus the yield (Tiwari and Tiwari, 1981). The dwarf wheat varieties have a larger root system (Bapu et al., 1971; Welbank and Taylor, 1971).

Although the degree of root system development is unrelated to plant height, studies of Netis et al. (1985) on the root system of winter wheats show that the roots of semidwarf varieties generally reach deeper into the soil and are better developed than those of the tall varieties. Plants with poorly developed root system had the lowest yields. Yield was related not only to total root weight, but also to the depth of the root penetration into the soil (*see* Chapter on 'Root type' in this volume).

The extent of the root system, and particularly the depth of rooting, is probably most important when water and nutrient supplies are limiting. The dwarf pearl millet cultivars have greater root : shoot ratio suggesting better stability under rainfed dry regions (Tongoona et al., 1984). Breeding for more extensive root system or increasing root length of the high yielding dwarf varieties is possible (*see* Chapter on 'Root type' in this volume.)

Growth and Development

Reduction in the height of dwarf barley plants is due to reduced cell number (Blonstein and Gale, 1984). Dwarfing genes exert their influence on plant height by inhibiting cell division and not by inhibiting cell elongation. Birnberg et al. (1987) concluded that the soybean variety T210 is unable to carry out GA-promoted cell division and that GA_3 had no effect on cell number in T 210 but increased cell numbers in Lincoln by 53 per cent. Later studies of Gale et al. (1986) with the lines of wheat varieties April Bearded and Maris Huntsman near isogenic for the dwarfing genes Rht_1, Rht_2 and Rht_3 indicated that the genes reduce plant height predominantly by reducing cell expansion.

A study of the internode development of rice Taichung 65 and its dwarf near-isogenic line T 65-d_2 showed that the rate of increase in cell prolification and elongation was lower in the dwarf line (Kamijima et al., 1985).

The main effect of the dwarfing gene is to increase the partitioning of assimilates to the developing ear, resulting in more fertile florets per spikelet and hence higher grain yields. The main yield contributing character 'grain number per ear' is increased due mainly to increased spikelet fertility as a result of greater assimilate availability. Ali et al. (1978) also observed that dwarf barley plants partitioned a greater share of assimilates to the grain than the tall plants. Also the dwarf rice varieties are less sensitive to low temperature-induced sterility than the closely related tall varieties (Rutger and Peterson, 1979). (*see* Chapter on 'Rate and duration of grain filling' in this volume).

REFERENCES

Aharoni, Y., C.T. Phan and M. Spencer, Ethylene production as related to dwarfness and earliness in plants. *Can. J. Bot.* 51: 2243–46, 1973.

Akmuradov, Sh. (Inheritance of low growth in *Gossypium barbadense*). Tilrkmenistan SSR Ylymlar Akademijasynyn Habarlary, Biologik Ylymlaryn No. 3: 80–82, 1979.

Aleshin, E.P., N.V. Vorobev and T.P. Zhurba. (The physiological reasons for the different levels of salt resistance in rice varieties). Doklady Vsesoyuznoi Ordena Lenina i Ordena Trudovogo Krasnogo Znameni Akademii Self skokhozyaistvennykh Nauk Imeni V.I. Lenina. No. 8: 3–5, 1984.

Alderov, A.A. (Inheritance of short straw following the hybridization of durum wheats. Byulleten Vsesoyuznogo Ordena Lenian i Ordena Druzhby Narodov Instituta Rastenievodstva Imeni N.I. Vavilova. No. 89: 27–31, 1979a.

Alderov, A.A. (The genetics of plant height in some tetraploid wheat species). *Idem.* No. 94: 76–78, 1979b.

Alderov, A.A. (Degree of dominance in respect of straw length in first generation hybrids of tetraploid wheats). *Idem.* No. 93: 15–17, 1979c.

Ali, M.A.M, S.O. Okiror and D.C. Rasmusson. Performance of semidwarf barley. *Crop Sci.* 18: 418–22, 1978.

Aliev, D.A. and E.G. Kazibekova. (Architectonics and photosynthetic function in high yielding wheat). *Fiziol. Rastenii*, 24: 962–68, 1977.

Alimov, F.M. (Interrelation between flag leaf area and grain weight from the main ear in different genotypes of durum wheat). Marunzalar. *Az SSR Elmlar Akad.* 36: 79–83, 1980.

Allan, R.E. Yield performance of lines isogenic for semidwarf gene doses in several wheat populations *In*: Proc. Int. Wheat Genet. Symp., 6th Kyoto Japan, 28 Nov. –3 Dec., Plant Germ - Plasm Institute, Kyoto Univ., Kyoto, Japan. pp. 265–70, 1983.

Allan, R.E., O.A. Vogel. F_2 monosonic analysis of culm length in wheat crosses involving semidwarf Norin 10, Brevor 14 and Chinese Spring series. *Crop Sci.* 3: 538–40, 1963.

Allan, R.E. and O.A. Vogel and J.R. Burleigh. Length and estimated number of coleoptile parenchyma cells of six wheat selections grown at two temperatures. *Crop Sci.* 2: 522–24, 1962.

Allan, R.E., O.A. Vogel and J.C. Craddock. Comparative response to gibberellic acid of dwarf, semidwarf, and standard short and tall winter wheat varieties. *Agron. J.* 51: 737–40, 1959.

Allison-Creese, K., C. Duchene and I.C. Murfet. Internode length in *Pisum:* the response of *lena* scions grafted to *Na* stocks. Ann. Bot. 55: 121–23, 1985.

Anand, S.C. and H.S. Aulak. Inheritance of semidwarfism in Olsen's dwarf (*Triticum aestivum* L.). Wheat Information Services, Kyoto, No. 32: 14–17, 1971.

Ancheva, M. and M. Vasileva. (Genetical and biochemical study of pea mutants. I. Activity and isozyme composition of peroxidase in stem mutant forms). *Genetika i Selektsiya*. 16: 73–84, 1983.

Babadzhanyan, G.A.; G.A. Saakyan and Zh. G. Khachatryan. (Features of the inheritance of plant height in crosses of short and long- stemmed varieties of *Triticum aestivum*). *Biologicheskii Zhurnal Armenii*. 27: 25–33, 1974.

Bai, S.H. (Studies on inheritance and ecological variation of culm length and related characters in short rice varieties). Research Reports of the Office of Rural Development, *Crop.* 15: 129–68, 1973.

Bapu, N., N.S. Reddy and J. Venkateswarlu. Active root distribution of four wheat varieties. *In*: Proc. Symp. on Radiation and Radioisotopes in Soil Studies and Plant Nutrition, Bangalore, India, December, 1970. Department of Atomic Energy, Bombay, India. pp.397–407, 1971.

Bingham, J. Physiological objectives in breeding for grain yield in wheat. Proc. 6th Eucarpia Congress, Cambridge, 1971. pp.15–29, 1972.

Birnberg, P.R., R.F. Coldero and M.L. Brenner. Characterization of vegetative growth of dwarf soybean genotypes including a gibberellin-insensitive genotype with impaired cell division. *Am.J. Bot.* 74: 868–76, 1987.

Blonstein, A.D. and M.D. Gale. Cell size and cell number in dwarf mutants of barley (*Hordeum vulgare*). *In*: Semidwarf Cereal Mutants and Their Use in Cross Breeding. FI. IAEF-TECDOC 307. Proc. Res. Coordination Meeting, Joint FAO/IAEA Division, Davis, California, Aug. 1982. VAEA, Vienna. pp. 19–29, 1984.

Börner, A. and G. Melz. Resonse of rye genotypes differing in plant height to exogenous gibberellic acid application. *Archiv. fur Zuchtungsforschung.* 18 (2): 71–74, 1988.

Brittto, M.R. and D.C. Cassalett. (The inheritance and heritability of plant height and its correlation with other characters in wheat (*Triticum vulgare* L.). Revista Instituto Columbiana. Agropecunario. 10: 255–68, 1975.

Brooking, I.R. and E.J.M. Kirby. Interrelationships between stem and ear development in winter wheat : the effects of a Norin 10 dwarfing gene. Gai/Rht$_2$. *J. agric. Sci.,* U.K. 97: 373–81, 1981.

Chary, S.N. and J.K. Bhalla. Peroxidase activity and isozymes in induced dwarf and tall mutants of pigeonpea (*Cajanus cajan* (L) Mill). *Legume Res.* 6: 74–78, 1983.

Chauhan, V.S. and J.P. Tandon. Inheritance of plant height in a cross of two cold tolerant rice varieties. *Int. Rice Res. Newsl.* 8 (4) : 3–4, 1983.

Chanterean, J. and C. Etasse. (Breeding dwarf populations of pearl millet (*Pennisetum typhoides* Staff) in Niger.) *Agronomie Tropicale.* 31: 254–57, 1976.

Cheema, A.A. and M.A. Awan. Performance of semidwarf mutants of Basmati 370. *Int. Rice Res. Newsl.* 13: 20–21, 1988.

Cooper, R.L. Development of short statured soybean cultivars. *Crop Sci.* 21: 127–31, 1981.

Chaudhry, A.S. A genetic and cytogenetic study of height in wheat. Ph.D. Thesis, Univ. of Cambridge, U.K. 1973.

Chowdhary, R.K. A bold seeded dwarf mutant of cowpea. *Tropical Grain Legume Bull.* No. 27: 20–22, 1983.

Das, K. and H. Kumar. Breeding behaviour of a dwarf mutant in 'Duke of Albany' pea. *Indian J. agric. Sci.* 44: 655–56, 1974.

Dechev, D.I. (Inheritance of plant height in durum wheat). *Rasteniev dni Nauki.* 22 (2): 15–20, 1985.

Dobachev, Yu. V. and L.A. Germantsev. (Short stature and drought resistance in spring bread wheat). *In*: Genetika, Fiziologiya i selektsiya zernooykh kultur na yugovostoke, Saratov, USSR: 39–45, 1987.

Dwivedi, S. and D.P. Singh. Inheritance of dwarf plant type in black gram (*Vigna mungo* (L.) Hepper). *Theo. appl. Genetics.* 70: 337, 1985.

Edwards, I.B.; R.C. Frohberg and E. L. Deckard. Nitrate reductase activity and nitrogen translocation in conventional height and semidwarf spring wheat varieties. *Agron. Abstr.* ASA, Madison, USA. p. 32, 1973.

Fick, G.N. and C.D. Qualset. Genes for dwarfness in wheat, *Triticum aestivum* L. *Genetics.* 75: 531–39, 1973.

Filev, K. Inheritance of culm height and grain yield in durum wheat. *In*: Semidwarf Cereal Mutants and Their Use in Cross Breeding. II. IAEA, Vienna, Austria. pp. 79–90, 1984.

Fischer, I.E. Developmental morphology of the inflorescence in the hexaploid wheat cultivars with and without the cultivar Norin 10 in their ancestry, *Can. J. Pl. sci.* 53: 7–15, 1973.

Gale, M.D. and R.S. Gregory. A rapid method for early generation selection of dwarf genotypes in wheat. *Euphytica.* 26: 733–38, 1977.

Gale, M.D., C.N. Law and A.J. Worland. The chromosomal location of a major dwarfing gene from Norin 10 in new British semidwarf wheats. *Heredity* 35: 417–21. 1975.

Gale, M.D. and G.A. Marshall. Insensitivity to gibberellin in dwarf wheats, *Ann. Bot.* 37. 729–38, 1973.

Gale, M.D. and G.A. Marshall. The nature and genetic control of gibberellin insensitivity in dwarf wheat grain. *Heredity.* 35: 55–65, 1975.

Gale, M.D. and G.A. Marshall. The chromosomal location of Gai_1 and Rht_1 genes for gibberellin insensitivity and semidwarfism in a derivative of Norin 10 wheat. *Heredity.* 37: 283–9, 1976.

Gale, M.D., A.M. Salter and J. Hoogendoorn. The effect of dwarfing genes of wheat on cell size and cell numbers. *In*: Annu Rep. Pl. Breed. Inst., 1985, Cambridge, U.K. (1986). p. 69, 1986.

Gale, M.D. and S. Youssefian. Dwarfing genes in wheat. *In :* Progress in Plant Breeding 1. G.E. Russell (ed), Butterworths, London. pp. 1–35, 1985.

Glukhovtseva, N.I. (Problems in breeding spring wheat for lodging resistance). *In*: Selektsiya Korotkostebeln pshenits, Moscow, USSR, Kolos. pp. 105–11, 1975.

Gu, M.H. and L.H. Zhu. (Genetic analysis of dwarfing genes in japonica rice). *Hereditas*, China. 3(3): 20–23, 1981.

Gupta, U.S. Production potential of dwarf genomes. *In*: Crop Physiology, U.S. Gupta (ed), Oxford & IBH Pub. Co., New Delhi. pp. 374–407, 1978.

Guseinova, V.M. (Genetic analysis of plant height in winter bread wheat varieties). Izvestiya Akademii Nauk Azerbaidzhanskoi SSR, Biologicheskie Nauki, NO. 1: 56–60, 1987.

Hansel, H. The use of the short straw mutant C_pB132 in breeding high yielding durum wheats in Australia. *Mutation Breed. Newsl.* No. 13: 2–4, 1979.

Harvey, D.M. and J. Goodwin. The photosynthetic net carbon dioxide exchange potential in conventional and 'leafless' phenotypes of *Pisum sativum* L. in relation to foliage area, dry matter production and seed yield. *Ann. Bot.* 42: 1091–98, 1978.

Hedley, C.L. and M.J. Ambrose. Designing 'leafless' plants for improving yields of the dried pea crop. *Adv. Agron.* 34: 225–77, 1981.

Hoogendoorn, J. and M.D. Gale. The effect of dwarfing genes on heat tolerance in CIMMYT germplasm. *In*: Proc. Conf. Cereal Sec. EUCARPIA, Wageningen, Netherlands, 24–26 Feb., 1988. pp. 61–66, 1988.

Hu, M.L. Genetic analysis of semidwarfing and insensitivity of gibberellin GA_3 in hexaploid wheat (*Triticum aestivum*(L) em Thell). Ph.D. Thesis, Washington State University. 1974.

Igoshin, A.P. and V.A. Kumakov. (Photosynthesis of individual organs during the grain filling period in different spring wheat cultivars). Trudy po Prikladnoi Botanike, Genetike i Selektsii. 72 (2): 40–45, 1982.

Jalalyar, M.S. Inheritance of insensitivity to gibberellin and of semidwarfing in *Triticum turgidum*(L) *durum*, Desf. Ph.D. Thesis, Washington State University. 1974.

Jha, M.P. and K.M. Singh. Identification of chromosome(s) carrying dwarfing gene(s) in *Triticum aestivum* 'Tom Thumb'. *Z. Pflanzenzuchtung* 79: 69–71, 1977.

Jha, M.P. and M.S. Swaminathan. Identification of chromosomes carrying the major genes for dwarfing in the wheat varieties Lerma Rojo and Sonora 64. *Curr. Sci.* 38: 379–81, 1969.

Jia, J.L. et al., (Response of Chinese semidwarf wheat varieties to gibberellic acid and a preliminary study of sources of dwarfness). Cf. Zuowu Pinzhong Ziyuan. No. 2: 18–20, 1988.

Johnson, E.C., K.S. Fischer, G.O. Edmeades and A.E. Palmer. Recurrent selection for reduced plant height in lowland tropical maize. *Crop Sci.* 26: 253–60, 1986.

Kamijima, O., Y. Takenaka and H. Nagasawa. (Character expression in a dwarf isogenic line of rice. II. The action of the dwarfing gene d_2 on cell proliferation and elongation in the parenchyma of the third internode). Science Reports of Fac. of Agric., Kobe Univ. 16: 377–84, 1985.

Karami, E. D.R. Kreig and J.E. Quisenberry. Water relations and carbon[14] assimilation of cotton with different leaf morphology. *Crop Sci.* 20: 421–26, 1980.

Kazaryan, M.Kh. (Complex wheat hybrids with different hybrid dwarfness phenotypes). *Biol. Zhur. Armenii.* 39: 345–48, 1986.

Khan, M.A. and S. Tsunoda. Differences in leaf photosynthesis and leaf transpiration rates among six commercial wheat varieties of west Pakistan. *Jap. J. Breed.* 20: 344–50, 1970.

Khangildin, V.V., V.I. Nikonov and I.F. shayakhmetov. (The use of *Triticum dicoccum* Schube in breeding durum wheat). Tr. Bashkir. NII S.kh. No. 10: 63–70, 1977.

Khmelev, B.I. and A.M. (Rozvadovskii. Inheritance of stem length in pea). Tsitologiya i Genetika. 18: 101–05, 1984.

Kim, N.I. and G.M. Paulsen. Assimilation and partitioning of photosynthate and nitrogen in isogenic tall, semidwarf, and double dwarf winter wheats. *J. Agron. Crop Sci.* 156: 73–80, 1986.

Kirillova, E.N. and N.I. Bocharnikova. (The relationship between auxins and growth inhibitors in dwarf and vigorously growing tomato plants). Izves tiya Akademi Nauk Moldavskoi SSR, Biologicheskikh i Khimicheskikh Nauk. No. 5: 73–74, 1986.

Kiss, A. Dwarfing of the hexaploid triticale. Cereal Res. Commu. 1: 35–44, 1973.

Konzak, C.F., M. Sadam and E. Donaldson. Inheritance of linkage in *durum* wheat of semidwarfing genes with low response to gibberellin A_3. In: Proc. Symp. on Genetics and Breeding *durum* wheat., Bari, Italy, May, 1973. pp. 29–40, 1973.

Kulshrestha, V.P. and S. Tsunoda. The role of Norin 10 dwarfing genes. *In:* Proc. 5th Int. Wheat Genetics Symp., New Delhi, Feb., 1978. pp. 988–94, 1981.

Kumar, C.R.A. and S.R. Rangasamy. Combining ability of dwarf lines of *indica* rice. *Oryza.* 21: 218–24, 1984.

Kumar, C.R.A. and S.R. Rangasamy. Diallel analysis for plant height in rice. *Egyptian J. Genet. Cytol.* 15: 223–30, 1986.

Kumar, H. A genetic study of plant height in garden pea. *Egyptian J. Genet. Cytol.* 5: 257–61, 1976.

Law, C.N., A.J. Worland and K.E. Stebbens. An effect of temperature on the fertility of wheat containing the dwarfing genes Rht_1, Rht_2 and Rht_3. *In:* Annu. Rep. Pl. Breed. Inst, 1984. Cambridge, U.K., (1985): 69–71, 1985.

Liang, G.H., E.G. Heyne, H.A. Cunnigham and R.E.H. Heiner. Peroxidase activities in sorghum and wheat differing in plant height. *Agron. Abstr.* Madison, U.S.A. p. 9, 1973.

Ling, T.N. and C.L.Li. Induction of short culm mutants for wheat improvement. *Mutation Breeding Newsletter.* No. 27: 8–9, 1986.

Lupton, F.G.H. Further experiments on photosynthesis and translocation in wheat. *Ann. appl. Biol,* 71: 69–79, 1972.

Lupton, F.G.H. and J. Bingham. Annual Rep. No. 34. Pl. Breed. Inst., Cambridge, England. 1970.

Lupton, F.G.H., J.Bingham, F.B. Ellis, B.T. Barness, K.R. Howse P.J. Welbank and P.J. Taylor. Root and shoot growth of semidwarf and taller winter wheats. *Ann. appl. Biol.* 77: 129–44, 1974.

Mardzilovich, M.I. (Inheritance of straw length in hybrid spring wheat plants). Vestsi A.N. BSSR, Ser, biyal. n. No. 3: 43–49, 1979.

Matsunaga, K., K. Maruyama and H. Ikehashi, Semidwarfism and varietal response to exogenous gibberellic acid in rice. *Oryza.* 18: 208–13, 1981.

McIntosh, R.A. Catalogue of gene symbols for wheat. *In:* Proc. 5th Int. Wheat Genetics Symp., New Delhi. S. Ramanujam (ed), Indian Soc. Genet. Pl. Breed., IARI, New Delhi. pp. 1299–1309, 1979.

Mellado, Z.M. (Response to nitrogen and phosphorus of tall and semidwarf wheat varieties. I. Variation in grain yield and its components). *Agricultura Tecnica.* 47: 152–59, 1987.

Meredith, W.R. Jr. Influence of leaf morphology on lint yield of cotton – enhancement by the subokra trait. *Crop Sci.* 24: 855–57, 1984.

Meredith, W.R. Jr. and R. Wells. Normal versus okra leaf yield interactions in cotton. I. Performance of near isogenic lines from bulk populations. *Crop Sci* 26: 219–22, 1986.

Meredith, W.R. Jr. and R. Wells. Subokra leaf influence on cotton yield. *Crop Sci.* 27: 47–48, 1987.

Mishra, R. and C.N. Chaubey. Modified mass selection for dwarf plant height in a composite mixture of sunflower. *Indian J. agric. Sci.* 57: 385–90, 1987.

Mohamed, S.N., A.W. Mohamed and R. Veeraswamy. Studies on the inheritance of certain plant characters in red gram (*Cajanus cajan* (L) Millsp.). *Madras agric. J.* 62: 64–65, 1975.

Morales, C. and G.L. Greene. Growth inhibitors in a radiation induced dwarf bean mutant. *Turrialba.* 22: 168–72, 1972.

Morris, R., J.W. Schmidt and V.A. Johnson. Chromosomal location of a dwarfing gene in Tom Thumb wheat derivative by monosomic analysis. *Crop Sci.* 12: 247–49, 1972.

Nalepa, S. and L. Pochaba. (Inheritance of culm length and stem length above the topmost node in hybrids between the winter wheat Kijow 2163/65 and the spring wheat Olsen Dwarf). Biuletyn Instytutu Hodowli i Aklimatyzacji Roslin. No. 41: 3–9, 1980.

Naskidshvili, P.P. (Obtaining short strawed forms as a result of the hybridization of wild and cultivated wheat species). Bull. Acad. Sci. Georgian SSR. 93: 683–88, 1979.

Nasr, M.I., S.H. Hassanein, A.K.A. Selim and M. El Hadeedi. Inheritance of dwarfness in five vulgare wheat crosses. *Alexandria J. agric. Res.* 24: 77–81, 1976.

Netis, I.T., P.P. Borovik, R.V. Morozov and S.A. Zaets. (Features of the development of the root system of semidwarf and tall varieties of winter wheat). *Oroshaemoe Zemledelie* No. 30: 28–31, 1985.

Ono, K. and F. Takaiwa. (Genetic analysis of homozygous mutations in regenerated rice plants). *Jap. J. Genet.* 58: 675–76, 1983.

Park, K.Y. Hormone content of normal and tall soybeans (*Glycine max* (L) Merr.). *Diss. Abstr. Int.B.* 38: 2989B, 1978.

Pearman, I., S.M. Thomas and G.N. Thorne. Effect of nitrogen fertilizer on photosynthesis of several varieties of spring wheat. *Ann. Bot.* 43: 613–21, 1979.

Piech, J. Monosomic and conventional genetic analysis of semidwarfism and grass-clump dwarfism in common wheat. *Euphytica* (suppl.). 1: 153–70, 1968.

Potts, W.C. and J.B. Reid. Internode length in *Pisum*. III. The effect of interaction of the *Na/na* and *Le/le* gene differences on endogenous gibberellin-like substances. *Physiol. Pl.* 57: 448–58, 1983.

Pylnev, V.V., A.V. Nefedov and M.G. Zakharin. (Efficacy of selecting short forms of winter wheat with a long coleoptile on the basis of stem structure). Nauchnotekhnicheskii Byulleten vsesoyuznogo Selektsionnogeneticheskogo Instituto. No. 3: 10–14, 1986.

Pyke, K.A. and R.M. Leech. Variation in ribulose 1-5-biphosphate carboxylase content in a range of winter wheat genotypes. *J. exp. Bot.* 36 (171) : 1523–29, 1985.

Quisenberry, J.E. Inheritance of plant height in cotton. II. Diallel analysis among six semidwarf strains. *Crop Sci.* 17: 347–50, 1977.

Radford, B.J. Effect of constant and fluctuating temperature regimes and seed source on the coleoptile length of tall and semidwarf wheats. *Aust. J. exp. Agric.* 27: 113–17, 1987.

Rao, M.V.P. The chromosomal location of a major gene for semidwarfism in the wheat variety Kalyan Sona. *Curr. Sci.* 49: 119–20, 1980.

Rao, S.A., M.H. Mengesha and C.R. Reddy. New sources of dwarfing genes in pearl millet (*Pennisetum americanum*). *Theo. appl. Genet.* 73: 170–74, 1986.

Ray, L.L. Breeding varieties for the broadcast method of cotton production. Proc. 17th Ann. Cotton Impr. Conf. pp. 89–92, 1965.

Ray, L.L. Breeding cotton varieties for narrow row production. Proc. Beltwide Cotton Prod. Res. Conf. p. 57, 1970.

Reddi, M.V. and E. G. Heyne. Inheritance of plant height and kernel weight in two wheat crosses. *Indian J. Genet.* 30: 109–15, 1970.

Rudenko, M.I. (Inheritance of plant height on crossing short-strawed and tall varieties of spring wheat). Sb. tr. NIISkh. tsenter. r-nov Nechernozemn, Zony No. 47: 140–46, 1979.

Rutger, J.N. and M.L. Peterson. Cold tolerance of rice in California. *In*: Report of a rice cold tolerance workshop. Los Banos, Legunna, Philippines, IRRI. and pp. 101–04, 1979.

Santos, J.B. and R. Vencovsky. (Genetic control of some components of plant height in French bean). *Pesquisa Agropecuaria Brasileria.* 21: 957–63, 1986.

Saxena, O.P. and G. Singh. Hormonal and enzymatic studies on dwarfism in wheat. *Narendra Deva J. Agric. Res.* 1: 79–87, 1986.

Schertz, K.F., D.T. Rosenow; J.W. Johnson and P.T. Gibson. Single Dw_3 height gene effects in 4- and 3- dwarf hybrids of *Sorghum bicolor* (L) Moench. *Crop Sci.* 14: 875–77, 1974.

Sears, E.R. The aneuploids of common wheat. *Missouri Agric. Exp. Sta. Res. Bull.* 572: 59, 1954.

Sharma, D. and H.C. Malik. Inheritance of dwarfing in wheat crosses involving standard varieties and a wheat-*Agropyron* translocation line. *Indian J. Genet. Pl. Breed.* 35: 100–08, 1975.

Sharma, G.S. and R.B. Singh. Inheritance of plant height and spike length in dwarf spring wheat. *Indian J. Genet. Pl. Breed.* 36: 173–83, 1976.

Shindin, I.M. and G.S. Karacheva. (A breeding and genetic analysis of plant height in spring wheat hybrids in the Far East). *Sib. vestn. s. kh. nauki.* No. 3: 28–34, 1979.

Singh, B.D., R. P. Singh and R.B. Singh. Endogenous gibberellins and amylase activity in tall and dwarf strains of rice (*Oryza sativa*). *Experientia.* 37: 363–65, 1981.

Singh, B.D. and Y. Singh. Dwarf mutants of rice (*Oryza sativa*) with partial block· in gibberellin utilization. *Biochem. Physiol. Pflanzen.* 117: 789–91, 1982.

Singh, V.P., E. A. Siddiq and M.S. Swaminathan. Mode of inheritance of dwarf stature and allelic relationships among various spontaneous and induced dwarfs of cultivated rice, *Oryza sativa* L. *Theo. appl. Genetics.* 55: 169–76, 1979.

Skorik, V.v. (Recurrent selection for short straw in winter rye). Selektsiya i Semenovodstvo, Ukrainian SSR No. 61: 21–30, 1986.

Song, Y., J. Harada and T. Tanaka. Gibberellin response of *japonica-indica* hybrids in Korean rice cultivars, *Int. Rice Res. Newsl.* 6 (3): 3, 1981.

Stelmakh, A.F. and G.P. Bondar. (Study of internode length in the semidwarf variety INIA 66 by means of monosomic analysis). Nauch-tekhn. byul. Vses. selektsgenet, in-ta No.1: 47–53, 1980.

Subbiah, B.V., J.C. Katyal, R.L. Narasimham and C. Dakshinamurti. Preliminary investigations on root distribution of high yielding wheat varieties. *Int. J. appl. Radiation and Isotopes.* 19: 385–90, 1968.

Sunderman, D.W. Seedling emergence of winter wheats and its association with depth of sowing, coleoptile length under various conditions, and plant height. *Agron. J.* 56: 23–25, 1964.

Tanaka, A. Efficiency of respiration. *In*: Rice Breeding. Int. Rice Res. Sta., Manila, Philippines. p. 738 and pp. 483-98, 1972.

Thorne, G.N. Distribution between parts of the main shoot and the tillers of photosynthate produced before and after anthesis in the top three leaves of main shoots of Hobit and Maris Huntsman winter wheat. *Ann. appl, Biol.* 101: 553–59, 1982.

Tietz, A. (Dwarfism and abscisic acid). *Biochem. Physiol. Pflanzen.* 174: 499–503, 1979.

Tiwari, J.P. and D.K. Tiwari. Root productivity of irrigated and rainfed wheat. *JNKVV Res. J.* 15: 5–11, 1981.

Tripathi, P. and R.N. Misra. Diallel analysis for plant height in dwarf and semidwarf rice. *Oryza.* 23: 27–31, 1986.

Tongoona, R. S.C. Muchena and J.K. Hendriz. The effect of dwarfing genes on root development in pearl millet inbreds and hybrids. *Zimbabwe J. Agric. Res.* 22: 67–83, 1984.

Tsilke, R.A. and I.I. Gerasimenko. (Genetic control of growth rhythm in plants of spring bread wheat under drought. I. F_1 and F_2 hybrids of short-strawed with medium strawed and tall varieties). *Genetika,* USSR. 23: 325–35, 1987.

Tupitsyn, N.Y. (The nature of dwarfness in wheat). Selektsiya i Semenovodstvo, USSR No. 3: 24–25, 1985.

Turner, F.T., C.C. Chen and C.N. Bollich. Coleoptile and mesocotyl lengths in semidwarf rice seedlings. *Crop Sci.* 22: 43–46, 1982.

Varenitsa, E.T., B.I. Sendukhadze, G.V. Kochetygov and E.A. Rogozhina. (Study of winter wheat hybrids from crosses of dwarf and short-strawed forms with long strawed hardy varieties). Sb. nauch tr. NII s. kh. tsentr.r-nov. Nechernozemn, Zony. No. 44: 39–52, 1978.

Vasilenko, I.I. (Breeding high yielding short-strawed wheat varieties). Vestnik Seleskokhozyaistvennoi Nauki No. 10: 102–07, 1977.

Vahiduzzaman, M. and M.S. Ahmad, Inheritance of plant height in seven crosses of rice. *Cereal Res. Commu.* 8: 527–32, 1980.

Waldia, R.S. and V.P. Singh. Inheritance of dwarfing genes in pigeonpea. *Indian J. agric. Sci.* 57: 219–20, 1987.

Wang, Y.c., X.Z.Xue, G.S. Tang and Q.Y.Wang. (Monosomic analysis of height in the wheat variety Albian i.) *Acta Agronomica sinica* 8: 193–98, 1982.

Welbank, P.J. Root growth of different species and varieties of cereals. *J.Sci. Food Agric.* 25: 231–32, 1974.

Welbank, P.J. and P.J. Taylor. Root growth of wheat varieties. Rothamsted Exp. Sta. Report. 1970: 95–97, 1971.

Wojcieska, U. and M. Slusarczyk. (The distribution of the products of photosynthesis in the stems of long and short-strawed winter wheats). *Acta Agrobotanica* 28: 263–73, 1975.

Worland, A.J. and C.N. Law. The genetics of hybrid dwarfing in wheat. *Z. Pflanzenzuchtung* 85: 28–39, 1980.

Worland, A.J. and S. Petrovic. The gibberellic acid insensitive dwarfing gene from the wheat variety Saitama 27. *Euphytica,* 38: 55–63, 1988.

Zeven, A.C. Tom Pounce Blanc and Tom Pounce Barbu Rouge, two *Triticum aestivum* sources of very short straw. *Wheat Information Service.* 29: 8–9, 1969.

CHAPTER 5

SHORTENING GROWTH DURATION: EARLINESS

INTRODUCTION

Early maturity and photoperiod insensitivity are highly desirable characteristics in grain crops, often extending their area of adaptation, permitting more than one crop per season and enabling them to mature grain in arid regions where later maturing cultivars would fail. Early maturing cultivars are also important with a view to reducing water and insecticide costs. The yield of such early maturing plants is generally low on per plant basis but is often higher on per unit area basis as the early plants become smaller and can be planted more densely (Lanza and Dionigi, 1974; Dionigi, 1975; Sharma and Ahmad, 1979; Costa, 1978; Zelenskii and Parkhomenko, 1980; Kitten and Schuster, 1980). One of the reasons of decrease in per plant yield is shortening in the duration of the reproductive phase, that is, flowering to physiological maturity, together with reduction in their vegetative phase. Our effort should be to cut down the vegetative phase with no reduction in the grain/seed-filling period for maximum partitioning and grain filling (Discussed in Chapter 6).

Shortening growth duration has several advantages, for example, escaping drought hazards occurring at the time of maturity (Kotova and Ryabtseva, 1980; Srivastava, 1983; Sanlescu et al., 1986; Gangadharan, 1986), escaping late arising incidence of pests (Kitten and Schuster, 1980) and diseases (Narula, 1983). Above all, short duration crops fit better in multiple or relay cropping systems aimed at obtaining maximum production per annum. Early forms of pigeonpea suit better in low rainfall areas and soils with poor moisture retention (Sharma et al., 1980). Efforts in this direction have been going on since long, but the breeders conversed on this important aspect of crop improvement only about two decades ago.

Samokhvalov et al. (1975) emphasized that in the early varieties of several crops (maize, spring and winter wheat, spring barley, oats, *Pannicum* millet, peas, sunflower), germination rate, accumulation of dry matter, photosynthesis, enzyme activity and absorption capacity of roots take place at a more rapid rate and for a shorter time than in their late forms. Dormancy in potato is also broken 30 to 40 days earlier in early and mid-early varieties than in the late ones (Becka, 1975).

For judging earliness of maturity in peanuts of four maturity groupings (very early, early, mid-season and late), Baily and Bear (1973) described four parameters to be studied, namely: (i) from sowing to the opening of the first flower, (ii) from opening of the first flower to the opening of a given number of flowers (15 to 30 per cent), (iii) from a flower's opening to seed maturity in the pod formed from this flower, and (iv) from maturity to the weakening of the gynophore. Using 12 varieties of cotton, Singh et al. (1978) emphasized that the most important components of earliness are number of days from sowing to first flowering, number of days from sowing to boll opening, position of the first fruiting node, and number of sympodia. The first two characters were said to have direct effects on earliness and the last were said to have mainly indirect effects. Pisarenko et al. (1985) determined that the relationship between earliness and water requirement of irrigated maize hybrids. Water requirement of mid-early and mid-season hybrids was longer (3374 to 4845, and 3432 to 4759 m^3/ha) and was higher for the mid-late and late season hybrids (3685 to 5389 and 4049 to 5452 m^3/ha).

VARIABILITY

Much is known about the variability in growth/maturity duration of all crops and specific examples need no mention. In short, out of 530 wheat forms studied by Rudenko and Dmitriev (1980), 1 per cent plants proved very early, 4 per cent early, 14 per cent mid-early, 70 per cent mid-season and 11 per cent late. Of 3,500 entries in the USDA world wheat collection, 350 were very early and photothermal insensitive (Qualset and Puri, 1974). Data on days to flowering and duration of heading (interval from start to completion of flowering) of early season rice (*aus*) varieties are available in a tabular form (Mitra and Biswas, 1977). Days to flowering ranged from 60 to 66 days to 123 to 129 days and the duration of heading from four to six days to 19 to 21 days. Teleutsa (1980) screened 369 varieties of soybean; their growth period ranged from 71 to 150 days. In a study of the growth period of 17 soybean varieties, Musorina and Natochieva (1986) reported that genotype accounts for 79.4 per cent and environment accounts for 20.6 per cent variation. These are just a few examples of three important food crops. Some varieties mature only in half the duration of the others but they could be as high as or slightly less yielders than their late counterparts. Thus there is scope of obtaining a yield to almost the same level, if not more, in a shorter period of time; thus avoiding the late arising hazards and allowing more crops to be harvested from the same land.

SELECTION

We have seen that so much variability in time of maturity between different crop varieties/populations occurs that it is possible to select early types and further improve the productivity of these early but less productive types. On this premise, much work has been done. For instance, most maize varieties from the tropic and subtropic regions exhibit excessive vegetative growth and delayed flowering when grown under the long days of temperate regions. If tropical and temperate zone varieties are crossed and early generation selection for fast maturity in the derived population is made, adapted temperate zone genotypes may be developed. Troyer and Brown (1972) selected three late maize synthetics for early silking by sibmating over a six generation period, and then compared their performance in the seventh cycle at three plant densities for two years. By selecting for early flowering 1 t/ha yield increase, 7.2 cm plant height decrease, 5.2 cm ear height decrease, and 1.8 days earliness in flowering was achieved for every cycle of selection (Tables 5.1 and 5.2).

Table 5.1: Agronomic performance of maize by cycles of selection. Three synthetics, three densities and two seasons. (After Troyer and Brown, 1972).

Cycle	Days to silk	Plant height (cm)	Ear height (cm)	Yield (q/ha)
0	77.9	338	155	54.6
1	77.8	325	150	54.2
2	74.4	318	147	56.8
3	72.4	310	137	60.2
4	70.7	307	135	59.8
5	69.5	307	132	60.2
6	67.3	295	124	60.7
Average/cycle	1.8	7.2	5.2	1.0

Table 5.2: Agronomic performance of seven late maize synthetics during six cycles of selection (After Troyer and Brown, 1976).

Cycle	Days to silk	Plant height (cm)	Ear height (cm)	Yield (q/ha)	Ears/100 plants
0	74.1	312	140	59.8	92
1	71.4	307	135	62.9	97
2	69.4	300	127	64.2	100
3	68.4	295	127	68.9	104
4	66.8	292	124	66.5	107
5	65.5	282	114	65.2	103
Average/cycle	1.7	6.0	4.2	1.1	2.2

The decrease in silking time and increase in ears per plant indicate that selection for early flowering during stress is effective in adapting materials to high plant densities. Flowering date responds readily to selection, and selection for early flowering is an effective way to adapting late maize varieties for cultivation in 'short season' areas. Improvements achieved per cycle of selection in various plant characters of maize are summarized in Table 5.3.

Avey et al. (1982a) carried out selection from a wheat population composed of 39 crosses among 12-, two-, three- and fourway hybrids involving 16 varieties of diverse flowering date. Selection in the first cycle acted mainly on major genes which showed some non-additive effects while selection in the other cycles acted on minor genes governed by additive and/or additive × additive effects. Selection for earliness resulted in increased culm length and the number of ears per plant but a reduction in grain yield (Lee, 1984). Path coefficient analysis indicated that grain yield was positively correlated with the number of

Table 5.3: Improvements achieved per cycle of selection in maize

Plant characters	Improvements achieved	Authority
Decrease in silking	1.8	Troyer and Brown, 1972.
time (days)	1.7	Troyer and Brown, 1976.
	0.6 – 1.2	Troyer, 1976.
	1.8 – 2.2	Fonturbel and Ordas, 1981.
Decrease in plant height (cm)	7.2	Troyer and Brown, 1972.
	6.0	Troyer and Brown, 1976.
	7.0	Troyer, 1976.
Decrease in ear height (cm)	5.2	Troyer and Brown, 1972.
	4.2	Troyer and Brown, 1976.
Grain yield (q/ha)	1.0	Troyer and Brown, 1972.
	1.1	Troyer and Brown, 1976.
	3.4	Troyer, 1976.
Ears per 100 plants	2.2	Troyer and Brown, 1976.

ears per plant, the number of grains per ear and 1000-grain weight in early populations. Skovmaand and Fox (1983) obtained 14 days earliness in maturity by selecting early segregates from F_2 and F_3 populations of triticale. Some of the triticale selections were earlier and yielded up to 15 per cent more than the early wheat cultivars Sonalika and Inia 66. However, crossing rye selected for earliness with early wheats did not give early triticales. Srivastava (1981) selected three strains of rice with a high yield potential (CR 289-1008, CR 289-11414 and CR 289-1208) from the cross AR 12422 × ARC 12751, which flowered in 70, 75 and 77 days, respectively.

Much more spectacular results were reported by Aylmer and Walsh (1979) with broad bean (*Vicia faba*). One cycle of mass selection for earliness in a mixed population of 33 accessions reduced the time to ripening by six days and the height by 18 cm without affecting yield. Earliness in maturity resulted from flowering at a lower node and at an earlier date without extension of the reproductive phase. They concluded that mass selection in normal intermediate populations is likely to be the more rewarding approach in breeding for earliness.

In French bean, selection for early maturity from F_2 is effective (CIAT, 1978a). Rapid progress was obtained by using a modified bulk-pedigree method. A few intermediate segregates were as early as or earlier than the determinate sources of early maturity used as parents and showed higher yields and greater stability.

Of late, some fast selection criteria have been developed. For instance, in rice, Chomoneva et al. (1984) determined peroxidase activity in the roots and above ground organs on the fifth and the tenth day of development in four varieties differing in earliness. The early varieties had a higher activity than the late ones. Also the analysis of DNA extracted from three-day old seedlings of cotton (*Gossypium barbadense*) varieties S 6030, 5904 I and 8763 I revealed varietal differences in the content of 5-methyl cytosine, with early variety S 6030 having the highest content and the late 8763I the lowest (Guseinov et al., 1985). The root cells of the early variety, 5904 I, divided and increased in size more rapidly and had a higher rate of elongation and a more active protein synthesizing system than those of the other varieties (Yurtaeva, 1986). Yurtaeva suggested that a study of root cells would facilitate early screening for growth period in breeding work.

In cucurbits, selection for earliness should be carried out at the bud stage of the female flowers, followed by removal of all late flowering plants. Hybridization between selected early forms is effective. Shmanaeva and Lebedeva (1987) report that under controlled climatic conditions it is possible to evaluate earliness in seven-day old seedlings of varieties and hybrids according to the length of the hypocotyl, due to high positive correlation between hypocotyl length and leaf area at the transplanting stage, a large leaf area leading to early yields. Seedlings with long hypocotyls (8 to 10 cm) flowered and fruited five to 29 days earlier than those with short hypocotyls (1.5 to 4.5 cm).

Prudnikova et al. (1987) developed a formula for determining earliness in potato by spectrofluorographic analysis of leaves for chloprophyll and protochlorophyll contents when taken from the field and then again after a period in the dark. The breeding forms evaluated exhibited characteristic levels of protochlorophyll accumulation in the dark. The greatest differences between genotypes were observed at field emergence, and the leaf samples were taken at this stage.

GENETIC STUDIES

Several genes are known to control earliness in flowering/fruiting/maturity of different crops. They may have dominant, additive, additive × additive or epistatic effects. For instance, in wheat, Heakal (1975) indicated that at least 14 chromosomes influence heading date. Ten genes have comparatively a major effect and four have a minor effect on days to ear emergence (Halloran, 1975). However, Pirasteh (1974) reports that earliness is controlled by three genes with dominant epistasis of one of the genes.

Genetic variation in the date of ear emergence is mainly additive, highly heritable and under polygenic control (Paroda et al., 1972; Edwards et al., 1976). Dominance and intermediate effects are most important; additive effects significant in a few hybrids, and epistasis has a significant effect (Botezan et al, 1978, Table 5.4). The genotypic variances were estimated and partitioned into additive, dominance and additive × additive components (Avey, 1981; Avey et al., 1980, 1982b). Additive effects were significant in the cross Aobachomugi × Double Crop, additive and dominance effects were significant for Aobachomugi × P 6879 and additive × additive effects were significant for Double Crop × P 6879. Dominance may be of some consequence in specific crosses, but in general, the fixable additive and additive × additive components are of primary importance.

In general, in the cereal crops, the heading date has high heritability (Bhatt, 1972; Johnson and Frey, 1967). According to Takahashi and Yasuda (1971) several distinct physiological complexes affect the time of heading. These include photoperiodic responses and vernalization requirements, as well as other factors which are not well understood. In wheat, additive gene action accounts for a large amount of the variation for days to heading (Crumpacker and Allard, 1962; Amaya et al., 1972; Bhatt, 1972) but dominance also is important (Biffen, 1905; Crumpacker and Allard, 1962). Epistasis reported in several studies (Amaya et al., 1972; Crumpacker and Allard, 1962) appears environment and cultivar specific.

Pinthus (1963) reported that in wheat, earliness in spike initiation is dominant over lateness but the short period from initiation to heading is dominant over long period. Of the eight loci found to control heading date in wheat progeny, seven show dominance for earliness and one shows dominance for lateness (Avey and Ohm, 1979).

Table 5.4: Inheritance behaviour of growth duration in different crop species.

Crop species	Inheritance behaviour	Authority
Wheat	Overdominant early	Lubmin, 1974.
		Kolomiets, 1979.
	Dominant early	Pinthus, 1963.
		Postrigan, 1976.
		Kolomiets, 1979.
	Intermediate early	Anwar and Chowdhry, 1967.
		Akhmetov and Fedorov, 1974.
		Lubmin, 1974.
		Botezan et al., 1978.
		Kolomiets, 1979.
		Al-Saheal, 1985.
		Singh et al., 1987.
	Dominant late	Lubmin, 1974.
		Kolomiets, 1979.
	Over-dominant late	Lubmin, 1974.
Barley	Partially dominant early	Andriyash, 1983.
	Dominant early	Goldenberg, 1979.
	Intermediate early	Prokofeva et al., 1980.
	Dominant late	
Rice	Overdominant late	Jun et al., 1986.
Triticale	Dominant early	
	Intermediate early	Georgieva, 1980.
	Dominant late	
Maize	Partially dominant early	Fachs, 1973.
		Daniel et al., 1974.
		El-Ghawas et al., 1978.
Broad bean	Dominant early	Garmash, 1977.
Soybean	Partially dominant early	Gilioli et al., 1980.
Pigeon pea	Partially dominant early	Dahiya and Satija, 1978.
Cotton	Dominant early	Al-Rawi and Kohel, 1969.
		Wilson and Wilson, 1976.
Sunflower	Over-dominant early	Zazharskii, 1978.
Tomato	Partially dominant early	Johnson and Hernandez, 1980.
	Dominant early	Egiyan and Lukyanenko, 1979.
Cucumber	Partially dominant early	Miller and Quisenbery, 1976.
Musk melon	Partially dominant early	Chadha and Nandpuri, 1978.

Following reciprocal crosses involving early (E), midseason (M) and late (L) wheat varieties, complete dominance of earliness was seen in the $E \times E$ crosses, partial dominance of earliness in the $E \times M$ crosses, values intermediate between parents in most $M \times E$ crosses, partial dominance of earliness in the $E \times L$ crosses and partial dominance of lateness in the $L \times E$ crosses (Savchenko, 1983). In the F_2, transgressive forms earlier than the early parent segregated, together with forms equalling the early parent forms intermediate in growth period between the parents and forms equal to or later than the late parent. The highest percentage of early transgressive forms was obtained from $E \times E$ crosses in both F_2 and F_3. When the F_1 and F_2 of a diallel cross involving eight parents were scored for date of ear emergence (Al-Saheal, 1985; Singh et al., 1987), earliness was partially dominant over lateness. The expression of date of emergence was largely governed by additive gene effects, but there was also evidence for non-additive gene effects. The F_1 showed heterosis towards earliness. Estimates of heritability were high. Singh et al. (1987) report that a single major gene governs the time of ear emergence, earliness being partially dominant over lateness, with high narrow sense heritability.

In barley, there is a partial dominance for early flowering, 10 to 30 per cent dominant and 70 to 90 per cent recessive alleles were obtained (Fischer, 1975). Heading date is controlled by a polygenic system with additive effects (Tuberosa et al., 1986) while the flowering time is controlled by both major and minor genes (Sharaan et al., 1982). Frequency distribution of parental, F_1, F_2 and backcross generations indicated control of heading date by three major loci with recessive and duplicate dominant epistasis accounting for most of the phenotypic variability (Gallagher et al., 1987). Recessive gene interaction resulted in plants that were about 19 to 35 days earlier than later counterparts with the dominant allele present. Of the 78 hybrid combinations made by Bugaev and Lukyanova (1984), 56 per cent were as early as or earlier than their parents, 23 per cent were intermediate and 21 per cent were later maturing. If the maternal parent was early, the hybrid progeny was more likely to be early or earlier. The early progenies were higher yielding than the late ones (Tuberosa et al., 1986). The photosensitive recessive gene ea_k, located on chromosome five controls extreme earliness in Japanese barley varieties while the recessive gene ea_c controls extreme earliness in Chinese varieties (Yaruda and Hayashi, 1981). Compared with Ea_k or Ea_c lines, lines carrying the recessive genes were about two weeks earlier. Gene ea_k increased the number of ears per plant and ea_c increased the 1000-grain weight.

Studies of McKenzie et al. (1978) indicated that in rice early heading is controlled by a single partially dominant gene while Mahmoud et al. (1984) indicated additive, dominance and dominance × dominance effects. Prasad et al. (1984) further indicated that earliness in flowering is controlled by three dominant genes. The gene E^a advanced panicle initiation by nine days while the effect of E^b was slightly weaker, but both genes also shortened the period from panicle initiation to heading by three to six days (Tsai, 1974). Thus E^a advanced heading by 13 to 14 days and E^b by 11 to 12 days. The earliness genes E^a, E^b and E^γ, E^x (radiation-induced) are all at the same locus and promoted heading in T 65 by a mean of 10 days (Tsai, 1980).

Early flowering in rice is dominant to late both under short and long day environments, but the varieties flowering early in one environment were late flowering in the other (Ganashan and Whittington, 1976). Their data indicated a digenic control of early flowering in short day with complementary interaction.

In maize, additive and additive × additive gene effects for the number of days to silking have been observed (Daniel, 1972; Sharma et al., 1972). Daniel observed occurrence of four partially dominant loci. Earliness was partially dominant over lateness and the additive gene effects and epistasis were important (El-Ghawas et al., 1978). In combinations of early × midseason varieties, length of the vegetative phase was controlled by a minimum of two or three genes, and in early × late combinations by four or five genes. Three to four times as many genes were involved in the control of the period up to flowering (Fachs, 1973). The earlier parent was partially dominant.

Kumar et al. (1982) performed combining ability and component analyses of data on days to heading and days to maturity of pearl millet from a 15 × 15 diallel cross. Although additive and non-additive genetic variance was important for both traits, non-additive variance predominated. They concluded that array means can be used for selection purposes.

In dry peas (*Pisum sativum*), days to flowering and maturity are under the control of additive gene action, although substantial dominance has also been observed for maturity (Kumar and Das, 1975; Dobias, 1974; Ranalli, 1982; Venkateswarlu, 1982). Although Venkateswarlu showed the importance of both additive and non-additive gene effects for number of days to flowering, additive gene effects predominated. Since narrow sense heritability is high and additive gene action predominates selection of suitable parental material may result in rapid genetic advance for earliness (Kumar and Das, 1975). However, Ranalli (1982) has shown that lateness was partially dominant over earliness and the maternal effects were significant. Earliness is controlled by a single recessive gene (Kaul, 1980).

In pigeonpea, flowering time is controlled by at least three major genes (Byth et al., 1983). Venkateswarlu and Singh (1982) indicated predominance of additive gene effects for the number of days to first flower opening. They suggested that crosses involving parents with diverse maturities and high specific combining ability for earliness may give early maturing high-yielding segregates.

In chickpea, Gupta and Ramanujam (1974) observed that days to flowering were conditioned by additive or non-additive gene actions. But in cowpea, flower initiation is governed by additive genetic variance which is highly heritable (Tikka et al., 1976). The variety Pusa Phalguni which displays complementary gene action for earliness was more useful in breeding than EC 16938 which owes its earliness to an accumulation of dominant genes. In *Phaseolus*, earliness in two of the three donors was controlled by a single dominant gene with no maternal effects, while in the third parent it was controlled by two or more genes with evidence of partial dominance and maternal effects (CIAT, 1978b).

In case of mustard, appreciable additive gene action on the flowering pattern was noted by Yadav et al. (1976). Olivieri and Parrini (1979) indicated that in rapeseed, earliness was dominant over lateness. Additive and dominance effects were present and a single gene appeared to condition the traits. However, Pal and Singh (1981) recorded additive and non-additive genetic components for a number of days to first flower and 50 per cent flowering in rapeseed. For the number of days to seed maturity, only non-additive gene action was observed. In another oil seed crop, sesame, overdominance was detected (Yermanos and Kotecha, 1980) and selection for flowering time could be very effective. In sunflower, however, length of the growth period from sowing to inflorescence initiation is controlled polygenically while the period from inflorescence initiation to the beginning of floret opening shows non-genic dominance of longer duration (27 days) over shorter duration (two days) (Machacek, 1979).

Flowering duration in cotton is controlled by additive gene action (Chinnadurai et al., 1973). They suggested that for combining earliness with high yield, MCU 5 should be crossed with K 3400 or KK 1543. Popov (1986) indicated polymeric gene interaction for earliness with isomeric interaction accounting for families exceeding parental limits. Pulatov (1986) observed high combining ability for earliness and developed early lines with economically useful traits.

In crosses between early forms of potato, 43.3 per cent of the hybrid progeny were early while in crosses between late forms it was only 3.4 to 3.7 per cent (Nazar, 1983). Earliness combined with other economically useful characters was found in 10.2 to 24 per cent of hybrids. In tomatoes, additive and dominance effects were significant for the number of days to most growth stages, and additive gene variance was greater than dominance variance (Li, 1976; Gibrel, 1983). Also Li recorded evidence for cytoplasmic effects on earliness. Dhaliwal and Nandpuri (1986) crossed the early tomato varieties Balkan and Cold Set with the main season Punjab Chhuhara. Both additive and non-additive genetic effects were significant for days to first ripe fruit, weight and number of early fruits and total early fruit yield. Genetic control of early yield appeared to be simpler in the cross involving Balkan than in that involving cold Set. Transgressive segregation for earliness was observed in the F_3, and exploitation of this by inbreeding, or by intercrossing and restricted recurrent selection, is recommended.

In cucumber, dominance effects are important in the inheritance of time to flowering (Miller, 1975). Further, earliness and yield in water melon and melon are controlled by genes with additive effects (Prosvirnin, 1978).

In addition to the gene action described above, further data on combining ability, heritability and heterosis for earliness are summarized in Table 5.5, and on loci controlling earliness in wheat are given in Table 5.6.

Table 5.5: Some studies on combining ability, heritability and heterosis for earliness in different crop species.

Crop species	Character studied	Authority
	Combining ability for Earliness	
Maize	GCA, high	Daniel et al., 1974
Rice	GCA, high in varieties Kanchi Karuna and Krishna	Ranganathan et al., 1973
	SCA, high in Karuna × 7447 and Kanchi × Krishna	
Pigeonpea	GCA, high	Venkateswarlu and Singh, 1982
Rapeseed	GCA and SCA, high	Olivieri and Parrini, 1979
Cotton	GCA, high in F_2 and F_3	Pulatov, 1986
	SCA, high in F_1	Radwan and El-Zahab, 1974
	Heritability of Earliness	
Wheat	Broad sense, high — 84.1%	
	Narrow sense — 68.3%	Al-Saheal, 1985
Barley	Broad sense in F_1 - F_2 (0.799 and 0.997)	Sharaan et al., 1982
	Narrow sense in F_1 - F_2 (0.879 and 0.920)	
Peas	Narrow sense, high	Kumar and Das, 1975
		Venkateswaralu and Singh, 1982
Cucumber	Broad sense, high	Miller, 1975
	Heterosis for Earliness	
Maize	11 lines, marked heterosis	Daniel, 1972
Barley	In hybrids from ecologically and geographically distinct varieties	Abramova and Kharionovskaya, 1976
Tomato	35.1% heterosis (Both parents were early but differed markedly in morphological characters and origin)	Oganesyan, 1971
		Kurganskaya and Agentova, 1975
	V 729 × Cross 525	
	Cross 525 × Sort 123	
	Sort 123 × Podarok 105	Egiyan and Lukyanenko, 1979

Table 5.6: Loci controlling earliness in wheat.

Loci	Description	Authority
Chromosome 2D	Carrier of the major gene with dominant epistasis effect	
Ch. 2B and 4B	Carrier of two other genes	Pirasteh, 1974
Ch. 5A, 7A, 7B, 3D and 6D	Carry minor modifying factors	
Ch. 5A	Affect ear emergence in Chines Spring	Snape et al., 1976
Ch. 5A and 5D	Affect ear emergence	Law et al., 1976; Ilina, 1981
Ch. 5A, 5D, 3B and 6D	Affect ear emergence	Dylenok and Yatsevich, 1978
Ch. 3A and 3D	Affect ear emergence	Khatyleva and Shavyalukha, 1978
Ch. 5D, 3B, 5A, 5B, 2D and 1D	Affect ear emergence to heading period in var. Avrora	
Ch. 5D, 5A, 4B, 2B, 5B and 7D	Affect ear emergence to heading period in var. Conche × *Agropyron elongatum*	Ganeva and Bochev, 1987

BREEDING WORK

F_1 hybrids from a cross between early and late wheat varieties were closer to the earlier parent in heading date (Abdel and Mokhie El' din, 1975). In the F_2 most of the segregates were closer to the late parent. $F_1 - F_3$ reciprocal hybrids from the cross San Pastore × Bezostaya (early × midearly), Akerman 804 × PPKh 186 (midearly × late) and San Pastore × Triamph (early × early) were studied by Boyadzhieva (1976). The crosses between early and midearly gave intermediate hybrids closer to the early parent. Crosses between midearly and late forms gave intermediate hybrids closer to the late parent. The earliest hybrids were obtained by crossing the two early varieties, which headed two to four days before their parents. Similarly, the variety Carmen obtained from Australia 57-79 × Bezostaya 1 is seven to eight days earlier than the early parent Bezostaya 1 (Ceapotu; 1976).

The study on P_1, P_2, F_1, F_2, BC_1 and BC_2 by Botezan et al. (1978) on nine varieties of wheat indicated that the F_1 hybrids tended to be earlier than the parents and that the BCs of the F_1 to the early parent produced early hybrids and those of the late parent late hybrids. F_2 values were intermediate between those of the F_1 and the BC parent. Wheat varieties Sonalika, Sonora 64 and Hira are good combiners for earliness, Hira × Sonora 64 being the best under high fertility and Hira × Sonalika being the best under low fertility (Jatasra and Paroda, 1980). Similarly, Amaya et al. (1972) studied the P_1, P_2, F_1, F_2, F_3, BC_1 and BC_2 of four crosses between seven durum wheat varieties for heading date (Table 5.7). The F_1 hybrids were intermediate or as early as the early parent.

Table 5.7: Mean performance of the parental , F_1, F_2 and F_3 bulks BC_1 and BC_2 generations for heading date in four crosses of durum wheat (After Amaya et al., 1972).

Generation	Cross 1 Longdon (P₁) × 61–130(P₂)	Cross 2 Akmolinka(P₁) × 61–130(P₂)	Cross 3 Cappelli(P₁) × Leeds (P₂)	Cross 4 Akmolinka(P₁) × Wells (P₂)
	Date headed, days from July 1			
P_1	14.8	19.5	20.7	20.2
P_2	13.4	13.8	14.4	15.9
F_1	13.4	14.7	17.3	17.6
F_2	13.6	14.9	16.1	16.9
F_3	13.6	15.5	16.1	16.5
BC_1	14.1	16.2	18.6	18.7
BC_2	13.4	14.2	15.4	16.4

In the crosses between ecologically and geographically distinct barley varieties, 5 to 22 per cent segregates showed transgression for earliness (Abramova and Khariouovskaya, 1976). The *Avena sterilis* lines CI 8077 and PI 295932 are later than *A. sativa* lines Garland, Portal, Jaycee, Proker and X 1047-3 by two to 30 days. The F_1 plants from each of six crosses were earlier than *A. sativa*, ranging from two days for Jaycee × CI 8077 and X 1047-3 × PI 295932 to 16 days for Jaycee × CI 8077. All crosses produced varying numbers of F_3 lines which were earlier than *A. Sativa* (Lyrene and Shands, 1975).

Cross (1978) developed a short duration yellow dent line ND100 from the cross W129 × W128. Dunn and Beck (1981) made crosses between op × op and broad-based synthetic × white inbred maize varieties. Some of the strains obtained are three weeks earlier than the currently available strains. An interesting observation was made by Menzi (1981). Of the two dent inbreds, CH 54 invariably conferred greater earliness on its progeny than did W 401, although W 401 is itself earlier than CH 54.

Rozvadovskii (1973) suggested the possibility of selecting early genotypes of dry peas from the segregating F_2 population from a cross between early and late varieties. When late × early and late × midearly varieties were crossed, growth duration was intermediate in the F_1 and transgressive segregation occurred in the F_2. In the F_3, four groups were found: (a) earlier than the early parent, (b) as early as the early parent, (c) close to the late parent, and (d) later than the late parent (Kalaidzhieva, 1978). Groundnut varieties Gangapuri and MH₂ (both Valentia types), Chico (Spanish type) and Robut 33-1 (Virginia bunch type) were crossed as donors of earliness with four Spanish bunch type varieties in a line × tester fashon (Basu et al., 1986). Chico was the best general combiner for days to 50 per cent flowering, days to maturity and number of mature pods per plant. Basu et al. recommended Chico, Robut 33-1 and Gangapuri as parents for breeding early maturing, high yielding varieties. Several pigeonpea hybrids produced were earlier in maturity than their early parent T 21 and produced more seed per day of growth than the higher yielding parent (Gupta et al., 1983). From a study on black seeded tropical *Phaseolus* varieties, scientists at Colombia (CIAT, 1978b) have concluded that selection for earliness could be practiced from the F_2 using a combined pedigree-bulk method. They observed transgressive segregation for days to physiological maturity.

Garlyev (1979) made reciprocal crosses between early cotton varieties (S 4727 × S 5306), late varieties (133 × Acala 4-42), and early and late forms (S 4727 × Acala 4-42). F_1 hybrids between early forms ripened six to eight days earlier than their parents and F_2 hybrids three days earlier. F_1 and F_2 hybrids between late forms equalled their parents in ripening date or were one to three days earlier. Narayanan et al. (1987) performed three cycles of disruptive crossing and selection from F_2 onward. Growth period to maturity was reduced by 45, 35 and 20 days in three different hybrids.

Kurganskaya and Agentova (1975) studied 370 tomato hybrids and observed that 23.7 per cent were superior to the standard variety in early and total yield, 33.7 per cent only in total yield and 20.2 per cent only in early yield. The remaining hybrids were close to one of their parents in both total and early yields. In the number of days from emergence to flowering the hybrids tended to be closer to the early parent, but early flowering did not always bring a high early yield. In order to obtain F_1 populations with higher early yields than the parental varieties, a parent which bears a large number of early fruits should be crossed with one which bears heavy fruits (Lopez-Rivares and Cuartero, 1985). In potatoes, the promising hybrids obtained by crossing the early variety Vostok with the late Oliev, exceeded the approved varieties in early and total yield (Zhuk, 1980).

Mutation Breeding

Mutations caused by radiation or chemical mutagens have often resulted in desirable populations. Efforts have also been made to produce short duration varieties through mutagenesis. Hsieh and Chang (1975) treated seeds of five japonica rice varieties with X-rays and thermal neutrons. In general, mutants showed wider variation in the number of days to heading, photoperiod sensitivity and thermosensitivity than the original varieties. When rice variety Kwangluai 4 was γ-irradiated, three homozygous early maturing mutants were obtained (Gao, 1981) which were backcrossed to Kwangluai 4. The F_1 and F_2 data indicated that all the three mutant lines resulted from simple recessive gene action. Mallick and Bairagi (1979) treated the grains of dwarf rice IR 8 with ethylmethane sulphonate and selected a mutant plant in M_2. In replicated trials of the M_5 and M_7 generations, the mutant line was significantly earlier to flower. Carmona (1986) derived several mutants which matured four to 27 days earlier than the parent varieties by gamma irradiation of the semidwarf rice varieties. The mutant 410MU30 matured 10 days earlier than its parent but equalled in yield.

Sixty-one early maturing mutants were obtained from the barley variety Chikurinibaraki 1 (Ukai and Yamashita, 1979). Of these, 31 were obtained after treatment of seeds with γ-rays, five after treatment with thermal neutrons and 25 after treating the seeds with ethyleneimine, ethylmethane sulphonate or N-nitroso-N-methylurea. Chemical mutagens were less efficient at inducing mutation for earliness than were γ-rays and thermal neutrons. Goldenberg (1975) studied the inheritance of earliness in induced early mutants of barley. The early mutants M 93 and Mari were crossed. The mutant gene of M 93 was epistatic over that of Mari, resulting in F_2 segregation of nine late, three early and four very early plants as revealed by F_3 progeny tests. Epistasis contributed to 40.26 per cent of the genetic effects while 32.58 per cent resulted from additive effects and 27.15 per cent from dominance. Shevtsov (1985) obtained the mutant 54 M 17 from the barley cv. Regia after grain treatment with N-nitroso-N-ethylurea. Two of the lines selected showed extreme earliness of heading, good yield, and high lodging resistance.

Sun et al. (1986) derived the mutant wheat Longfumei 1 in the F_4M_3 following treatment of the F_1 of X in 3 × Lio with thermal neutrons, which is very early to mature (70 to 75 days from emergence), has 15 to 18.2 per cent protein, 0.38 to 0.40 per cent lysine and yields 3 to 4.5 t/ha.

Treatment of maize grains of line W23 with 0.1 per cent solution of 1,4-diazoacetylbutane produced mutants maturing 12 to 15 days earlier than the initial form (Morgan et al., 1972). However, light-impulse irradiation of the undeveloped tassels gave rise to a form ripening 27 days earlier than the normal (Balaur, 1972). Hanna and Burton (1985) treated seeds of pearl millet cv. Tift 23B with ethylmethane sulphonate to give e_1 and with thermal neutrons to give e_2 early mutations. Both mutations are controlled by single recessive genes. Plants with e_1e_1 and e_2e_2 genotypes flowered 49 and 38 days after planting, respectively. Tift 23B flowered after 76 days.

The early mutants EM_1 and EM_2, obtained by X-irradiated seeds of dry pea var. Bonneville with a 5 Krad dose were respectively 40 and 31 days earlier than their parent and also had higher seed yield and higher protein content (Kaul, 1980; Kaul et al., 1979). Reciprocal crosses and backcrosses of these mutants with their parent indicated that earliness in each was controlled by a single recessive gene. In nine soybean varieties treated with γ-rays, the frequency of early mutants in the M_2 was 0.28 per cent. The range of variation in early maturity in the M_3 was greater than in M_2 (Lu, 1981).

New short forms of cotton (*Gossypium barbadense*) with very early ripening were produced by means of radiation mutagenesis combined with intervarietal crosses (Tyaminov, 1980). The same mutant TL2 obtained by treatment of seeds of the variety Neelum with ethylmethane sulphonate was earlier and shorter. Although its 1000-seed weight was lower, its seed yield and oil content were superior (Nayar, 1979).

PHYSIOLOGICAL BASIS OF EARLINESS

In a study of 26 wheat varieties, Zelenskii et al. (1979) noted that the earlier varieties generally had a higher photochemical activity. An inverse correlation was found between duration of the period, emergence to heading and photochemical activity. But the story of total leaf area is different. Late varieties generally have higher leaf area and the early varieties lower (Gamzikova and Gudinova, 1981), but among the recently bred varieties midseason and early varieties have a higher leaf area.

During the early stages of development all cotton varieties had a fairly high photosynthetic rate but the early ones generally had a higher rate throughout the growing period (Vodogreeva, 1976). Azizkhodzhaev (1979) has also confirmed that the photosynthetic rate and photochemical activity during flower bud formation and flowering were higher in the early varieties than in the midseason and late varieties. An analysis of photosynthetic activity in the early cotton variety 1306 DV and the midlate variety 108 F at the stage of three to four true leaves, flower bud formation, the onset of flowering, maximum flowering and the onset of ripening indicated that the reasons for a higher rate of photosynthesis in 1306 DV may be the higher rate of mesophyll differentiation and a greater number of chloroplasts in 1306 DV than in 108 F at the 10-leaf stage (flower bud formation) (Nazarov, 1979). The early variety also had a more rapid rate of chlorophyll synthesis.

In soybean, LAI is correlated with days from sowing to seed formation while yield is correlated with days from seed formation to maturity (Paul et al., 1979). As LAI is not correlated with grain yield, Gordon et al. (1982) explored on the photosynthetic efficiency and found that the early maturing varieties had higher photosynthetic rates than the later maturing varieties, but the former showed a rapid decline in rate after attaining the peak, while the later maturing varieties maintained their photosynthetic activity longer. Du et al. (1982) noted that the photosynthetic rate in 49 different varieties varied from 11 to 40 mg $CO_2/dm^2/h$. In improved varieties, Hill reaction activity was 66.6 per cent greater and photosynthetic rate was 60 per cent greater than in the best parent. They observed that photosynthetic rate was positively correlated with yield in varieties of certain maturity groups, although apparently not in very early or very late varieties.

Out of the eight varieties and 46 hybrids of potato, eight were selected having higher rates of photosynthesis. Though seven of these belonged to the late group, the hybrid VL 292/71 having the highest rate of photosynthesis belonged to the midearly group (Zrust and Smolinkova, 1977).

In addition to the generally higher rate of photosynthesis in the early maturing varieties, some of them have more efficient nutrient absorption capacity and possess rapid fertilization rate. For instance, the comparative cytoembryological studies of Oganesyan (1975) on tomato varieties and hybrids revealed a direct correlation between earliness and fertilization rate. In a variety with a short growth period, fertilization was more rapid than in the late one; the hybrid was intermediate between the parents. Further, Bruetsch (1977) observed that the shallow root system in the early maize genotypes was responsible for efficient nutrient (P, Fe, Mg) uptake and utilization as compared to the late genotypes. This made the early maize lines more responsive to applied phosphorus, even without tillage, while the late lines required conventional tillage.

Hoogendoorn (1984) linked earliness in wheat to photoperiod and/or vernalization insensitivity. Most of the accessions from low latitude regions reached ear emergence rapidly owing to their insensitivity to photoperiod and vernalization with earliness factor accelerating ear emergence (Hoogendoorn, 1985).

A study of six soybean varieties differing in growth period duration revealed that up to flowering the earliest varieties had the highest rate of photosynthetic surface area formation (Kuzmin, 1986). Among the varieties studied, net photosynthetic production was highest during the period from flowering to the beginning of pod formation. The early flowering families out-yielded the late flowering ones by 7, 28 and 14 per cent in maturity groups I, II and III (Lin and Selson, 1988). Short early flowering types had the highest mean yield and gave the highest yielding individual families in all the three maturity groups. The early selections gave grain yields equal to or greater than those of the late selections (Innes et al., 1985). Mean weight per grain of the early selections was greater than in the late selections.

MODELS FOR PREDICTING FLOWERING/GROWTH DURATION

Using multiple regression technique, the photothermal models of rice growth duration from sowing to heading for three major cultivars (early, medium and late) were established based on the crop data of 12 representative cultivars grown at different locations throughout China and the meteorological data taken from the same locations with the same periods. Making a comparison between the modelling method and the traditional temperature summation method, the former reduced the errors by about three days for early rice, six to nine days for medium rice and 18 to 20 days for late rice cultivars. Dua and Garrity (1988) further attempted on developing a model for predicting flowering in rice varieties. They evaluated data from 43 experiments conducted at 23 irrigated sites in Asia, Africa and South America (lat. $0^o 9'$ S to $37^o 16'$ N) for relationships between mean temperature, photoperiod and flowering date. Cultivars IR 9729-67-3 (very early), IR 36 and BG 35-2 (early) and Taichung Sen Yu 285 (medium) were tested for best fit to generalized thermal or photoperiod models. A simple model using accumulated thermal units accurately estimated the rate of progress to flowering in IR 36. Best fit models included a quadratic photoperiod term that accounted for day length effects. Models for the other cultivars gave similar performances.

REFERENCES

Abdel, K.S. and M. Mokhie El'Din. (Study of the length of the growth period in winter wheat hybrids in the first and second generation). Nauch tr. Ukr. s-Kh. akad. 146: 38–40, 1975.

Abramova, Z.V. and A.I. Kharionovskaya. (Inheritance of length of growth period and yield components in barley after crosses of varieties differing in ecological and geographical origin). Nauch. tr. Sev. Zap. NII s. Kh. No. 35: 52–66, 1976.

Akhmetov, A.Z. and P.F. Fedorov. (The inheritance of plant height and length of growth period in F_1 hybrids of winter wheat). Kazaksyan anyl saruasylyk gylymynyn habarsysy No. 3: 32–34, 1974.

Al-Rawi, K.M. and R.J. Kohel. Diallel analysis of yield and other agronomic characters in *Gossypium hirsutum* L. *Crop Sci.* 9: 779–83, 1969.

Al-Saheal, Y.A. Diallel analysis of date of spike emergence in eight spring wheat crosses. Cereal Res. Commu. 13:11–18, 1985.

Amaya, A.A., R.H. Bush and K.L. Lebsock. Estimates of genetic effects of heading date, plant height and grain yield in durum wheat. *Crop Sci.* 12: 478–81, 1972.

Andriyash, N.V. (Inheritance of heading date in winter bread wheat hybrids). Nauchno-teknnicheskii Byulleten Vsesoyuznogo Ordena Lenina i Ordena Druzhby Narodov Nauchnoissiedovatelskogo Instituta Rastenievodstva Imeni N.I. Vavilova. No. 134:11–13, 1983.

Anwar, A.R. and A.R. Chowdhry. Heritability and inheritance of plant height, heading date and grain yield in four spring wheat crosses. *Crop Sci.* 9: 760–61, 1969.

Avey, D.P. Genetic control, breeding and prediction of maturity in winter wheat. *Diss. Abstr. Int. B.* 42: 18B, 1981.

Avey, D.P. and H.W. Ohm. Genetic model for days to heading in F_2 populations from a set of winter wheat cultivars. *Agron. Abstr.* Madison, Wis, U.S.A. pp. 54–55, 1979.

Avey, D.P., H.W. Ohm and F.L. Patterson. Advanced generation analysis of the genetic control of earliness in three crosses of winter wheat. *Agron. Abstr.* Madison, Wis, U.S.A. p. 48, 1980.

Avey, D.P., H.W. Ohm; F.L. Patterson and W.E. Nyquist. Three cycles of simple recurrent selection for early heading in winter wheat. *Crop Sci.* 22: 908–12, 1982a.

Avey, D.P., H.W. Ohm, F.L. Patterson and W.E. Nyquist. Advanced generation analysis of days to heading in three winter wheat crosses. *Crop Sci.* 22: 912–15, 1982b.

Aylmer, J.M. and E.J. Walsh. An evaluation of two approaches to breeding for earliness in field beans (*Vica faba* L). *Irish J. Agric. Res.* 18: 253–61, 1979.

Azizkhodzhaev, A. (Rate of photosynthesis and photochemical reactions in cotton varieties differing in earliness). Tr. VNII selection i semenovod. Khlopchatuika No. 17: 38–43, 1979.

Bailey, W.K. and J.E. Bear. Components of earliness of maturity in peanuts (*Arachis hypogaea* L.). *J. Am. Peanut Res. Ed. Assoc.* 5: 32–39, 1973.

Balaur, N.S. (On a light-induced very early form of maize in Meldavia). *In*: Dokl-1-i Vses. nauchteckhn. knof. po. vozobuovlyaemyn. istochnikam energii, 3. Moscow, USSR. *Energiya.* pp. 71–72, 1972.

Basu, M.S., M.A.Vaddoria, N.P. Singh and P.S. Reddy. Identification of superior donor parents for earliness through combining ability analysis in groundnut (*Arachis hypogaea.* L.). *Ann. Agric. Res.* 7: 295–301, 1986.

Becka, J. (A study of dormancy of potato varieties with growth periods of different lengths). *Rostlinna Vyroba.* 21: 161–75, 1975.

Bhatt, G.M. Inheritance of heading date, plant height and kernel weight in two spring wheat crosses. *Crop Sci.* 12: 95–98, 1972.

Biffen, R.W. Mendel's laws of inheritance and wheat breeding. *J. Agric. Sci.* 1: 4–48, 1905.

Botezan, V., V. Moldovan, and M. Snciu. (Studies in the inheritance of earliness and breeding for this characteristic in wheat). *Probleme de Genetica Teoretica si Aplicata.* 10: 299–330, 1978.

Boyadzhieva, D. (Possibilities of producing intervarietal hybrids of *Triticum aestivum* L. with an early growth period). *Genetika i Selektsiya.* 9: 488–92, 1976.

Bruetsch, T.F. Physiological factors affecting the differential uptake and accumulation of phosphorus by long and short season maize genotypes. *Diss. Abstr. Int. B.* 37: 5472B–5473B, 1977.

Byth, D.E., K.B. Saxena, E.S. Wallis and C.Lambrides. Inheritance of phenology in pigeonpea (*Cajanus cajan*). *In*: Proc. Aust. Pl. Breed. Conf., Adelaide, 14–18 Feb. pp. 136-38, 1983.

Bugaev, V.D. and M.V. Lukyanova. (Inheritance of vegetative period in F_1 and F_2 hybrids of spring barley). Nauchnotekhnicheskii Byulleten Vsesoyuznogo Ordena Lenina i Ordena Druzhby Narodov Nauchno-issledovatelskogo Instituta Rastenievodstva Imeni N.I. Vavilova. No. 138: 22–26, 1984.

Carmona, P.S. Early maturing induced mutants in high yielding rice varieties of Rio Grande do Sul State (Brazil). *Mutation Breed. Newsl.* No. 28: 6–7, 1986.

Ceapotu, N. (New varieties and hybrids of agricultural plants bred in Romania). *Probleme de Genetica Theoretica si Aplicata* 8: 93–121, 1976.

Chadha, M.L. and K.S. Nandpuri. Mode of inheritance of earliness in muskmelon (*Cucumis melo* L.). *Indian J. Hort.* 35: 123–26, 1978.

Chinnadurai, K., S.R. Shree Rangasamy and P.M.Menon. Genetic analysis of duration of flowering in relation to yield and halo length in *Gossypium hirsutum* L. *Genetica Agraria.* 27: 396–409, 1973.

Chomoneva, T, Z.Z. Li, V. Kapchina, T. Nikolov, V. Vasev and I. Chilikov. (Peroxidase activity in seedlings of early and late, short and tall rice varieties). *Fiziol. Rast.* 10(3): 28–37, 1984.

CIAT (Centro International de Agricultura Tropical, Colombia). French bean. p. 64, 1978a.

CIAT. *Phaseolus.* Colombia, 1978b.

Costa, J.A. Study of plant population and row spacing on several soybean genotypes. *Diss. Abstr. Int. B.* 38(12): 5684 B, 1978.

Cross, H.Z. ND100 and ND 300 new corn inbred lines for producing early maturing hybrids. North Dakota Farm Research. 36(2): 25–29, 1978.

Crumpacker, D.W. and R.W. Allard. A diallel cross analysis of heading date in wheat. *Hilgardia.* 32: 275–318, 1962.

Dahiya, B.S. and D.R. Satija. Inheritance of maturity and grain yield in pigeonpea. *Indian J. Genet. Pl. Breed.* 38: 41–44, 1978.

Daniel, L. (Biometrical and genetical analysis of flowering time in maize). Lednice na Morave, Czechoslavakia; Mendeleum Institute. 1972.

Daniel, L, I. Bajtay and J. Deak. Investigation of the inheritance of flowering time in a diallel cross of inbred lines of sweet corn (*Zea mays* L.) Conv. saccharata Koern. Zoldsegtermesztesi Kutato Intezet Bulletinje. 9: 87–98, 1974.

Dhaliwal, M.S. and K.S. Nandpuri. Genetics of earliness in tomato. *Ann. Biol.* 2: 72–76, 1986.

Dionigi A (The character 'earliness' in cereals). *Genetica Agraria.* 29: 135–50, 1975.

Dobias, A. (The use of diallel analysis for the investigation of the inheritance of quantitative characters in agricultural crops). Acta Facultatos Rerum Naturalium Universitatis Comenianae. *Genetica.* 5: 53–65, 1974.

Du, W.G., Y.M. Wang and K.H. Tan. (Differences in photosynthetic activity between soybean varieties (strains) and relationships with yield). *Acta Agronomica Sinica* 8: 131–35, 1982.

Dua, A.B. and D.P. Garrity. Models for predicting rice flowering. *Int. Rice Res.Newsl.*13(2): 7–8, 1988.

Dunn, G.W. and D. Beck. Selection for an early ornamental flint corn. *Maize Genetics Coop. Newsl.* No. 55: 42, 1981.

Dylenok, L.A. and A.P. Yatsevich. (Monosomic analysis of some quantitative characters in the spring wheat Pitic 62. II. Length of the main ear, number of spikelets per ear, ear compactness and heading date). *In* : Genetika produktivnosti s. kh. Kultur. Minsk, Belorussian SSR. Nauka i tekhnska. pp. 67–73, 1978.

Edwards. H., H. Ketata and E.L. Smith. Gene action of heading date, plant height and other characters in two winter wheat crosses. *Crop Sci.* 16: 275–77, 1976.

Egiyan, M.E. and A.N. Lukyanenko. (Inheritance of earliness in first generation tomato hybrids). Trudy po Prikladnoi Botanike, *Genetike i Selektsii.* 65(3): 10–14, 1979.

El-Ghawas, M.I., H.A. Khalil and M.A. Abo-Elfadhl. (Inheritance of earliness in maize). Research Bull. Fac. of Agric., Ain Shams Univ., No. 806: 16, 1978.

Fachs, A. (Biometrical investigations of the inheritance of the organogenetic developmental phases in maize). Acta Universitas Agriculture, Brno, A. 21: 335–41, 1973.

Fischer, V. (Studies in the genetics of flowering date in barley.) *Z. fur Pflanzenzuchtung.* 74: 71–76, 1975.

Fonturbel, M.T. and A. Ordas. (Mass selection for earliness in two populations of maize). *Genetica Iberica* 33: 225–35, 1981.

Gallagher, L.W., M. Belhadri and A. Zahour. 1987. Interrelationships among three major loci controlling heading date of spring barley when grown under short day lengths. *Crop Sci.* 27: 155–60, 1987.

Gamzikova, O.I. and L.G. Gudinova. (Some indices of photosynthesis in wheat varieties differing in duration of growth period). *In:* Selekts. i semenovod. zern. Kultur. v sib. Novosibirsk, USSR. pp. 54–59, 1981.

Ganashan, P. and W.J. Whittington. Genetic analysis of the response to day length in rice. *Euphytica.* 25: 107–15, 1976.

Ganeva, G. and B. Bochev. (Genetic control of the main morphological, biological and economic characters in useful varieties of *Triticum aestivum* L. III. Earliness). *Genetika i Selektsiya.* 19: 473–82, 1987.

Gangadharan, C. CR 666, the 60-day rice strains. *Int. Rice Res. Newi.* 11(3): 4–5, 1986.

Gao, M.W. Genetic analysis of the earliness of the early maturing mutants in indica rice. *Mutation Breed. Newsl.* No. 17: 9, 1981.

Garlyev, G. (Early ripening and other economically useful characters in hybrids from intervarietal crosses). *In:* Vozdelyoanie s-kh. Kulture. Ashkhabad, Turkman SSR. pp. 18–24, 1979.

Garmasn, E.S. (Inheritance of growth period duration in hybrids of *Vicia faba*). *In:* 3-i s'ezd, Vses. o-va genetikov i selektsionerov im. N.I. Vavilova, Leningrad, 16–20 maya 1977 g. 1. Genet. i selektsiya rast. Tez. dokl. Leningrad, USSR. p. 111, 1977.

Georgiva, I. 1980. (Inheritance of earliness in hybrids between different forms of triticale). *Genetika i Selektsiya.* 13: 413–9.

Gibrel, G.F. The heritability of earliness, fruit size, yield and high processing quality in tomatoes as measured by early generation. *Diss. Abstr. Int. B.* 43: 3085 B-3086 B, 1983.

Gilioli, J.L., T. Sediyama, T.C. Silva, M.S. Ries, and J.T.L. Thiebant. (Inheritance of number of days to flowering and to maturity in four natural soybean (*Glycine max* (L) Merril) mutants). *Revista Ceres.* 27: 256–69, 1980.

Goldenberg, J.B. Utilization of induced early mutants to study the inheritance of earliness in barley. *In:* 3rd Int. Barley Genet. Symp., July 7–12 pp. 3–39, 1979.

Goldenberg, Kh. (Inheritance of earliness in barley). *Genetika i Selektsiya.* 12: 153–58, 1979.

Gordon, A.J., J.D. Hesketh and D.B. Peter. Soybean leaf photosynthesis in relation to maturity classification and stage of growth. *Photosynthesis Research* 3: 81-93, 1982.

Gupta, S.C., L.J. Reddy and D.G. Faris. Early maturing pigeonpea hybrids. *Int. Pigeonpea Newsl.* No. 2: 19–20, 1983.

Gupta, V.P and S. Ramanujam. Genetic architecture of yield and its components in chickpea. *In*: S. Ramanujam and R.D. Iyer (eds). Breeding Researches in Asia and Oceania. Proc. 2nd General Congr. Soc. for Advancement of Breeding Researches in Asia and Oceania Session XII. Improvement of grain legumes. Indian Soc. Genet. Pl. Breed., New Delhi. pp. 793–99, 1974.

Guseinov, V.A., E.I. Yurtaeva and K.G. Tretyakov. (Study of the structural organization of DNA in seedling roots of cotton varieties of different maturity groups). *Uzbekistan Biologija Zurnali* No. 1: 61–62, 1985.

Halloran, G.M. Genetic analysis of time to ear emergence in hexaploid wheat, *Triticum aestivum*, using intervarietal chromosome substitution lines. *Can. J. Genet. Cytol.* 17: 365–73, 1975.

Hanna, W.W. and G.W. Burton. Mutations for early maturity in pearl millet. *Mutation Breed. Newsl.* No. 26: 3–4, 1985.

Heakal, M.Y. F_2 monosomic analysis of heading date in *Triticum aestivum* L. *Egyptian J. Genet. Cytol.* 4: 476, 1975.

Hoogendoorn, J. Variation in time of ear emergence of wheat (*Triticum aestivum* L.): physiology, genetics and consequences for yield. Thesis, Wageningen Agricultural Univ., Netherlands. p. 123, 1984.

Hoogendoorn, J. The physiology of variation in the time of ear emergence among wheat varieties from different regions of the world. *Euphytica.* 34: 559–71, 1985.

Hsieh, S.C. and T.M. Chang. Radiation induced variations in photoperiod sensitivity, thermo-sensitivity and the number of days to heading in rice. *Euphytica.* 24: 487–96, 1975.

Ilina, L.B. (Inheritance of duration of the emergence-heading period and the effects of monosomy for some chromosomes. 1.) Rast i produktivnost rast. Dept. 3135–81, 130–40, 1981.

Innes, P., J. Hoogendoorn and R.D. Blackwell. Effects of differences in date of ear emergence and height on yield of winter wheat. *J. agric. Sci.*, U.K. 105: 543–49, 1985.

Jatasra, D.S. and R.S. Paroda. Genetic architecture of ear emergence in wheat. *Haryana Agric. Univ. J. Res.* 10: 34–40, 1980.

Johnson, C.E. and T.P. Hernandez. Heritability studies of early and total yield in tomatoes. *HortScience.* 15: 280, 1980.

Johnson, G.R. and K.J. Frey. Heritabilities of quantitative attributes of oats (*Avena* sp.) at varying levels of environmental stress. *Crop Sci.* 7: 43–46, 1967.

Jun, B.T., S.Y. Cho and K.Y. Chang. (Studies on the inheritance of quantitative characters in rice. 7. Analysis of days to heading in the F_1 and F_2 generations of a diallel cross). Res. Rep., Rural Dev. Admin., Korea Republic, Crop. 28(1): 55–62, 1986.

Kalaidzhieva, S. (Inheritance of growth period on hybridization of different pea varieties). *Rasteniev dni Nauki* 15: 26–35, 1978.

Kaul, M.L.H. Radiation genetic studies in garden pea. 2. Non-allelism of early flowering mutants and heterosis. *Z. fur Pflanzenzuchtung.* 84: 192–200, 1980.

Kaul, M.L.H; R.Garg and A. Jain. Radiation genetic studies in garden pea. 6. Interaction of two early mutant genes. *Sci. & Cult.* 45: 81–83, 1979.

Khatyleva, L.U. and T.A. Shavyalukha. (Genetic control of ear formation in spring wheat). Vestsi A.N. BSSR Ser. biyal. n. No. 3: 14–20; 137, 1978.

Kitten, W.F. and M.F. Schuster. (Impact of insect resistant and early maturing varieties of cotton on insecticide applications and production practices in Mississippi). Abstr. 86. Agric. and For. Exp. Sta., U.S.A. 1980.

Kolomiets, L.A. (Duration of growth period in intervarietal hybrids of winter wheat). *In*: Voes. shkola molod. uchenykh. i spertsialistov po teorii i prakt. selektsii rast, 1979. Tez. dokl. Moscow, USSR, pp. 15–16, 1979.

Kotova, G.P. and M.T. Ryabtseva. (Breeding early maize hybrids of the intensive type). *Kukuruza.* No. 5: 27–28, 1980.

Kumar, H. and K. Das. Genetics of flowering and maturity time in garden pea. *Indian J. Genet. Pl. Breed.* 35: 17–20, 1975.

Kumar, P., R.L. Kapoor, S. Dass and S. Chandra. Genetics of days to heading and maturity in pearl millet. *Haryana Agric. Univ. J. Res.* 12: 282–86, 1982.

Kurganskaya, N.V. and M.V. Agentova. (Earliness of heterotic hybrids of tomato). Genetika i selektsiya rast. i zhivotnykh v Kazakhstane. Alma Ata, Kazakh. SSR; Kainar. pp. 40–43, 1975.

Kuzmin, M.S. (Formation of photosynthetic surface area and photosynthetic production in soybean plants). *In*: Biologiya, Selektsiya i genetika soi Novosibirsk, USSR (1986): 125–34, 1986.

Lanza, F. and A. Dionigi. (Correlation between earliness and yield in maize.) Annali dell. Institute Spearimentale Agronomico. 5: 383–403, 1974.

Law, C.N.; A.J. Worland and B. Giorgi. The genetic control of ear emergence time by chromosomes 5A and 5D of wheat. *Heredity.*36: 49–58, 1976.

Lee, B.H. (Efficiency of male sterile-facilitated recurrent selection for earliness in wheat breeding). Res. Rep., Office Rural Dev., S. Korea. *Crops.* 26 (2): 61–77, 1984.

Li, S.C. Genetic studies of earliness and growth stages of *Lycopersicon esculentum* Mill. *Diss. Abstr. Int. B.* 36: 3143 B, 1976.

Lin, M.S. and R.L. Selson. Effect of plant height and flowering date on seed yield of determinate soybean. *Crop Sci.* 28: 218–22, 1988.

Lopez, Rivares, P. and J. Cuartero. (Genetics of the components of early yield in tomato). *Genetica Iberica.* 37: 25–38, 1985.

Lu. Z.T. (Studies on mutation breeding for early maturity in soybean). Yuanzineng Nongye Yingyong, No. 2: 5–10, 1981.

Lubmin, A.N. (The inheritance of duration of the period germination-heading in wheat). *Selektsiya i Semenovodstvo.* No. 2: 27–28, 1974.

Lyrene, P.M. and H.L. Shands. Heading dates in six *Avena sativa* L. × *A. sterilis* L. crosses. *Crop Sci.* 15: 359–60, 1975.

Machacek, C. Investigation of the inheritance of earliness in sunflower (*Helianthus annuus* L.). *Genetika a Slechteni.* 15: 225–32, 1979.

Mahmoud, A.A., S.M.A.Sayyed, A.H.Fayed, M.A.A. Latif and M.A. Ismail. (Genetic consequences of the transfer of induced mutations in rice (*Oryza sativa* L.) II. Days to heading and tillering capacity). *Indian J. Genet. Pl. Breed.* 44: 533–37, 1984.

Mallick, E.H. and P. Bairagi. An early maturing and fine grain mutant induced in rice. *Sci. & Cult.* 45: 124–25, 1979.

McKenzie, K.S., J.E. Board, K.W. Foster and J.N. Rutger. Inheritance of heading date of an induced mutant for early maturing rice (*Oryza sativa* L.). *SABRAO J.* 10: 96–102, 1978.

Menzi, M. (Aspects of ripening in several selected inbred lines of maize in incomplete diallel crosses). Mitteilungen fur Schweizerische Landwirtschaft 29: 188–95, 1981.

Miller, J.C. Jr. Genetic studies on time of flowering as a component of earliness in cucumber. *HortScience* 10: 319, 1975.

Miller, J.C. Jr. and J.E. Quisenberry. Inheritance of time to flowering and its relationship to crop maturity in cucumber. *J. Am. Soc. hort. Sci.* 101: 497–500, 1976.

Mitra, D. and S. Biswas. Variability in heading among aus rice varieties in West Bengal, India. *Int. Rice Res. Newsl.* 2 (5): 5, 1977.

Morgan, V.V., I.P. Chuchmii and V.S. Boreiko. (An early maturing maize mutation induced by 1, 4-diazoacetylbutane). *Tsitologiya i genetika* 6: 419–21, 1972.

Musorina, L.I. and N.N. Natochieva. (Variation in growth period). *Maslichnye Kultury.* No. 4: 37–38, 1986.

Narayanan, S.S., P. Singh, V.V. Singh and S.K. Chauhan, Disruptive selection for genetic improvement of upland cotton. *Indian J. agric. Sci.* 57: 449–52, 1987.

Narula, P.N. Short duration pigeon peas escape leaf blight disease in late seedings in Bihar. *Int. Pigeonpea Newsl.* No. 2: 50–51, 1983.

Nayar, G.G. EMS-induced high-yielding, early mutant in linseed (*Linum usitatissimum* L.). *Curr. Sci.* 48: 214–16, 1979.

Nazar, S.G. (Inheritance of earliness in hybrid potato populations). *Kartoplyarstvo* No. 14: 11–14, 1983.

Nazarov, R.S. (Rate of $^{14}CO_2$ absorption in cotton varieties differing in earliness). *Subtropicheskie Kultury* No. 3: 142–43, 1979.

Oganesyan, R.P. (The occurrence of heterosis in tomato backcross). *Izv. s-kh.* n. No. 4: 27–29, 1971.

Olivieri, A.M. and P. Parrini. Earliness of flowering in winter and summer rape seed. *Cruciferae Newsl.* No. 4: 22–23, 1979.

Pal, R. and H. Singh. Diallel cross analysis of maturity traits in rape seed. *Haryana Agric. Univ. J. Res.* 11: 36–44, 1981.

Paroda, R.S., V.P. Singh and A.B. Joshi. Genetics of ear emergence in wheat (*Triticum aestivum* L.). *Indian J. agric. Sci.* 42: 653–56, 1972.

Paul, M.H., C. Planchon and R. Ecochard. (Study of the relationships between leaf development, growth cycle and yield in soya). *Annales de l'Amelioration des Plantes.* 29: 479–92, 1979.

Pinthus, M.J. Inheritance of heading date in some spring wheat varieties. *Crop Sci.* 3: 301–04, 1963.

Pirasteh, B. Monosomic analysis of photoperiod response in wheat. *Diss. Abstr. Int. B.* 34: 640 B, 1974.

Pisarenko, V.A., D.R. Iokich and E. Ya, Grigorenko. (Water requirements of irrigated maize hybrids in earliness). *Oroshaemoe Zemledelie.* No. 30: 42–44, 1985.

Popoe, P. (The genetics of the "hybridogenic" complex.) *Khlopkovodstvo* No. 7: 16–17, 1986.

Postrigan, V.F. (Inheritance of duration of growth period in wheat). *In:* Sozdanie novykh gibridov i sortov kukuruzy i ozinioi pshenitsy. Dnepropetrovsk, Ukrainian SSR. pp. 85–87, 1976.

Prasad, G.S.V., M.V.S. Sastry, T.E. Srinivasan and M.B. Kalode. Inheritance of tolerance to rice stem borer, *Scirpophaga (Tryponyza) incertulas* (Walker), and its association with plant habit and maturity period. *Indian J. agric. Sci.* 54: 352–55, 1984.

Prokofeva, E.V., I.F. Loshak and E.M. Esimbaeva, (Analysis of the inheritance of growth period in F_1 and F_2 hybrids of spring barley). *In:* Selekts, semenovod. i sortov. agrotekhn. zern. Kultur i mnogolet. tra sev. Zap. nechernozem. zony, Leningrad, USSR. pp. 69–74, 1980.

Prosvirnin, V.I. (Genetic control of quantitative characters in watermelon and melon). *In:* Genet, Kolichestv. priznakov s.-kh. rastenii, Moscow, USSR; Nauka. pp. 246–52, 1978.

Prudnikova, I.V., L.K. Sukhover, T.K. Tishkevich, R.A. Chkanikova, L.A. Makhanko and A.P. Makhanko. (A method for determining earliness in breeding forms of potato). USSR Patent (1987) A.s. 1266487. 1987.

Pulatov, M. (Inheritance of earliness in *Gossypim barbadense* hybrids). *Khlopkovodstvo* No. 5: 23–24, 1986.

Qualset, C.O. and Y.P. Puri. Sources of earliness and winter habit in durum wheat. *Wheat Information Service* No. 38: 13–15, 1974.

Radwan, S.R.H. and A.A.A. El-Zahab. Diallel analysis of some agronomic characters in *Gossypium barbadense*. *Z. fur Pflanzenzuchtung* 72: 291–304, 1974.

Ranalli, P. (Genetic and environmental effects in the control of earliness in peas (*Pisum sativum* L.) for processing). *Rivista di Agronomica.* 16: 392–95, 1982.

Ranganathan, T.B., P. Madhava Menon and S.R. Sree Rangasamy. Combining ability of earliness and yield in dwarf varieties of rice). *Madras Agric.J.* 60: 1134–38, 1973.

Rozvadovskii, A.N. (Inheritance of the length of the growth period in intervarietal hybrids of pea).

In: Novoe v issled. po sakhar. svekle i zern, kulturam, Kiev, Ukrarian, SSR. pp. 155–60, 1973.

Rudenko, M.I. and V.E. Dmitriev. (Choice of initial material for breeding spring wheat for earliness in the Krasnoyarsk region). Trudy po Prikladnoi Botanike. *Genetike i Selektsii* 68: 44–48, 1980.

Samokhvaalov, G.K., Yu.F. Zatsepa and N.I. Sherstnyuk. (The physiological characteristics of early and late plants). *Vestn. Kharkov, unta, biologiya.* 7: 62–67, 1975.

Saniescu, N.N., G. Ittu, C.Tapu and P.Veadu. (Observations on the performance of some wheat and triticale cultivars under drought conditions). *Probleme de Genetica Tloretica si Aplicata.* 18(1): 1–16, 1986.

Savchenko, D.I. (Inheritance of earliness in winter wheat hybrids). *Selektsiya i Semenovodstvo*, USSR, No. 9: 10–14, 1983.

Sharaan, A.N., T.M. Sabet and H.A. Hussein. Diallel analysis of barley induced early mutants. 1. Inheritance of earliness. Res. Bull., Fac. of Agric. Ain. Shams Univ. Egypt. No. 1945. 20, 1982.

Sharma, D., S.S. Baghel and S. Kumar. Inheritance of the number of days to silking and seed size in maize (*Zea mays* L.) *Indian J. agric. Sci.* 42: 998–1001, 1972.

Sharma, D., L.J. Reddy, J.M. Green and K.C. Jain. *In*: Proc. Int. Workshop on Pigeonpeas Vol. I. ICRISAT. pp. 71–81, 1980.

Sharma, J.C. and Z. Ahmad. Genetics of yield and developmental traits in bread wheat. *Indian J. agric. Sci.* 49: 299–306, 1979.

Shevtsov, V.M. Early ripening winter barley mutant. *Mutation Breed. Newsl.* No. 26: 1–2, 1985.

Shmanaeva, T.N. and A.T. Lebedeva. (Analysis of cucumber variety populations for earliness). Doklady Vsesoyuznoi Ordena Lenina i Ordena Trudovogo Krasnogo Znameni Akademii Sekokok-hozyaistvennykh Nauk Imeni V.I. Lenina (1987). No. 7: 19–21, 1987.

Singh, D.P., I.P. Singh, S.Seth, B.Chhabra and A.P. Tyagi. Association analysis of earliness in upland cotton. *Indian J. agric. Sci.* 48: 516–18, 1978.

Singh, V.P., R.S. Rana, M.S. Chaudhary and A.S. Redhu. Genetic architecture of ear emergence in bread wheat. *Indian J. agric. Sci.* 57: 381–84, 1987.

Skovmand, B. and P.N. Fox. Selection and testing for early hexaploid triticales. *In: Agron. Abstr.*, Madison, Wis, U.S.A. pp. 80–81, 1983.

Snape, J.W., C.N. Law and A.J. Worland. Chromosome variation for loci controlling ear emergence time on chromosome 5A of wheat. *Heredity.* 37: 335–40, 1976.

Srivastava, D.P. Very early maturing strains in rice. *Curr. Sci.* 50: 831, 1981.

Srivastava, D.P. CR 289–1008 and CR 289–1208-very early maturing rices. *Int. Rice Res. Newsl.* 8: 5, 1983.

Sun, G.Z., Y.C. Ghen, Z.W.Wan, Y.X. Zhang and Z.M. Shang. Super early maturing wheat variety Longfumei No. 1. *Mutation Breed. Newsl.* No. 27: 8, 1986.

Takahashi, R. and S. Yasuda. Genetics of earliness and growth habit in barley. Int. Barley Genetics Symp. Proc. 2nd (Pullman, Wash, 1970). pp. 388–408, 1971.

Teleutsa, A.S. (Growth period and yield of soybean in the central zone of Moldavia). Bull. Vsesoyuznogo Ordena Lenina i Ordena Druzhby Narodov Inst. Rastenicvodstva Imeni N.I. Vavilova. No. 97: 64–68, 81, 1980.

Tikka, S.B.S., R.K. Sharma and J.R. Mathur. Genetic analysis of flower initiation in cowpea (*Vigna unguiculata* (L) Walp). *Z. fur Pflanzenzuchtung* 77: 23–29, 1976.

Troyer, A.F. Selection for early flowering in corn. III. (18 F_2 populations). *In: Agron. Abstr.* Madison, Wis, U.S.A. p. 65, 1976.

Troyer, A.F. and W.L. Brown. Selection for early flowering in corn. *Crop Sci.* 12: 301–04, 1972.

Troyer, A.F. and W.L. Brown. Selection for early flowering in corn. Seven late synthetics. *Crop Sci.* 16: 767–72, 1976.

Tsai, K.H. (Effects of the earliness gene *E* on development and adaptability in the genetic background of the rice varieties Taichung 65). *J. Agric. Assoc.* China. No. 87: 1–20, 1974.

Tsai, K.H. (Genetic studies on earliness genes of rice, using isogenic lines). *J. Agric. Assoc.* China. No. 110: 1–22, 1980.

Tuberosa, R., M.C. Sanguineti and S. Conti. Divergent selection for heading date in barley. *Pl. Breed.* 97: 345–51, 1986.

Tyaminov, A.R. (Development of earliness in intensive varieties of fine-fibred cotton). Tr. VNII selektsii i semenovd. Khlopchatnika No. 18: 97–101, 1980.

Ukai, Y. and A. Yamashita. (Early barley mutants induced by radiation and chemical agents. I. Frequency of induced and a brief description of the characteristics of the mutants). *Jap. J. Breed.* 29: 255–67, 1979.

Venkateswarlu, S. Genetic analysis of flower initiation in *Pisum sativum* L. *Madras Agric.J.* 69: 461–66, 1982.

Venkateswarlu, S. and R.B. Singh. Combining ability for earliness in pigeonpea. *Indian J. Genet. Pl. Breed.* 41: 252–54, 1982.

Vodogreeva, L.G. (Photosynthetic rate and earliness in *Gossypium barbadense*). Sb. nauch rabot. Turkm. NII selektsii i semenovodstva tonkovoloknist. Khlopchatnika. No. 14: 93–99, 1976.

Wilson, F.D. and R.L. Wilson. Breeding potentials of non-cultivated cottons. III. Inheritance of date of first flower. *Crop Sci.* 16: 871–73, 1976.

Yadav, T.P., .V.P. Gupta and H. Singh. Inheritance of days to flowering and maturity in mustard. *SABRAO Journal.* 8: 81–83, 1976.

Yaruda, S. and J. Hayashi. (Effect of two genes for extreme earliness for yield and its components in barley). *Nogaku Kenkyu.* 59: 113–24, 1981.

Yermanos, D.M. and A. Kotecha. Gene action in flowering time of sesame, *Sesamum indicum* L. *Agron. Abstr.* Madison, Wis. U.S.A. p. 73, 1980.

Yurtaeva, E.I. (Earliness in cotton). *In*: Intensib. s.-kh. pr - va. Tez. dokl.Nauch. Konf., Tashkent, Uzbek SSR (1986): 35-37, 1986.

Zazharskii, V.T. (Heterosis and the inheritance of growth period duration in intervarietal sunflower hybrids). *Nauch. tr. Voronezh. s-kh. in-t.* 100: 80–88, 1978.)

Zelenskii, M.I., G.A. Mogileva and I.P. Shitova. (Duration of the period emergence - heading and the photochemical activity of the chloroplasts in wheat). *Selikokhozyaistvennaya Biologiya.* 14: 202–06, 1979.

Zelenskii, M.A. and A. Parkhomenko, 1980. (Yield of single early maize hybrids in the forest steppe of the Ukraine). Zhuk, V. Yu. (Breeding potato for early ripening). Kartoplyrstvo No. 11: 16–17, 1980.

Zrust, J. and A. Smolinkova. (Differences in assimilation rate in potato hybrids and some parental varieties). *Rostlinna Vyroba.* 23: 723–32, 1977.

CHAPTER 6

RATE AND DURATION OF GRAIN FILLING

Introduction
Variability in the duration and rate of grain filling
Assimilate partitioning
Grain size
Selection for duration and rate of filling period
Genetic and breeding studies
References

INTRODUCTION

Grain production in any crop depends on accumulation of dry matter in seeds. The amount of grain produced is a product of the number of seeds and their size, determined by the rate and duration of dry matter accumulation into them. Several authors have reported that under normal years, long duration crops have higher grain yields (Boyle et al., 1979; Wych et al., 1982), but in adverse years, especially with drought, and hot spells of wind during the crop ripening period, early maturing-short duration varieties outyield their long duration counterparts (Wych et al., 1982; *see* the chapter on Earliness). Now arises the question, which phase of crop growth is more responsible for higher grain yield: is it the long duration robust vegetative phase (source) (Aksel and Johnson, 1961; Gebeyehou et al., 1982a) or the long duration reproductive phase which is capable of attracting more assimilates (sink) (Hanway and Russell, 1968; Daynard et al., 1971; Peaslee et al., 1971; Spiertz et al., 1971; Cross, 1975; Daynard and Kannenberg, 1976). Many scientists working with several crops have supported that longer the duration of grain filling period, higher is the yield realized (Asana and Bagga, 1966; Syme, 1967; Johnson and Tanner, 1972; Shehata et al., 1976; Wein and Ackah, 1978; Boote et al., 1979; Boerma and Ashley, 1988). However, Eastin (1972) found that short vegetative period and a long grain filling period resulted in high yield in sorghum in Nebraska, whereas a long vegetative period and a short grain filling period were required for grain yield under Texas conditions. Thus the optimum length of the two growth periods depends upon the environment, particularly temperature (Krenzer and Moss, 1975; Bruckner and Frohberg, 1987). Wiegland and Cuellar (1981) have shown that in wheat with each °C increase in mean daily air temperature, there was a reduction of 3.1 days in the duration of grain filling. Similar data have also been reported by other workers (Chinoy, 1974; Asana and Williams, 1965; Ford et al., 1976; Chowdhury and Wardlaw, 1978). With the reduction in filling duration, kernel weight also decreased by 2.8 mg/kernel/°C (Wiegland and Cuellar, 1981).

Canopy photosynthesis and seed-fill period have been shown to be positively related to seed yield in soybean. Boerma and Ashley (1988) compared 20 soybean genotypes. The genotypes consisted of seven recently developed cultivars, seven plant introductions, which had been previously identified as having high seed yield, three high photosynthetic cultivars, and three low photosynthetic cultivars. Five canopy apparent photosynthesis measurements were taken during the reproductive stage of development. When averaged over maturity groups, the recently developed cultivars were 5 per cent higher in canopy apparent photosynthesis than the low photosynthetic cultivars and 13 per cent higher than the plant introduction. The recently developed cultivars and high photosynthetic cultivars had four to five days longer seed-fill period than plant introduction or low photosynthetic cultivars (Table 6.1). The recently developed cultivars averaged 12 per cent higher in seed yield than the low photosynthetic cultivars and 6 per cent higher than plant introductions. Partial correlation coefficients with the effect of maturity removed were positive between seed yield and canopy apparent photosynthesis, seed-fill period, and the product of canopy apparent photosynthesis and seed-fill period. The results indicate that high photosynthetic capacity and long seed-fill period were associated with high seed yield in this diverse group of genotypes. However, some other scientists have reported that it is not the duration of grain filling (seed-fill) period which is important, rather the rate at which grains are being filled (Asana and Williams, 1965; Bingham, 1967; Nass and Reiser, 1975; Brocklehurst, 1977; Singh et al., 1977; Jones et al., 1979; Bruckner and Frohberg, 1987). A crop variety with a shorter grain filling period but a faster filling rate can give as much or even higher grain production than the others having longer filling period but slower filling rate (Nass and Reiser, 1975; Daynard and Kennenberg, 1976). The rate of grain filling depends on the partitioning ability of the crop species or variety which is not necessarily dependent on the amount of source available. Our experiences with the high yielding dwarf genomes have also shown that source is seldom a limiting factor (Gupta, 1978); we know well by now that the major portion of assimilates for grain filling comes from the local photosynthate in the upper strata of the canopy. Still a certain minimum vegetative growth is essential to support the bigger ear, as the triple gene dwarf wheats have not yielded more than the double gene dwarfs (Gupta, 1978). In order to avoid the late season drought and hot spells of wind we can definitely go for the short season varieties. But we should not sacrifice the length of the reproductive phase, as in most cases it is related to grain yield, rather we must also try to increase the rate of grain filling (Park, 1979). For achieving this, the rate of assimilate partitioning has to be hastened and the period lengthened, as far as possible.

Genotypes vary for both grain filling rate and duration, but increasing temperature during grain filling tend to stop grain growth prematurely and to hasten physiological maturity. Rate, but not duration of grain filling, is correlated with kernel weight (Bruckner and Frohberg, 1987). Increasing kernel weight by extension of the grain filling period does not appear to be a promising strategy for increasing grain yield in these environments. Simultaneous selection for grain filling rate and high kernel weight is possible without lengthening grain filling duration, and selection for high grain filling rate through selection for high kernel weights is possible (Bruckner and Frohberg, 1987). High rate and short duration of grain filling appear to contribute to increased stress tolerance in these geno-types, but only one of the 20 genotypes investigated had both characteristics (Bruckner and

Table 6.1:   Mean plant and seed traits for four soybean genotype groups over two years (After Boerma and Ashley, 1988).

Genotype groups	Canopy apparent photosynthesis ($\mu mol\ CO_2\ m^{-2}s^{-1}$)	Seed yield (t/ha)	Seed fill period (days)	Seed weight (mg/seed)	Plant height (cm)
LPC	19.6	3.24	37.9	139	127
HPC	21.0	3.72	42.7	156	128
RC	20.8	3.62	42.1	156	133
PI	18.3	3.41	38.4	171	116

LPC = Low photosynthetic cultivars
HPC = High photosynthetic cultivars
RC = Recently developed cultivars
PI = Plant introduction

Frohberg, 1987). High grain filling rates with short to medium grain filling durations appear to be desirable objectives in environments in which the growing season frequently is shortened because of severe stress.

After seed number has been determined, cereal grain yields become proportional to kernel weight (Wiegand and Cuellar, 1981), which is a function of the rate and duration of grain filling. Rate of grain filling, which is dependent upon the number of endosperm cells is formed during the first two weeks after anthesis (Brocklehurst, 1977) and increases only moderately with increasing temperature, but duration of grain filling has a strong negative response to increasing temperatures (Gallagher et al., 1976: Sayed and Gadallah, 1983; Sofield et al., 1977; van Sanford, 1985; Wardlaw et al., 1980; Wiegand and Cuellar, 1981; Wych et al., 1982). High temperatures during the grain filling period of wheat impose major limitations on kernel weight and grain yield through reduction of grain filling duration (Sayed and Gadallah, 1983; Sofield et al., 1977; Wiegand and Cuellar, 1981). High temperatures enhance movement of photosynthate from the flag leaf to the ear, but do not necessarily increase grain filling rate, because increased respiratory losses of carbon occur at high temperatures (Wardlaw et al., 1980).

If grain filling duration is limited by high temperature, final kernel weight becomes proportional to grain filling rate under high temperature stress conditions (Wiegand and Cuellar, 1981). Wiegand and Cuellar (1981) suggested that genetic variability in grain filling rate should be searched for and exploited in wheat improvement programmes because genetic factors (cultivar) largely determine grain filling rate and environmental factors (temperature) largely determine grain filling duration. High rate and short duration of grain filling may contribute to higher kernel weights and yield in cultivars developed for short growing season environments (Nass and Reiser, 1975), environments prone to severe post anthesis stress (Sayed and Gadallah, 1983), and in early maturing wheat cultivars suitable for double cropping systems (van Sanford, 1985).

Alternative strategies for improvement of wheat productivity in environments prone to progressive, post-anthesis high temperature stress include lengthening of the grain filling period through earlier head emergence and flowering (Metzger et al., 1984; Wiegand and Cuellar, 1981) and identification and incorporation of heat stress tolerance mechanisms that would allow photosynthesis and grain growth to continue under high temperatures (van Sanford, 1985). He identified wheat genotypes that were relatively insensitive to high temperatures and maintained long grain filling duration under high temperature conditions. Although genetic variation for grain filling duration has been reported (Gebeyehou et al., 1982a, b; Jones et al., 1979; Metzger et al., 1984; Nass and Reiser, 1975; Sayed and Gadallah, 1983; Wych et al., 1982), longer grain filling duration has often been associated with early heading date (Metzger et al., 1984; Sayed and Gadallah, 1983, van Sanford, 1985; Wych et al., 1982).

Lanceolate leaflets of near isogenic lines of soybean, cv. Clark, consistently maintained a higher CO_2 exchange rate than the plants with normal leaflets, throughout the reproductive stages of development (Hsieh and Sung, 1986). The peduncle cross-sections of the semidwarf rice varieties Jaya, Palman and IR 8 contain more vascular bundles and have a greater phloem area than that of the tall Jhona 349, suggesting that the superior yields of the semidwarf types may be related to superior ability to translocate nutrients to the inflorescence during grain filling (Kaur and Singh, 1987).

Endogenous hormones may also have a role to play. Wheat var. HD 2009 with large grains is rich in auxins and Kalyan Sona, the highest yielding cultivar is rich in gibberellins (Bhardwaj and Verma, 1985). Foliar sprays with kinetin and IAA on rice have been found effective in increasing vascular bundle number and phloem area, respectively in the tall cv. Jhona 349 (Kaur and Singh, 1987). Both kinetin and IAA significantly increased grain yield and yield components.

Here in this chapter, I will review the amount of existing variability in these traits and the genetic and breeding efforts that have gone towards understanding this problem and improving the yielding ability of some of the grain crops.

VARIABILITY IN THE DURATION AND RATE OF GRAIN FILLING

Variations in the amount and rate of photosynthate formation (photosynthesis) have been reported (Sinclair, 1980; Buttery et al., 1981). Buttery et al. found a positive correlation between leaf photosynthesis during the reproductive phase and seed yield. Cultivar variation in canopy photosynthesis has also been reported (Egli et al., 1970; Wells et al., 1982). Harrison et al. (1981) found genotypic variation in canopy photosynthesis during the seed-fill period in progeny from two crosses. The heritability of canopy photosynthesis was 41 per cent and 65 per cent in two crosses for selection. The rate and duration of assimilate transport for grain filling is naturally dependent on photosynthate formation.

The period from anthesis to maturity in 156 lines of cowpea tested (Wien and Ackah, 1976, 1978) varied from 17 to 24 days and was positively correlated with seed yield and pod size. Variation in the duration of the seed filling period of soybeans has also been reported and a positive relationship has been found between the length of the seed filling period and yield (Hanway and Weber, 1971; Dunphy et al., 1979). The seed filling period ranging from 34 to 44 days and their seed yield ranged from 1,750 to 2,900 kg/ha (Zweifel et al., 1978). The high yielding varieties had a longer seed filling period. Further, Reicosky (1979) reported a wider range from 29 to 47 days. The varieties with longer seed filling period also had a larger seed size. Reicosky et al. (1982) reported further that soybean strains varied in the reproductive period (days from beginning of the bloom to normal green pods), seed filling period (days from beginning of seed fill to physiological maturity) and filling period estimate (days from beginning of seed fill to maturity). The seed filling period ranged from 18 to 54 days with a mean of 40 days. The strains with longer seed filling periods maintained seed growth at the end of seed filling for a longer time than strains with a shorter seed filling period. Lengths of reproductive periods and seed filling periods of five each of long seed filling period and short seed filling period-strains are give in Table 6.2.

Table 6.2: Reproductive development of five soybean strains with a long and five strains with a short seed filling period (After Reicosky et al., 1982)

Strains	Reproductive period (days)	Seed filling period (days)
	Long seed filling period	
PI 243.519	80	54
L 76.0166	84	53
PI 342.003	74	53
Columbus	78	51
PI 340.004	75	51
	Short seed filling period	
PI 200.548	48	28
PI 96.199	47	27
PI 200.482	46	26
PI 229.354	40	20
PI 261,469	37	18

Variations in the duration of seed filling period of inbred and hybrid maize lines have also been reported (Daynard et al., 1971; Carter and Poneleit, 1973; Poneleit and Egli, 1979). Cross (1975) measured the rate of fill and duration of grain filling period in a diallel of seven maize lines and found significant genotypic variation. In a 60 random S_1 families from a broad-based maize synthetic, significant variation was noted in the time to silking, time to physiological maturity, length of grain filling period, and grain yield (Tietz, 1979). The length of the grain filling period and the rate of accumulation during this period were correlated with the final grain weight. Boyle et al. (1979) noted this period to range from 31 to 43 days.

Variation in the duration of grain filling period of different lines of barley (Rasmusson et al., 1979), cultivars of wheat (Asana and Joseph, 1964; Stoy, 1965; Rawson and Evans, 1971; Simmons and Crookston, 1979) and rice (Tsunoda, 1964; Jones et al., 1979; Saini and Tandon, 1984) have been reported. Nass and Reiser (1975) reported data on vegetative and reproductive phases together with grain yield of 10 spring wheat cultivars (Table 6.3). From the data, percentage period spent under grain filling has been calculated. Grain filling index (the ratio of grain filling duration to number of days from emergence to maturity) of 44 true breeding lines was also calculated (Czaplewski, 1983) and was found to range between 9 and 11 per cent.

Not only the duration of grain filling period but also the rate of filling period are important variables. Lee et al. (1978) in a two-year experiment taking four varieties of soft red winter wheat showed a relationship between these traits and grain yield. In 1976, varieties Arthur, Hart, Pennoll and Redcoat exhibited yields of 2,455, 3,942, 4,637 and 4,259 kg/ha and filling periods of 28, 30, 30 and 31 days, and daily filling rates of 88, 131, 155 and 137 kg/ha, respectively. In 1977, Pennoll and Redcoat yielded 4,406 and 4,794 kg/ha and had filling periods of 25 and 32 days and daily filling rates of 176 and 150 kg/ha, respectively. In a study of 12 maize populations, Poneleit et al. (1978) showed the kernel

Table 6.3: Days taken under vegetative and reproductive phases and their proportion and grain yield of 10 spring wheat cultivars (Adapted from Nass and Reiser, 1975).

Cultivar	Planting to maturity (days)	Planting to anthesis (days)	Anthesis to maturity (days)	% period under grain filling	Grain yield (g/plot) 14 plants per plot
Ankra	104	60	44	42.31	520
Opal	102	60	42	41.18	491
Strong	102	60	42	41.18	427
Apu	92	54	38	41.30	388
Sonora 64	91	53	38	41.76	378
Nainari 60	102	60	42	41.18	372
Pitic 62	106	67	39	36.19	366
Selkirk	96	57	39	46.62	350
Ostka Hlopicka	100	64	46	46.00	349
Red Fife	106	64	42	39.62	304

growth rates to range from 5.5 to 12.8 mg/kernel/day and the effective filling period from 26 to 42.3 days. Positive correlations were shown between kernel growth rate and mean kernel weight and between effective filling period and potential yield, while negative correlations were found between kernel growth rate, kernel number per plant and potential yield, and between kernel number per plant and mean kernel weight.

Table 6.4: Length of grain filling period, grain filling rate and grain yield of five oat genotypes (Adapted from Wych et al., 1982).

Genotype I.D. number	Grain filling period (days)	Grain filling rate (g m^{-2}day^{-1})	Grain yield (g m^{-2})
1	35.0	8.4	293
2	33.3	9.8	325
3	30.0	11.8	354
4	29.0	15.3	441
5	25.5	17.7	450

Table 6.5: Length of the grain filling period, grain filling rate and grain yield of five soybean genotypes (After Egli et al., 1978).

Genotype	Grain filling period (days)	Grain filling rate (mg/seed/day)	Seed weight (mg/seed)
Kaurich	24.2	9.11	217
Williams	25.4	6.19	154
Cutler 71	25.6	6.13	156
Custer	22.2	6.00	129
Essex	32.2	3.72	114

N.B. Essex has the maximum grain filling period but the minimum seed weight, and Kaurich having maximum grain filling rate also has maximum seed weight (grain filling period is short).

Egli (1975) observed a range of 3.38 to 8.32 mg/seed/day in the filling rate of soybeans. The greater importance of grain filling rate, especially in areas with hot and/or dry weather during the late grain filling period, are clearly brought out from the data on five oat genotypes given in Table 6.4 and five soybean cultivars in Table 6.5.

ASSIMILATE PARTITIONING

Although for grain filling, maximum contribution comes from the local photosynthate, for example, from the awns, glumes, rachis and the flag leaf in cereals, and the leaf immediately above the cob in maize, and the pod wall and the leaf immediately below the pod in legumes; green stem also contributes sizable amount of assimilate for grain filling, especially in cereals (Gupta, 1978).

In the swede rape variety Oro, nearly 75 per cent of the assimilates from the topmost leaf are translocated to the siliqua. During intensive dry matter accumulation in the seeds, the respective net photosynthetic activities of seeds, siliqua walls and siliqua bearing stems are 2, 28 and 41 per cent respectively, relative to that of the leaves (100 per cent) (Thies and Brar, 1984). The respective CO_2 fixation capacities of the organs (total plant 100 per cent) are 49 per cent for the leaves, 25 per cent for the siliqua walls, 16 per cent for the siliqua bearing stems, 9 per cent for the leaf bearing stems and 1.1 per cent for the seeds.

Koomanoff et al. (1980) studied six winter wheat cultivars. At the anthesis, stems of early heading cultivars had a higher stored carbohydrate concentration, and gave lower dry matter yields when compared to the later heading cultivars. Nonstructural carbohydrate and dry matter yields decreased more (greater partitioning) in the stems than in the leaves of all cultivars. The contribution of flag leaf and awns to grain filling is more in tall than in dwarf wheats (Nefedov and Pylnev, 1984). The contribution of second leaf and green stem to grain filling is more in dwarf than in tall genomes (Babuzhina, 1983; Perez and Martinez-Carrasco, 1985). In early wheat cultivars, ears emerge sooner and thus the grain filling period becomes usually longer, and also since the ears photosynthesise and transport for grain filling, they yield more especially under stress environments (Watson et al., 1963). Nass and Reiser (1975) presented data on stem transport in relation to grain yield in 10 wheat genotypes (Table 6.6).

Table 6.6: Stem transport, harvest index(HI) and grain yield in 10 spring wheat genotypes (Adapted from Nass and Reiser, 1975).

Genotypes	Maximum stem weight (g/ plot of 14 plants).	Reduction in stem weight from maximum to harvest (g/plot of 14 plants)	Transport from stem (%)	HI (%)	Grain yield (g/plot of 14 plants)
Ankara	901	169	18.76	41.5	520
Opal	892	163	18.27	40.2	491
Strong	800	137	17.13	39.2	427
Apu	807	156	19.33	37.3	388
Sonora 64	711	182	25.60	41.7	378
Nainari 60	797	87	10.92	34.4	372
Pitic 62	858	156	18.18	34.4	366
Selkirk	809	174	21.51	35.5	350
Ostka Alopicka	1005	217	21.59	30.7	349
Red Fife	998	208	20.84	29.4	304

In the wheat variety Anza (Sturdy), only 15 per cent of the leaf assimilates formed at anthesis were remobilized and used in the grain, compared to 47 per cent in Stephens and 36 per cent in Hill 81 (Mkamanga, 1985). Two weeks later the ears, stems and top two leaves were the major photosynthetic organs, and 75 per cent of assimilates formed then were later transported to the ears. In Stephens and Hill 81, respectively, 91 and 89 per cent of the assimilates formed by the ears and stems after senescence of the leaves were used to make grain, compared with 79 per cent in Anza (Sturdy) and Yamhill. Stephens and Hill 81 gave the highest yields. Further, Swain et al.(1987) showed that out of the four rice varieties, Swarnaprabha and Ptb 10 were efficient in translocation of assimilates to the panicle while Co 41 and Adt 32 were efficient in photosynthesis but also showed greater respiratory losses.

An analysis of dry matter accumulation in maize showed that as grain development progressed the rate of grain filling began to exceed the rate of dry matter accumulation indicating a net redistribution of stored assimilates (Jurgens et al., 1978). In the droughted plants, redistribution continued after the cessation of net plant dry matter production and had occurred earlier, and was to a larger extent than in the controls. A ^{14}C translocation study confirmed that grain fill drew on stored photosynthate to a greater extent in droughted plants than in controls. Because of this redistribution of photosynthate, grain yield was related to the total dry matter accumulated by the plants during the growing season. Also in rice, van Dat and Peterson (1983) concluded that more carbohydrate partitioned from the long duration cultivars than from the short duration cultivars. As much as 21 per cent assimilate comes from the previously stored carbohydrates in the vegetative parts (Cock and Yoshida, 1972); higher percentage coming in late maturing cultivars than in early maturing cultivars (Yoshida and Ahu, 1968).

In soybean, just after the mid-pod-filling stage, when maximum pod dry weight was recorded in both short season Norman and full season Amsoy 71, a significant loss of accumulated pod dry matter occurred in Norman as the seed attained maximum weight (Thorne, 1979). This apparent redistribution of assimilates, assuming a 41 per cent efficiency of conversion, accounted for 12.7 per cent of the final seed yield per plant in Norman and only 1.8 per cent in Amsoy 71. Further, Gay et al. (1979, 1980) argued that the yield advantage of cv. Williams resulted partially from a longer seed filling period (12 per cent), which was associated with higher CO_2 uptake and N_2 fixation during the late filling period. However, the yield advantage for cv. Essex resulted from greater partitioning of photosynthate to the seed. By selecting for early partitioning of photosynthate to the reproductive parts, Grantz and Hall (1982) advocate cowpea varieties adapted to semi-arid environments which might be obtained. The decrease in straw weight of oats (partitioning) between heading and maturity was more common and drastic in the dry years (Wych et al., 1982).

GRAIN SIZE

Genetic variation in grain size (weight) has been attributed to differences in duration of grain filling (Asana and Bagga, 1966; Syme, 1967) or to variation in the rate of grain filling (Asana and Williams, 1965; Bingham, 1967; Brocklehurst, 1977) or to both sources (Stamp and Geisler, 1976). These also depend on floret position within the ear, cultivar characteristics and environmental temperature. Apical kernels on an ear of maize have shorter durations of fill than base kernels, but the rate of growth is the same for individual kernels in either position (Tollenaar and Daynard, 1978). There is no appreciable difference in the rate of seed growth in soybeans between seeds from the first and those from the last pods to develop on the plant (Egli et al., 1978); however, the size of seed from the last pods to develop is reduced owing to a shorter filling period. They used five cultivars of varying seed sizes : Kaurich 234 mg/seed; Williams 171 mg/seed; Coulter-71, 165 mg/seed, Custer 129 mg/seed and Essex 114 mg/seed. Kaurich had the highest seed growth rate (7.96 mg/seed/day), Essex the lowest rate (3.64 mg/seed/day) and the other cultivars had intermediate rates. Seed growth rates were relatively constant across the early and late pods. The duration of the effective filling period was less for the late pods on the intermediate cultivars and this resulted in slightly smaller seeds.

In cowpeas, Wien and Ackah (1978) found a significant positive relation between pod development period and seed size. However, the seed number per pod was negatively correlated with pod development period (Fig. 6.1).

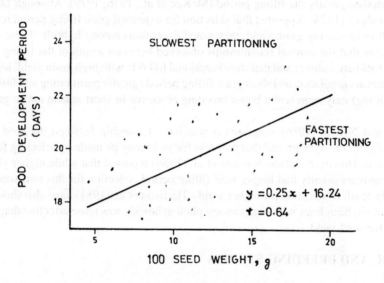

Figure 6.1: The relationship between 100-seed weight and pod development period for 65 cowpea lines (After Wein and Ackah, 1978).

SELECTION

The germ-plasm of crop plants (cereals and legumes) should initially be selected for the length of grain/seed filling period by measuring the time between flowering and 95 per cent mature grains/pod. Several workers have shown that substantial genetic variability exists in this trait and selection will be successful (Fakorede and Mock, 1978). A positive correlation has been found between the length of the grain formation period and 1000-grain weight of early and midseason winter wheat varieties but not in the late varieties (Oryuk and Bazalei, 1977). But in maize the early, medium and late varieties had the grain sizes of 26.5, 30.0 and 32.6 g/100 grain and their yields were 8.3, 8.7 and 9.6 t/ha, respectively, showing a relation between grain size and yield. By direct selection for larger grain size, varieties with larger filling period/faster filling rates are also selected (Poneleit and Egli, 1979). Ten wheat lines, selected for kernel weight from 100 lines screened, were intercrossed by Busch and Kofoid (1982) to form the initial C_0 population. Selection among S_1 plants for high kernel weight was used with approximately 22 plants selected in each cycle. The selection intensity varied from 1.5 to 2.9 per cent per cycle and averaged 2.2 per cent. Kernel weight increased 3 per cent per cycle from C_0 through C_2 as measured by the lines, and 7 per cent per cycle from C_1 through C_4 as measured by the populations. Genetic variance for most measured traits remained as high in the C_2 as in the C_0 population. Two cycles of selection resulted in lines with kernel weight higher than any line in the C_0 population. These results were further confirmed by the results of the population tests which indicated continued gain for kernel weight through cycles 3 and 4.

Multiple correlation and regression studies (McKee et al., 1979) of nine cultivars of oat indicated that low rates of grain filling and short LAD limited yield, suggesting that it would be easier to alter the rate of grain filling by selection for increased LAI and LAD than to lengthen greatly the filling period (McKee et al., 1976, 1979). Although Daynard and Kannenberg (1976) supported that selection for extended grain filling period in maize may result in increasing grain yield, they noted exceptions among hybrids. These exceptions indicate that the general relationships observed between length of the filling period and grain yield are indirect and that the exceptional hybrids with high grain yield, high rate of dry matter accumulation and short grain filling period (greater partitioning and faster rate of grain filling) may represent a better breeding objective in short season maize growing regions.

Smith and Nelson (1986) suggested a positive relationship between the seed filling period and yield of soybean, and that progress for yield may be made by selecting for seed filling period. However, Salado-Navarro et al. (1986) reported that while higher yielding soybean cultivars usually had longer seed filling period, selection for this trait would not necessarily result in selection for higher yields. Harrison et al. (1981) have also shown that selection of soybean lines with high canopy photosynthesis was more effective than direct selection for seed yield.

GENETIC AND BREEDING STUDIES

Several investigators have suggested that it should be possible to increase yield in grain crops by achieving an optimal duration for the vegetative and the grain filling periods of growth. Aksel and Johnson (1961) observed that long sowing-to-heading period tended to be associated with high yield in barley. However, Hanway and Russell (1969) and Daynard and Kannenberg (1976) reported a positive relationship between the length of the grain filling period and grain yield in maize. Further, Bingham (1967) concluded that both vegetative and grain filling periods are important for achieving high yields in wheat. He noted that the yield of grain is directly dependent on sink size, which is largely determined during the vegetative period, and on the photosynthetic capacity of the crop during the grain filling period. On the other hand, Nass and Reiser (1975) concluded that the length of the grain filling period was not important in determining yield in 10 wheat cultivars. Now arises the question whether higher grain yields can be obtained by optimizing the duration of vegetative and grain filling periods ? Keeping this in view, Rasmusson et al. (1979) selected nine barley cultivars representing three distinct types spending 25, 33 or 39 per cent of their growth cycle in grain filling. Estimates of heritability were high for the duration of the vegetative period. Estimates were low for the grain filling period when based on a single plot, but relatively high when based on the means of replicated plots.

Correlations between the two growth periods and between them and days to maturity suggest that selection aimed at optimizing the duration of the two growth periods and days to maturity would be hindered but not precluded by the associations.

If sizable heritable differences occur within a species in the duration of the vegetative and the grain filling periods, the opportunity exists for improving yield through altering the lengths of these periods in a breeding programme. Table 6.7 shows that it does.

Table 6.7: Duration of growth period of nine barley cultivars (Adapted from Rasmusson et al., 1979).

Cultivar/Length of grain filling period	Days in vegetative period	Days in grain filling period	Days to maturity	Grain filling period (% of total)
Short				
CI 5827	62.4	21.3	83.6	34.13
CI 5809	61.0	20.6	81.6	33.77
Intermediate				
Dickson	55.6	25.1	80.7	45.14
Manker	53.6	26.1	79.8	48.69
Larker	54.3	26.5	80.8	48.80
Long				
Primus	49.8	29.9	79.6	60.04
CI 6573	49.1	30.8	81.1	62.73
CI 5926	48.6	30.8	79.4	63.37
Vaughan	48.1	32.2	80.4	66.94

As expected, the proportion of days in grain filling was negatively correlated with the number of days in the vegetative period. Days to the vegetative period were negatively correlated with days in grain filling in four of five populations. It was found that short sowing to heading period was determined over a long period and that long heading-to-ripening period was dominant over a short period (Aksel and Johnson, 1961). In a given parent a long sowing to heading was associated with a short period and *vice versa*. These relations were broken, but very rarely in progenies. Rare breakdown of parental character associations among the progeny provided good evidence for close linkage.

Singh et al. (1977) investigated nine genetically diverse wheat varieties and their 36 one-way hybrids and found that the grain filling period was negatively associated with days to anthesis and was largely determined by additive gene effects. Varieties S308 and Kalyan Sona showed good general combining ability for grain filling period. But the rate of grain development appeared most important than during the grain filling period in determining yield differences. Narrow sense heritability estimates were high for the duration of heading, anthesis, grain filling and physiological maturity (Choi, 1984). Both additive and nonadditive gene action was present for all traits. It appeared that a shorter duration of grain filling along with a shorter lag period from heading to anthesis is important for high rates of grain filling.

Cross (1975) carried out a diallel analysis of duration and rate of grain filling of seven inbred lines of maize. The general combining ability effects for duration and rate of grain filling were larger than specific combining ability effects which suggested that simple selection techniques should be effective in changing these characters within the material studied. Bajaj and Phul (1982) studied the inheritance of span of maturity and harvest index in pearl millet. Analysis of data from a 10×10 diallel excluding reciprocals revealed a high heterotic response in several crosses for HI but not for length of the grain filling period, which tended to be longer in the parents than in their hybrids. Both traits were affected by additive and nonadditive gene action. Metz (1982) generated genetic variability for length of the seed filling and reproductive periods in soybean by crossing. Significant and heritable genetic variability for length of the seed filling and reproductive periods was observed among lines in both photoinsensitive determinate and photosensitive intermediate hybrid crosses, except for the length of the seed filling period in the indeterminate lines. Realized heritabilities for length of the reproductive period and length of the seed filling period ranged from 0.63 to 1.17 and 0.46 to 1.02, respectively. Seed filling and reproductive periods of F_4 lines could be lengthened by six days due to F_3 selection for these traits.

REFERENCES

Aksel, R and L.P.V. Johnson. Genetic studies on sowing to heading and heading to ripening periods in barley and their relation to yield and yield components. *Can. J. Genet. Cytol.* 3: 242–57, 1961.

Asana, R.D. and A.K. Bagga. Studies in physiological analysis of yield. 8. Comparison of development of upper and basal grains of spikelets of two varieties of wheat. *Indian J. Pl. Physil.* 9: 1–21, 1966.

Asana, R.D. and C.M. Joseph. Studies in physiological analysis of yield. 7. Effect of temperature and light on the development of the grain of two varieties of wheat. *Indian J. Pl. Physiol.* 8: 86–101, 1964.

Asana, R.D. and R.F. Williams. The effect of temperature stress on grain development in wheat. *Aust. J. Agric. Res.* 16: 1–13, 1965.

Babuzhina, D.I. (Characteristics of assimilate translocation in the forms of rye differing in height). Cf. *Plant Breed. Abstr.* 56: 6641, 1983.

Bajaj, R.K. and P.S.Phul. Inheritance of harvest index and span of maturity in pearl millet. *Indian J. agric. Sci.* 52: 285–88, 1982.

Bhardwaj, S.N. and V. Verma. Hormonal regulation of assimilate translocation during grain growth in wheat. *Indian J. exp. Biol.* 23: 719–21,1985.

Bingham, J. The physiological determinants of grain yield in cereals. *Agric. Prog.* 44: 30–42, 1967.

Boerma, H.R. and D.A. Ashley. Canopy photosynthesis and seed-fill duration in recently developed soybean cultivars and selected plant introductions. *Crop Sci.* 28: 137–40, 1988.

Boote, K.J., T.R. Zweifel, A.M. Akhanda and K.Hinson. Relationship of soybean yield to seed filling period. *In*: World Soybean Research Conf. I. Corbin, F.T. (ed), Westview Press. pp. 51–52, 1979.

Boyle, M.G., D.F. Enievel and J.C. Shannon. Assimilate partitioning to kernels of maize differing in maturity. *Agron. Abstr.* Am. Soc. Agron., Madison, Wis, U.S.A. p. 9, 1979.

Brocklehurst, P.A. Factors controlling grain weight in wheat. *Nature* 266: 348–49, 1977.

Bruckner, P.L. and R.C. Frohberg. Rate and duration of grain fill in spring wheat. *Crop Sci.* 27: 451–55, 1987.

Busch, R.H. and K.Kofoid. Recurrent selection for kernel weight in spring wheat. *Crop Sci.* 22: 568–72, 1982.

Buttery, B.R., R.I. Buzzell and W.I. Findlay. Relationship among photosynthesis rate, bean yield and other characters in field grown cultivars of soybean. *Can. J. Pl. Sci.* 61: 191–98, 1981.

Carter, M.W. and C.G. Poneleit. Black layer maturity and filling period among inbred lines of corn (*Zea mays* L.). *Crop Sci.* 13: 436–39, 1973.

Chinoy, J.J. Correlation between yield of wheat and temperature during ripening of grain. *Nature* 159: 442–44, 1947.

Choi, B.H. Nature of inheritance and association of time, duration and rate of grain filling and subsequent grain yield in crosses of winter and spring wheats (*Triticum aestivum* (L) em Thell). *In*: *Diss. Abstr. Int. B.* 44: 32608, 1984.

Chowdhury, S.I. and I.F. Wardlaw. The effect of temperature on kernel development in cereals. *Aust. J. Agric. Res.* 29: 205–23, 1978.

Cock, J.H. and S. Yoshida. Accumulation of ^{14}C-labeled carbohydrate before flowering and its subsequent redistribution and respiration in the rice plant. *Proc. Crop Sci. Soc.* Japan 41: 226–34, 1972.

Cross, H.Z. Diallel analysis of duration and rate of grain filling of seven inbred lines of corn. *Crop Sci.* 15: 532–35, 1975.

Czaplewski, S.J. Duration of the grain filling period and grain yield in barley. *Diss. Abstr. Int. B.* 43: 24108, 1983.

Daynard, T.B. and L.W. Kannenberg. Relationships between length of the actual and effective grain filling periods and the grain yield of corn. *Can. J. Pl. Sci.* 56: 237–42, 1976.

Daynard, T.B., J.W. Tanner and W.G. Duncan. Duration of the grain filling period and its relation to grain yield in corn, *Zea mays* L. *Crop Sci.* 11: 45–48, 1971.

Dunphy, E.J., J.J. Hanway and D.E. Green. Soybean yields in relation to days between specific developmental stages. *Agron. J.* 71: 917–20, 1979.

Eastin, R.A. Photosynthesis and translocation in relation to plant development. *In*: Sorghum in Seventies N.G.P. Rao and R.L. House. (eds) Oxford & IBH Pub. Co., New Delhi. pp. 214–46, 1972.

Egli, D.B. Rate of accumulation of dry weight in seed of soybeans and its relationships to yield. *Can.J. Pl. Sci.* 55: 215–19, 1975.

Egli, D.B., J.E. Leggett and J.M. Wood. Influence of soybean seed size and position on the rate and duration of filling. *Agron. J.* 70: 127–30, 1978.

Egli, D.B., J.W. Pendleton and D.B. Peters. Photosynthetic rates of three soybean communities as related to carbon dioxide levels and solar radiation. *Agron. J.* 62: 411–14, 1970.

Fakorede, M.A.B. and J.J. Mock. Changes in morphological and physiological traits associated with recurrent selection for grain yield in maize. *Euphytica* 27: 397–409, 1978.

Ford, M.A., I. Pearman and G.N. Thorne. Effect of variation in ear temperature on growth and yield in spring wheat. *Ann. appl.* Biol. 82: 317–33, 1976.

Gallagher, J.N., P.V.Biscoe and B.Hunter. Effects of drought on grain growth. *Nature* 264: 541–42, 1976.

Gay, S., D.B. Egli and D.A. Reicosky. Physiological basis for yield differences in selected soybean cultivars. *In*: World Soybean Res. Conf. II. F.T. Corben (ed) Westview Press. p. 59, 1979.

Gay, S., D.B. Egli and D.A. Reicosky. Physiological aspects of yield improvement in soybeans. *Agron. J.* 72: 387–91, 1980.

Gebeyehou, G., D.R. Knott and R.J. Baker Relationships among durations of vegetative and grain filling phases, yield components, and grain yield in durum wheat cultivars. *Crop Sci.* 22: 287–90, 1982a.

Gebeyehou, G., D.R. Knott and R.J. Baker. Rate and duration of grain filling in durum wheat cultivars. *Crop Sci.* 22: 337–40, 1982b.

Grantz, D.A. and A.E. Hall. Earliness of an indeterminate crop, *Vigna unguiculata* (L) Walp., as affected by drought, temperature, and plant density. *Aust.J. Agric. Res.* 33: 531–40, 1982.

Gupta, U.S. Production potential of dwarf genomes. *In*: Crop Physiology (ed) U.S. Gupta, Oxford & IBH Pub. Co., New Delhi. pp. 374–407, 1978.

Hanway, J.J. and W.A. Russell. Dry matter accumulation in corn (*Zea mays* L.) plants, comparisons among single cross hybrids. *Agron. J.* 61: 947–51, 1969.

Hanway, J.J. and C.R. Weber. Dry matter accumulation in eight soybean (*Glycine max* (L) Merrill) varieties. *Agron. J.* 63: 227–30, 1971.

Harrison, S.A., H.R. Boerma and D.A. Ashley. Heritability of canopy apparent photosynthesis and its relationships to seed yield in soybeans. *Crop Sci.* 21: 222–26, 1981.

Hsieh, W.L. and F.J.M. Sung. CO_2 exchange rate and translocation during productive growth in soybean differing in leaf morphology. *J. Agric. Assoc.* China 135: 25–33, 1986.

Johnson, D.R. and J.W. Tanner. Comparisons of corn (*Zea mays* L.) inbreds and hybrids grown at equal leaf area index, light penetration, and population. *Crop Sci.* 12: 482–85, 1972.

Jones, D.B., M.L. Peterson and S.Geng. Association between grain filling rate and duration and yield components in rice. *Crop Sci.* 19: 641–44, 1979.

Jurgens, S.K., R.R. Johnson and J.S. Boyer. Dry matter production and translocation in maize subjected to drought during grain filling. *Agron. J.* 70: 678–82, 1978.

Kaur, J. and G. Singh. Hormonal regulation of grain filling in relation to peduncle anatomy in rice cultivars. *Indian J. exp. Biol.* 25: 63–65, 1987.

Koomanoff, N.E., D.H. Smith and J.R. Welsh. Partitioning of nonstructural carbohydrates and dry matter of six winter wheat cultivars. *Agron. Abstr.* Am. Soc. Agron, Madison, Wis, U.S.A. p. 86, 1980.

Krenzer, E.G. and D.N. Moss. Carbon dioxide enrichment effects upon yield components in wheat. *Crop Sci.* 15: 71–74, 1975.

Lee, H.J., G.W. McKee and D.P. Knievel. Rate of fill and length of the fill period in soft red winter wheat. *Agron. Abstr.*, Am. Soc. Agron.,Madison, Wis, U.S.A. p. 11, 1978.

McKee, G.W., H.J. Lee and D.P. Knievel. Rate of fill and duration of the filling period in nine cultivars of spring oats. Agron. Abstr., Am. Soc. Agron., Madison, Wis, U.S.A. p. 73, 1976.

McKee, G.W., H.J. Lee, D.P. Knievel and L.d. Hoffman. Rate of fill and length of the grain fill period for nine cultivars of spring oats. *Agron. J.* 71: 1029–34, 1979.

Metz, G.L. Length of reproductive period and degree of photoperiod sensitivity in soybean (*Glycine max* (L) Merr). *Diss. Abstr. Int. B.* 43: 13248, 1982.

Metzger, D.D., S.J. Czaplewski and D.C. Rasmusson. Grain filling duration and yield in spring barley. *Crop Sci.* 24: 1101–05, 1984.

Mkamanga, G.Y. Assimilate production and partitioning in wheat. *Oiss. Abstr. Int. B.* 46: 1393B, 1985.

Nass, H.G. and B. Reiser. Grain filling period and grain yield relationships in spring wheat. *Can. J. Pl. Sci.* 55: 673–78, 1975.

Nefedov, A.V. and V.V. Pylnev. (Role of leaves of particular insertion heights and of awns in grain filling in different winter wheat genotypes). Nauchno-Tekhnicheskii Byulleten Vsesoyuznogo Selektsionno-geneticheskogo Instituta.2: 11–14, 1984.

Oryuk, A.P. and V.V. Bazalei. (Inheritance of growth period in winter wheat hybrids under irrigation). *Selektsiya i Semenovodstvo.* 1: 17–19, 1977.

Park, K.Y. Seed development and germination of soybeans at various filling stages. *In :* World Soybean Res. Conf. II. F.T. Corbin (ed), Westview Press. p.71, 1979.

Peaslee, D.E., J.L. Ragland and W.G. Duncan. Grain filling period of corn as influenced by phosphorus, potassium and the time of planting. *Agron. J.* 63: 561–63, 1971.

Perez, P. and R. Martinez-Carrasco. (Physiological causes of differences in yield between tall and dwarf varieties of winter wheat). *Anales de Edafologia y Agrobiologia.* 43: 1491–1502, 1985.

Poneleit, C.G. and D.B. Egli. Kernel growth rate and duration in maize as affected by plant density and genotype. *Crop Sci.* 19: 385–88, 1979.

Poneleit, C.G., D.A. Reicosky and D.B. Egli. Dry weight accumulation rate and effective filling period duration on genetically variable maize populations. *In: Agron. Abstr.*, Madison, Wis, USA. p. 59, 1978.

Rasmusson, D.C., I. McLean and T.L.Tew. Vegetative and grain filling periods of growth in barley. *Crop Sci.* 19: 5–9, 1979.

Rawson, H.M. and L.T. Evans. The contribution of stem resources to grain development in a range of wheat cultivars of different height. *Aust. J. Agric. Res.* 22: 851–63, 1971.

Reicosky, D.A. Cultivar variation in the length of the grain filling period. *In:* World Soybean Res. Conf. II F.T. Corbin (ed), Westview Press.

Reicosky, D.A., J.H. Orf and C.G. Poneleit. Soybean germ-plasm evaluation for length of the seed filling period. *Crop Sci.* 22: 319–22, 1982.

Saini, J.P. and J.P. Tandon. Variation of ripening periods among rice genotypes. *Int. Rice Res. Newsletter.* 9: 4–5, 1984.

Salado-Navarro, L.R., T.R. Sinclair and K.Hinson. Yield and reproductive growth of simulated field grown soybean. I. Seed filling duration. *Crop Sci.* 26: 966–70, 1986.

Sayed, H.I. and A.M. Gadallah. Variation in dry matter and grain filling characteristics in field cultivars. *Field Crops Res.* 7: 61–71, 1983.

Shehata, A.H., S.I. Salama and M.A. Khalifa. Estimates of maturity and their implications to yield improvement in maize. *Egyptian J. Genet. Cytol.* 5: 248–56, 1976.

Simmons, S.R. and R. Kent Crookston. Rate and duration of growth of kernels formed at specific florets in spikelets of spring wheat. *Crop Sci.* 19: 690–93, 1979.

Sinclair, T.R. Leaf CER from post-flowering to senescence of field grown soybean cultivars. *Crop Sci.* 22: 255–59, 1980.

Singh, V. P., B.N. Dahiya and R.K. Chowdhury. Genetic studies of grain filling period in wheat. *In:* A.K. Gupta (ed), Genetics of Wheat Improvement. II. Genetics of Economic Traits., Oxford & IBH Pub. Co., New Delhi. pp. 42–43, 1977.

Smith, J.R. and R.L. Nelson. Relationship between seed-filling period and yield among soybean breeding lines. *Crop Sci.* 26: 469–72, 1986.

Sofield, I., L.T. Evans, M.G. Cook and I.A. Wardlaw. Factors influencing the rate and duration of grain filling in wheat. *Aust. J. Pl. Physiol.* 4: 785–97, 1977.

Spiertz, J.H. J., B.A. Ten Hag and L.J.P. Dupers. Relation between green area duration and grain yield in some varieties of spring wheat. *Net. J. Agric. Sci.* 19: 211–22, 1971.

Stamp, P.H. and G. Geisler. Der Verlauf des Kornwachstums in Abhangigkeit von der Kornspwsition bei Zwei Sommerweizensorten. *Z. Acker Pflanzenbau.* 142: 264–74, 1976.

Stoy, V. Photosynthesis, respiration and carbohydrate accumulation in spring wheat in relation to yield. *Physiol. Plant* (suppl). 4: 1–125, 1965.

Swain, P., S.K. Nayak and K.S. Murty, 1987. Photosynthesis and translocation of ^{14}C photoassimilates among rice varieties. *J. Nuclear Agric. & Biol.* 16: 18–21, 1987.

Syme, J.R. Growth and yield of irrigated wheat varieties at several rates of nitrogen fertilizer. *Aust. J. exp. Agric. Anim. Husb.* 7: 337–41, 1967.

Thies, W. and G. Brar. (Photosynthesis, respiration and dry matter accumulation in ripening seeds of swede rape). *Berichte der Deutschen Botanischen Gesellschaft.* 97: 119–23, 1984.

Thorne, J.H. Assimilate distribution from soybean pod walls during seed development. *Agron. J.* 71: 812–16, 1979.

Tietz, R.L. The effect of the rate and duration of dry matter accumulation on the grain yield of maize. *Diss. Abstr. Int. B.* 40: 2474 B, 1979.

Tollenaar, M. and T.B. Daynard. Dry weight, soluble sugar content, and starch content of maize kernels during the early podsilking period. *Can. J. Pl. Sci.* 58: 199–206, 1978.

Tsunoda, S. Leaf characteristics and nitrogen response in the mineral nutrition of rice plants. *In:* Mineral Nutrition of Rice Plants. Proc. Symp. at the IRRI, Johns Hopkins Press, Baltimore, MO. pp. 401–18, 1964.

Van Dat, T. and M.L. Peterson. Performance of near isogenic genotypes of rice differing in growth duration. II. Carbohydrate partitioning during grain filling. *Crop Sci.* 23: 243–46, 1983.

Van Sanford, D.A. Variation in kernel growth characters among soft red winter wheats. *Crop Sci.* 25: 626–30, 1985.

Wardlaw, I.F., I. Sofield and P.M. Cartwright. Factors limiting the rate of dry matter accumulation in the grain of wheat grown at high temperatures. *Aust. J. Pl. Physiol.* 7: 387–400, 1980.

Watson, D.J., G.N. Thorne and S.A.W. French. Analysis of growth and yield of winter and spring wheats. *Ann. Bot. N.S.* 27: 1–22, 1963.

Wells, R., L.L. Schulze, D.A. Ashley, H.R. Boerma and R.H. Brown. Cultivar differences in canopy apparent photosynthesis and their relationship to seed yield in soybean. *Crop Sci.* 22: 886–90, 1982.

Wiegland, C.L. and J.A. Cuellar. Duration of grain filling and kernel weight of wheat as affected by temperature. *Crop Sci.* 21: 95-101, 1981.

Wien, H.C. and E.E. Ackah. Genetical and environmental effects of pod-filling time in cowpea (*Vigna unguiculata* (L) Walp). *In: Agron. Abstr.*, Madison, Wis, U.S.A. p. 97, 1976.

Wien, H.C. and E.E. Ackah. Pod development period in cowpeas. Varietal differences as related to seed characters and environmental effects. *Crop Sci.* 18: 791–94, 1978.

Wych, R.D., R.L. McGraw and D.D. Stuthman. Genotype × year interaction for length and rate of grain filling in oats. *Crop Sci.* 22: 1025–28, 1982.

Yoshida, S. and S.B. Ahu. The accumulation process of carbohydrates in rice varieties in relation to their response to nitrogen in the tropics. *Soil Sci. Pl. Nutr.* 14: 153–61, 1968.

Zweifel, T.R., K.J. Boote and K. Hinson. Relationship of yield to growth, filling period, and dry matter partitioning on eight soybean genotypes. *In: Agron. Abstr.*, Madison, Wis, U.S.A. p. 15, 1978.

CHAPTER 7

CROP IMPROVEMENT FOR HIGH HARVEST INDEX

INTRODUCTION

Harvest index (HI) is the ratio of economic yield (y-econ) to total plant biological yield (root weight not included) y-biol. Earlier, Engledow and Wadham (1923) used the term 'migration coefficient' and Nichiporovich (1960) 'coefficient of effectiveness' to represent the physiological capacity of the plant to mobilize photosynthate and translate it to organs of economic value. Since economic yield (y-econ) is only a fraction of total dry matter produced (y-biol), HI is less than unity, but some scientists express HI as percentage of the total produce.

Harvest index forms a useful measure of yield potential and is relatively easy to measure on a large number of plants. The importance of HI as a determinant of yield was gleaned from the trend in this character with crop improvement. Hayes (1971) examined HI of barley varieties released for productivity before the nineteenth century (Table 7.1). In spring oats, Lawes (1977) showed a high correlation between HI and grain yield of a range of varieties introduced between 1908 and 1962. Shims (1963) claimed that the improvement in grain yields in Australian oat cultivars has been due almost entirely to an increased HI without an increase in straw yield, or total dry matter production, when compared to old cultivars. Van Dobben (1962) compared old wheat cultivars evolved since the turn of the century with the leading wheats of the sixties and found a progressive increase in HI from 0.34 to 0.40. A comparison of nine wheat varieties (Syme, 1970) showed only a small range in y-biol of 11 to 12.5 t/ha while the grain yield (y-econ) varied from 2.9 to 4.9 t/ha with HI ranging from 0.24 to 0.39. Further, Syme (1972) studied 49 spring wheat genotypes and found that 72 per cent of the grain yield variability in the field could be estimated on the basis of HI values from single plants grown in a glass house. The high yielding semidwarf wheat cultivars have an improved grain/straw ratio (HI) over tall cultivars and show a change in HI from 0.32 to 0.38 (Vogel et al., 1963). Gent and Kiyomoto (1985) reported that the yield advantage of the modern wheat cultivar houser over the older one, is due to a greater HI rather than increased photosynthesis.

Table 7.1: The change in harvest index of barley varieties in the U.K. (Adapted from Hayes, 1971).

Variety	Year of release	Harvest index
Spratt	Pre-1900	0.401
Plumage Archer	1914	0.487
Kenia	1933	0.550
Proctor	1953	0.535
Zephyr	1966	0.571

The main factor responsible for poor grain yield in legumes is their lower HI and not photosynthetic limitation or total dry matter production. Full genetic potential of legumes cannot be realized unless HI is improved. For achieving this, the morphological frame of legumes had to be reconstructed in such a way that the dry matter accumulated is more efficiently partitioned between grains and vegetative parts. In screening for high HI in legumes, the consideration is to be given to increasing the relative proportion of effective pods per plant. Such other physiological attributes as high photosynthetic efficiency, LAD sink/ source size, photosynthate partitioning, local photosynthate, structure and position of pods, low photorespiration, etc., are expected to increase productivity and thus HI. In pigeon pea, total dry matter production is negatively associated with HI (Khapre and Nerkar, 1986). Reduction in plant height lowers the dry weight of vegetative parts and thus the photosynthate saved is increasingly transported to grains, resulting in higher HI. Though HI is highly correlated with grain yield, yield improvement through HI is faster due to its high coefficient of variability, high heritability and high expected genetic advance (Thankural et al., 1979). Harvest index exerts maximum positive direct effect (70.56 per cent) on wheat yield/m^2 followed by number of spikes/m^2 (48.33 per cent) (Kumbhar et al., 1982).

Cultivars with high HI make the most efficient use of nutrients. Analysis of data recorded on eight wheat and seven barley varieties showed that the weight of N, P, K, Ca, and Mg taken up for each kg of grain produced was the least in varieties with high HI. In other varieties much of the nutrients applied are used for the vegetative material (Popovic, 1981). When the proportion of nutrients utilized for grain production is calculated, the semdiwarf rice variety, IR30, which has a very high HI (0.867), used a greater proportion of nutrients in the production of grain than straw (Oyedokum, 1979).

Harvest index is one of the best criterion for determining yield stability. Sharma and Singh (1983) indicted that in wheat, HI has the greatest direct effect on grain yield followed by biological yield and tillers/plant. The first two were more important in the normal environment while HI and tillers/plant were more important under stress conditions. While selecting cultivars of *Vigna mungo* for high HI, Yadav et al. (1983) reported that out of 31 cultivars examined, 18 were stable under three environments, but only three of them had high HI. Thus varieties like H70-3, KMu3 and H70-11 possessing stability and high HI should prove promising.

Genotype × environment interaction for HI in millet was investigated for 80 genotypes in six environments, including a range of moisture stress environments (Sagar et al., 1984). Four hybrids showed high HI and comparatively greater stability than other hybrids.

VARIABILITY

Sufficient genetic variability in HI in different varieties of almost all crop plants has been recorded; cereals (Singh and Stoskopf, 1971), lentils (Singh, 1977), dry beans, soybeans (Buzzell and Buttery, 1977) and many more (Table 7.2). Chandler (1969) reported that new high yielding rice cultivars have a HI of 0.47 to 0.57 while the traditional rice plant has a HI of 0.23 to 0.37. Oyedokum (1979) reported that in south-western Nigeria, the semidwarf rice variety, IR30, showed a HI as high as 0.867 when grown in pots at different levels of nitrogen. Breeding of higher yielding wheat cultivars has also resulted in the increase of HI from 0.34 to 0.40 (Van Dobben, 1962).

Table 7.2: Variability (range) of HI recorded in different varieties of some crop plants.

Crop species	Range of HI		Authority
Barley	0.401	to 0.571	Hayes, 1971
Wheat	0.24	to 0.39	Syme, 1970
	0.32	to 0.38	Vogel et al., 1963
	0.34	to 0.40	Van Dobben, 1962
Rice	0.23	to 0.57	Chandler, 1969
Sorghum	0.08	to 0.41	Goldsworthy, 1970
Soybean	0.20	to 0.64	Luftensteiner,1981
	0.41	to 0.68	Dadson et al., 1983
Broadbean	0.30	to 0.78	Luftensteiner, 1981
Chickpea	0.218	to 0.380	Lal, 1976
Pigeonpea	0.089	to 0.578	Singh and Shrivastava, 1980
Chillies	0.397	to 0.694	Chhoukar et al., 1981

Lal (1976) examined 14 cultivars of chickpea and recorded that cv. C235 gave the lowest seed yield of 1.79 t/ha compared with the highest yield of 3.31 t/ha given by cv. 390, but the dry matter yield of both cultivars was similar. The HI ranged from 0.218 in C235 to 0.380 in 390 and the seed yield was positively correlated with HI. In cowpea, HI and yield per plant are significantly correlated and HI can be increased by selecting for increased 100-seed weight and the number of seeds per pod (Choulwar and Borikar, 1985).

Among the sorghum cultivars grown in Nigeria, the local cv. Farafara has a very high photosynthetic efficiency (y-biol - 24 t/ha) but a very poor photosynthate partitioning (y-econ - 1.95 t/ha) and thus a poor HI (0.08) (Goldsworthy, 1970). But the dwarf American hybrid, NK300, has a low photosynthetic efficiency (y-biol, 9.8 t/ha) but very efficient partitioning and thus HI (0.41 - five times of Farafara) and grain yield (4 t/ha). On the other hand, Samaru hybrid has an intermediate partitioning ability (HI 0.24 i.e. three times of Farafara) but y-biol is close to Farafara (19.8 t/ha - only 20 per cent less) and thus the grain yield achieved is (4.54 t/ha) which is higher than the high yielding dwarf American hybrid, NK300.

SELECTION

Breeders generally plant the segregating/ advance generation test lines under isolation for grain yield evaluation. When the space-selected plants are sown closely as a crop, they generally behave differently. Fisher and Kertesz (1976) and Sharma and Singh (1983) showed that HI of spaced plants is superior to grain yield of spaced plants, and thus can not give a reliable prediction of grain yield in large plots at optimum plant population. Syme (1972) reported that nearly 72 per cent of the grain yield variability of a group of 49 spring wheats could be estimated from their HI values obtained from single plots. Harvest index shows less variation in different environments than biological yield or grain yield. Thus HI appears to be a more suitable criterion for selection than its components.

In wheat, Austin (1980) argues that there is scope for increasing yield up to 20 per cent by selecting for greater HI. The limit to which HI can be increased is considered to be around 60 per cent (Austin et al., 1980). Hence a cv. with a low HI would indicate that further improvements in partitioning of biomass would be possible. On the other hand, a cultivar with a HI between 50 and 60 per cent would probably not benefit by increasing the HI.

Selections for high and low HI were made in the F_2 from three wheat crosses (Niehaus, 1981). High HI selections produced higher yielding F_4 lines than the low HI selections. High yielding plants selected from this design produced higher yielding F_4 lines than random selections from the same cross. Selecting the best 10 per cent of F_2 to F_4 gave F_5 lines that out-yielded random selections by 53 per cent for DX39 × Mexico 8156 and 5 to 23 per cent for Olympic × DX39 (When et al., 1982). Kramer (1984) compared HI and grain yield per plant as criteria for selecting yield in monoculture in two trials involving 15 spring wheat cultivars grown at plant densities ranging from 6.25 to 400/m². Heritability for HI was higher than for grain yield/plant. Harvest index was found to be a good selection criterion. Harvest index showed positive heterosis due to dominance (Khalifa and Al-Saheal, 1984). They suggested that selection for HI should be practiced in segregating populations.

In barley, Sanguineti and Tuberosa (1984) studied the HI of the main culm of P_1, P_2, F_1, F_2 and F_4 generations of four single crosses. In three crosses, low narrow sense heritability values indicated that main culm HI should be used as a selection criterion in advanced generations only. However, in the fourth cross, narrow sense heritability was high enough for selection to start in the F_2. Nevertheless, Balkema-Boomstra and Mastebrock (1984) concluded that early generation selection for HI in barley cannot replace selection for yield.

Improvement by selection for HI in grain yield of oats has been recorded (Howey, 1982). However, Kotvics (1981) concluded that selection for high yield must be based on both HI and seed yield per plant. Harvest index with its high heritability and high GCA variance provides better opportunity for selection than its other two components. Harvest index is positively correlated with grain yield but negatively correlated with biological yield (Singh and Stoskopf, 1971; Chaudhary et al., 1977; Thakural et al., 1979). Thakural et al. (1979) further observed that grain yield and biological yield were in turn, uncorrelated. Thus an improvement in grain yield through indirect selection on HI appears possible without affecting the biological yield. Harvest index shows greater constancy over environments (normal and stress) than its components (Sharma and Singh, 1983). It means, any prediction made on the basis of these parameters regarding genetic gain through selection will have more validity for the HI than for the biological yield and grain yield.

The use of HI as a selection criterion for grain yield appears more feasible among inherently taller than shorter backgrounds. Harvest index is considered more reliable at high population densities than at low densities (Nass, 1980; Tanno et al., 1985; Li, 1986). Also, HI is a better indicator of yield potential at medium fertility than at low or high fertility levels (Khurana and Yadav, 1986). Harvest index in soybean is more stable to plant competition, photoperiod and drought and thus is a better selection criterion than high biomass (Osman, 1985).

Kertesz (1984) suggested that cereal breeders might be able to increase HI from the current highest values of abut 0.50 to 0.60; when combined with an increase in biological yield, this would give a genetic gain in grain yield of 20 to 25 per cent. Harvest index is considered a promising selection criterion for yield (because of high genetic variation, moderate sensitivity and positive correlation with grain yield), but not perfect, especially in early segregating generations (because of the effects of heterosis, competition and segregation). Early generation selection for increased biological yield while maintaining the highest possible HI, or for parallel increases in HI, and growth rate, is recommended. Kulshrestha (1985) has suggested that future breeding in dwarf wheats should aim at increasing total biomass and at combining high 1000-grain weight and high grain weight per spike while maintaining a high number of grains per ear.

GENETIC STUDIES

High HI is partially dominant over low HI (Bhatt, 1976). The genetic action governing expression of HI is largely additive, but in some crosses nonadditive action has been noted. High parent heterosis for HI was absent but midparent heterosis was observed in all crosses. Estimates of heritability and genetic advance were moderate to high (Bhatt, 1976). Sharma (1981) studied a diallel cross amongst nine wheat varieties, in two environments, for HI and other characters. Additive genetic effects predominated for height, days to heading and HI, whilst non additive genetic effects predominated for the number of tillers and grain yield. Heritability estimates were high for height, the number of days to heading, ear length and HI. In the normal environment, UP368 had good GCA for height, number of tillers, grain yield and HI; in the stress environment, WH147 and WG377 had good GCA for HI. Path analysis showed that the HI accounted for much of the genetic variability for grain yield. Wheat varieties Haruhikari and Pitic 62 showed consistently high HI values among the 33 cultivars studied (Tanno and Gotoh, 1983). Estimated heritability values for HI ranged from 57 to 95 per cent. Of 10 single crosses generated by crossing five genetically diverse wheat varieties in all possible combination, seven showed a preponderance of additive and dominance gene effects for inheritance of HI (Dhinsa and Bains, 1987). The crosses WL711 × Sonalika and HD2009 × Sonalika showed digenic and trigenic types of epistasis together with additive and dominance effects. Duplicate epistasis was noticed in HD2009 × Sonalika.

Sharma and Singh (1983) analysed the HI and its two components, biological yield and grain yield of wheat. The GCA variance was significant only in case of HI, while SCA variances were significant for all the three traits. Higher GCA/SCA ratio and the estimates of heritability in narrow sense were observed for HI than those for the other two traits. The general predictability ratio were near unity for the former, while its estimates were low and far from unity for the latter two components.

Sharma (1986) reports that selection for high and low HI in the F_3 generation of three wheat populations was effective in producing progenies with high and low HI. Heritability estimates for HI ranged from 0.44 to 0.60. In a diallel cross of seven parents, both general and specific combining ability were significant for HI, grain yield and biomass yield. Also, selection for high HI resulted in shorter plants with earlier heading dates and lower biomass yield (Sharma and Smith, 1986).

In the F_1 of four rice hybrids, there was a positive heterosis for HI (Roy and Smetanin, 1984). In the F_2, one hybrid showed inbreeding depression, while in two others there was an improvement in HI. Nonallelic gene interaction effects determined HI in the F_2. It was suggested that selection for HI would be effective due to its high coefficient of genotypic variation, heritability and genetic advance.

In oats, Lawrence (1974) and Rosielle and Frey (1977) observed that HI shows primarily additive gene action, although this is complicated by the occurrence of negative skewness (Rosielle and Frey, 1977). Crosses between parents both of which had high HI tended to show small phenotypic variation for HI.

Bajaj and Phul (1982) observed a high heterotic response in several pearl millet crosses for HI. Harvest index was affected by additive and nonadditive gene action, but in chickpea, non additive gene effects between HI and maturity duration (Bajaj et al., 1983) were observed. Variety G543 was a good combiner for both characters. In lentils, both GCA and SCA effects were important for HI and grain yield (Sandhu et al., 1981). Maximum heterosis over the better parent reached 29.1 per cent for HI and 22 per cent for seed yield. Genetic information about the combining ability of parents, and the nature of gene effects involved in the inheritance of HI, would be of immense value to breeders in the choice of suitable parents for improving HI leading to higher productivity.

BREEDING EFFORTS / STRATEGIES

In the quest of increasing productivity, earlier scientists might have thought of developing more vigorous plant forms. But they subsequently realized that this does not help; y-econ does not increase proportionately with y-biol and the plants become reproductively inefficient. Supplying excess water (Tables 7.3, 7.4; Fig 7.1) and/ or nitrogen resulted in overvegetativeness, reduced fructification and increased lodging and self-shading. This is more so in legumes than in cereals. Many grain legumes have a tendency of over-vegetativeness if raised under optimum growing conditions. If the growing buds, for instance of chickpea (Misra and Sangal, 1958; Deshmukh and Mitkari, 1959) and soybeans (Greer and Anderson, 1965; Bauer et al., 1976) are nipped during the preflowering stage, excessive vegetative growth is cut down, branching favoured, LAI increased, number of pods/plant and grain yield are increased, and thus the HI. In indeterminate soybeans, attenuating apical dominance by nipping apices mechanically or chemically (TIBA spray) during early flowering, enhanced seed set and yield by 10 to 15 per cent (Greer and Anderson, 1965). Weber (1955) found that 50 per cent defoliation of soybean at the preflowering stage did not result in any yield reduction. In a good soybean crop; only 41 to 67 per cent of the LAI is effective (defined as the per cent of the LAI that intercepted 90 per cent of the sunlight) (Sakamoto and Shaw, 1967). Fifty per cent reduction in LAI by debranching did not reduce grain yield (Beuerlein et al., 1971). Similar experiments do not seem to have been conducted with other legumes or cereals.

Table 7.3: The influence of nitrogen on the growth and HI of two wheat cultivars (Barley and Naidu, 1964).

Wheat cultivars/ Nitrogen levels	Biological yield (t/ha)	Grain yield (t/ha)	HI
Gabo, 0 kg N/ha	8.4	3.0	0.36
67kg N/ha	9.5	2.7	0.28
134kg N/ha	9.4	2.6	0.27
Bencubbin, 0 kg N/ha	8.5	2.0	0.24
67kg N/ha	8.6	1.8	0.21
134kg N/ha	9.7	1.3	0.14

Table 7.4: Influence of number of irrigations on growth and HI of wheat var. Site Cerros (Gupta, U.S. and P.R. Maurya, in press).

Number of irrigations	y-biol (t/ha)	y-econ (t/ha)	HI	Number of irrigations	y-biol (t/ha)	y-econ (t/ha)	HI
	←	1988	→		←	1989	→
8	8.1	3.8	0.470	7	7.2	3.6	0.500
9	7.7	3.9	0.506	10	7.4	3.8	0.513
12	10.8	4.6	0.416	13	8.8	4.1	0.466
16	11.7	4.5	0.384	17	9.2	4.3	0.468

Since nipping/ debranching in order to reduce over-vegetativeness in legumes is a costly, time- and labour-consuming process and requires recurring expenditure; the idea of developing vegetatively less vigorous, dwarf, determinate, less branching/tillering type was conceived. All these result in increased HI, that is, greater grain production per unit of nutrient or water supplied. It is concluded that the vegetative growth should be reasonably good but not to the extent of luxuriance, in fact, after the transition from vegetative to reproductive growth, partitioning should be so fast and efficient that vegetative organs get preempted. For achieving these, breeding for the following characters may be considered:

Dwarf Genomes

Increases in production potential of dwarf genomes has been discussed (Gupta; 1978). Decrease in plant height results in increased HI (Donald and Hamblin, 1976). The reduction in height, so evident in recent cultivars, appears to lead to an increased HI not only through a reduction in weight of vegetative parts but also through a direct contribution to grain production. The period of heavy allocation of assimilates to stem growth has been shown to coincide with the period of maximum growth of the ear. Shorter and lighter stems use less carbohydrate and this reduced competition may permit increased ear growth and grain number.

Dwarf rice cultivars have tended to have higher HI than the tall ones. The cultivars grown before the forties were tall, gave poor yield responses to applied N and had HI of 0.29 to 0.41, but the recently developed cultivars are shorter (ca. 70 cm), give yield responses to applied N and have HI around 0.52 (Ishizuka, 1969; Chandler, 1969). Yields have increased from about four to five t/ha.

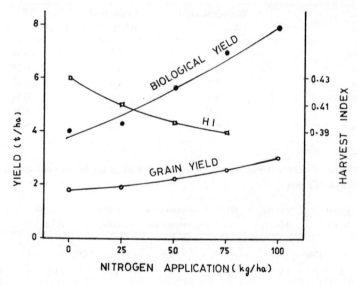

Figure 7.1. Influence of nitrogen level on growth and harvest index (HI) of wheat (After McNeal et al., 1971).

The grain yield and HI of wheat varieties with the dwarfing genes are consistently higher than those of tall varieties (Cleves Vargas, 1980). Short 2-gene semidwarf wheat lines have significantly higher HI values than their 1-gene dwarf and normal counter-parts (Allan, 1983). One-gene dwarf lines had higher HI values than their normal sibs. when calculated across tests and populations, the HI means of 2-gene dwarf, 1-gene dwarf and normal lines were 0.414, 0.380 and 0.316, respectively. The 2- and 1-gene dwarf doses increased HI by 31 and 20 per cent, respectively.

Shortening Crop Duration: Earliness

A crop with high HI but long maturity duration will not be economic, if the yield is calculated on per day basis, as against another with a little less HI but short maturity duration. Thus in addition to HI, shortening crop duration is a very important consideration in crop productivity and popularity (*see* Chapter five in this volume).

Short duration-photoinsensitive varieties start flowering sooner and do not produce as much vegetative organs, Efforts are made to shorten only the vegetative period with no reduction in the reproductive phase. Thus the HI of such short-duration cultivars is increased. Harper (1979) measured HI of 12 varieties of soybean from each maturity group I, II, III and IV, and reported that the varieties in early maturity groups had consistently higher HI than the varieties of the late maturity groups. Desai and Goyal (1981) recorded HI of some long and short duration sesame varieties (Table 7.5).

Table 7.5: Harvest indices of some long and short duration sesame varieties (Desai and Goyal, 1981).

Long duration varieties	HI	Short duration varieties	HI
IS 655	0.10	Gujrat Till	0.33
Purva 1	0.18	TC 25	0.34
4	0.21	Guari	0.25
		Vinayak	0.27

Determinate Habit

With the determinancy in growth habit, vegetative growth stops and thus the reproductive organs develop without any competition for metabolites, while in the indeterminate cultivars, both the growing vegetative buds and reproductive organs compete for the common pool of metabolites; thus the reproductive growth suffers and the HI falls down. A mean HI of 0.46 for indeterminate and 0.49 for determinate soybean varieties at a planting density of 50 plants /m² has been recorded (Baker et al., 1983). Dayde and Ecochard (1985) further confirmed that the determinate soybean varieties had higher HI. With 34 cowpea lines, Ojoma (1976) recorded higher HI for the determinate types. Further to this, Fernandez and Miller (1985) reported that determinate cowpea lines had HI three times higher than the indeterminate ones (*see* Chapter 3).

Uniculm/Nonbranching Types

Donald and Hamblin (1976) have clearly elucidated how the yield per unit area of the uniculm varieties should be higher than the multiculm varieties, as the secondary and tertiary spikes have lesser number of grains than in the spikes on the mother shoot. Thus, if the seed rate of uniculm varieties is increased to get the same number of spikes per unit land area, loss due to higher seed rate is richly compensated. In wheat, Merritt (1982) showed that selection of uniculm and oligoculm progeny of multiculm × uniculm crosses caused significant improvements in HI over those of either parent. Similarly, the mutant nonbranching Guar (*Cyamopsis tetragonoloba*) obtained by Swamy and Hashim (1980) having fewer and longer pods with more seeds per pod had a higher HI.

Population Tolerance

In order to obtain potential yield per unit area of land from the modern dwarf, early, determinate and nonbranching/non-tillering type of varieties with high HI, but occupying less space, it is necessary to maintain higher plant population and simultaneously incorporate population tolerance in them with no reduction in grain yield/ HI. Singh and Shrivastava (1980) suggested selecting for high HI under high density for obtaining high yielding lines. Walker and Fioritto (1984) have stressed that the total dry matter production and final plant population are as important as HI in determining the yield of determinate soybeans. With an increase in plant population density, y-biol continues to increase until lodging or other stress factors become limiting, but y-econ and HI show a decline at slightly lower population density. This density should be taken as maximum for the prevailing conditions. This becomes amply clear from the data in Table 7.6.

Table 7.6: Relationship between plant density, economic and biological yields (t/ha) and HI in cereals.

Cereal spp.	Plant population (1000/ha)						Authority
Wheat	35	70	1540	4470			
y-econ	1.7	2.5	2.3	1.8			Puckridge and
y-biol	4.8	8.1	8.9	7.4			Donald, 1967.
HI	0.358	0.304	0.262	0.251			
Sorghum	80	120	160	200			
y-econ	0.38	0.40	0.38	0.37			Gerakis and
y-biol	2.27	2.87	2.68	2.95			Tsangarakis,
HI	0.169	0.141	0.141	0.125			1969
Maize	15	23	29	39	59	117	
y-econ	3.4	4.5	5.1	4.8	4.7	3.7	Morrow and
y-biol	9.4	11.9	12.8	12.8	13.9	15.2	Hunt, 1891
HI	0.367	0.380	0.398	0.378	0.343	0.243	

PHYSIOLOGICAL ATTRIBUTES

Photosynthesis and Photosynthate Partitioning

Crossing parents having high overall photosynthetic efficiency (high biological yield) with parents having high HI (efficient partitioning) maximizes the probability of recombining 'yield genes' to give an optimal balance among the many interacting physiological processes that will give high economic yield (Hagemann et al., 1967). Sakharova (1980) reports that HI, which was 1.7 times higher in *Triticum aestivum*, than in T. boeaticum and *T. monococcum* respectively was also correlated with photophosphorylation activity at anthesis. Also in peas, Mahon (1982) reports that HI was generally greater in those selected for high carbon exchange rate.

Simplified growth analysis is primarily a procedure for improving selection of parents used in crosses aimed at breeding for higher yields. Though it does not directly ease the task of identifying high yielding varieties or progenies, its merit is in providing information whether each line, regardless of whether its yield is high or low, achieves yield largely through efficient partitioning.

Some evidence of increasing yield in *Phaseolus vulgaris* comes from the cross Redkote × Charlottetown, the respective parents having high biological yield with moderate economic yield and low HI, and low biological and economic yields combined with high HI (Wallace et al., 1976). The derived variety Redkloud gave 25 per cent higher yield than the maximum yields of Redkote. Redkloud matures in about 85 days while Redkote requires 105 days and is therefore much higher in economic yield per day, but it has the same biological yield per day. Furthermore, the Redkloud plant is smaller, has about 60 per cent as many leaves, leaf area and LAI. It has lower yield than Redkote at low (normal) populations and it achieves its higher yield only at higher plant population densities. Redkloud's high yield results from its high HI, which occurs because after flowering Redkloud partitions less of its photosynthate to continued vegetative growth, and more to seed.

In rice, after panicle initiation, growth of the vegetative organs such as tillers, new leaves and roots slows down and as a result, accumulation of available carbohydrates begins in the leaf sheath and culm base, sharply increases its amount during the two weeks before heading, and reaches its highest value at anthesis. Cook and Yoshida (1972) report that the carbohydrates stored before heading are efficiently translocated to the ear after anthesis; 68 per cent were translocated to the panicle, 20 per cent respired and 12 per cent remained in leaf sheaths and culms. The contribution of preheading storage to grain is variable but mostly between 20 to 40 cent (Matsushima and Wada, 1959; Yoshida, 1972). The rate of mobilization is greatest about one week after anthesis, reaching as high as twice the rate of total dry matter increase at that time (Monsi and Murata, 1970).

The ratio of photosynthetic capacity of a crop to its growth capacity during panicle initiation and heading is important, because, when this ratio is high, preheading storage is abundant, and when low, reserves are small (Murata, 1969). The HI is influenced by the amount of assimilates translocated to the ear during grain filling. If a large amount of starch and sugar remains in leaf sheaths and culms at harvest it is an indication that either translocation or storage is limiting.

Selection for greater diversion of photosynthate to seed production is a better approach for maximizing yield (Shibles and Weber, 1966). The aspects of selection for photosynthate partitioning (Snyder and Carlson, 1984) and crop improvement for efficient partitioning (Gupta, this volume) have been recently reviewed. Although genetic control of HI is an important aspect of differential partitioning of photosynthate, little information is available on the pattern of variation of this attribute in the segregating population following a cross.

Leaf Characters

Crop growth rates (CGR) for canopies of naturally erect-leaved varieties of rice (Hayashi and Ito, 1962) and barley (Tanner, 1969) are greater than those for canopies of droopy-leaved varieties, and such varieties can also be sown more densely. The modern more productive rice varieties have more erect leaves than the older varieties (Tanaka et al., 1966), and an association has been shown between high yield and erect leaves in barley, wheat and oat varieties (Tanner et al., 1966; Angus et al., 1972). At high LAI, canopies with more vertically inclined leaves have a higher photosynthetic rate than those with horizontal leaves, at least under clear skies with sun at high elevations, because of reduced light saturation of the upper leaves and more uniform distribution of light throughout the canopy. More vertical leaves may therefore be of significant advantage to crop, and Watson and Witts (1959) suggested that more inclined leaves constitute one of the important differences between wild and cultivated sugar-beet. Such varieties have high HI.

Frequently there is a negative relation between leaf size and photosynthetic rate, as in wheat, rice and maize (Hanson, 1971). The concept of larger leaf expansion or early canopy close up is fast changing. In fact, plants with smaller and more vertically disposed leaves allowing greater light penetration into the lower canopy strata and thus raising the CO_2 compensation point of the bottom leaves is preferred. Such plants can be sown more densely in order to obtain higher yields. Recent development of 'leafless' peas (Hedley and Ambrose, 1981) and okra and superokra type of cotton varieties (Kirby, 1977; Kirby et al., 1980) have proved beyond doubt that luxuriant leaf development can no longer be considered desirable.

Flower/Pod Abscission

Greatest abscission takes place after fertilization and most of this occurs during early stages of embryo development. Abscised buds are frequently the last buds to differentiate at a node, whereas pods are among the earliest to differentiate. Abscission is disproportionately greater at lower nodes/ branches. Grain legumes, in general, shed numerous flower buds, flowers and young pods even under optimum agronomic conditions. Duc and Picard (1982) recorded variability in flower and pod abscission among different genotypes of *Vicia faba*. Flower drop was determined by the young pod : flower ratio. Young pod drop was measured by the total pods : young pod ratio. The pod : flower ratio was negatively correlated with the number of nodes above the last podded node on a stem, and positively correlated with the ratio of number of podded nodes : number of flowering nodes. This may indicate an indirect way of selection for high pod per flower ratio, that is, by selecting for types having a limited growth above the last podded node and a pod set well distributed over all flowering nodes. By selecting genotypes with less abscission and higher pod set, grain production per plant will increase and thus the HI. The variability in flower abscission in some temperate and tropical grain legumes has been reported in Table 7.7.

Table 7.7: Per cent flower abscission in some temperate and tropical grain legumes.

Legume species	Flower abscission (%)	Authority
Temperate legumes		
Phaseolus vulgaris	80–85	Webster et al., 1975
	52–76	Subhadrabandhu et al., 1978
	45–80	Binnie and Clifiord, 1981
Vicia faba	85	Soper, 1952
	50	Lawes, 1974
	24–94	Duc and Picard, 1982
Pisum sativum	34	Meadley and Milbourn, 1970
Tropical legumes		
Glycine max	83	Van Schaik and Probst, 1958
	20–80	Hardman, 1970
Vigna radiata	47–61	Kaul et al., 1976
Vigna mungo	32–46	Kaul et al., 1976
Vigna unguiculata	40–56	Kaul et al., 1976
	54	Ojehomon, 1970
Gajanus cajan	80–90	Narayana and Sheldrake, 1976
Cicer arietinum	80–90	Saxena and Sheldrake, 1976

Spraying *V. unguiculata* plants during flowering with 1 per cent solution of sugar, monoammonium phosphate or diammonium phosphate decreased flower abscission and increased pod weight. Of the three, monoammonium phosphate was most effective. Spraying with propyzamide-44 after flowering prevented flower and pod drop in soyabean (Bae et al., 1975). Varieties of *V. unguiculata* resistant to flower bud abscission were selected after spraying with 400 ppm ethiphon one week before the start of flowering (Wien, 1979). Application of ethiphon during pod filling caused a greater proportion of older forms to be dropped from types susceptible to flower bud abscission than from types resistant to flower bud abscission. Further, Okelana and Adedipe (1982) observed that ethiphon enhanced HI in *V. unguiculata* cv. New Era. They concluded that ethiphon could be used for controlling pod abscission and for improving seed production in varieties such as New Era in which excessive vegetative growth is associated with reduced seed yield.

Basic investigations on the existence of genotypic variation in respect of flowerpod shedding are needed. An understanding of the cause whether it is due to limitation of nutrients, hormonal imbalance, light and/ or temperature dependence, or some other environmental influence, is needed urgently.

Growth Retardation

In species and varieties where genetic dwarfs have not been developed or further dwarfism is desired, growth retardants have played a useful role. For instance, CCC (2-chloroethyl trimethyl ammonium chloride) is used extensively on wheat and rye in Western Europe to prevent lodging and enhance productivity. CCC treated plants have a shorter and thicker stem, better root system, and shorter, broader and upright leaves (Birecka, 1967). Grain number per ear is often increased, whereas the grain size is reduced. The number of tillers and ears per unit area are also increased. De Vos (1968) obtained high wheat yields by reducing row width in combination with CCC. Retarded plants permit the use of higher doses of fertilizer and frequent irrigation. Dwarfed plants with increased productivity do have higher HI.

Bokhari and Youngner (1971a,b) obtained reduction in plant height and large increases in the number of tillers and grain yield of wheat and barley (Tables 7.8 and 7.9). These

Table 7.8: Effect of CCC on tillering of uniculum barley and grain yield (Bokhari and Youngner, 1971 a).

Molar CCC concentrations	Mean number of tillers/plant (after 42 days)	Mean number of ear bearing tillers/plant	Mean grain weight/plant (g)
10^{-1}	29.5	15.7	36.0
10^{-2}	18.0	12.7	19.9
10^{-3}	8.3	6.0	11.9
10^{-4}	8.7	5.0	9.0
10^{-5}	4.7	3.3	6.6
10^{-6}	4.0	3.0	4.7
10^{-7}	2.0	1.7	4.4
0	1.0	1.0	2.1

increases were maintained when seeds obtained from CCC treated wheat plants (retardized) were sown (Table 7.9).

Table 7.9: Growth and grain yield of untreated progeny of CCC treated (soil drench) Ramona-50 spring wheat plants (Bokhari and Youngner, 1971 b).

Molar CCC treatment of parents	Mean number of tillers/plant	Mean height (cm)	Mean grain weight/plant (g)
10^{-1}	16.33	31.66	18.06
10^{-2}	12.66	41.10	8.58
10^{-3}	9.66	43.66	7.02
10^{-4}	7.66	46.10	5.18
10^{-5}	6.33	51.26	4.09
0	3.00	˙71.60	1.84

On legumes, TIBA appears effective, and has successfully improved yield of soybean. Lodging is reduced, pod set increased, maturity enhanced and the grain yield increased (Wittwer, 1971). Yields of other legumes, for example, chickpea have also increased by TIBA sprays (Sinha and Ghildyal, 1973).

Growth retardants have been used in an attempt to limit excessive vegetative growth of cotton. Singh (1970) applied CCC to three cultivars of *Gossypium hirsutum* and one cultivar of *G. arborium*, in Punjab (India), where excessive growth was a problem. Application of CCC, 70 to 80 days after planting, retarded growth and increased the number of bolls per plant, boll weight and lint yield, and thus the HI. Sprays of 40 ppm CCC increased yield of the *G. hirsutum* cultivar by 18 to 15 per cent and sprays of 160 ppm CCC were required to increase the yield of *G. arborium* by 15 to 34 per cent.

In sugarcane, growth retardants favour sugar accumulation and regrowth after harvest. They permit close plantation, shorten the period between main crop harvest and ratoon, prolong the life of the plantation and permit the use of higher doses of fertilizers with no lodging (Papadakis, 1978).

In potato, application of growth retardants before tuber sprouting produces shorter, stronger, and more numerous sprouts; the crop grows denser, tuberization begins earlier and leaf senescence is delayed. Stimulation in tuberization by CCC and B9 application was recorded by Gifford and Moorby (1967) and Humphries and Dyson (1967). Ethylene has also been shown to promote tuberization (Catchpole and Hillman, 1969; Garcia-Torres and Gomez-Campo, 1972). In crops that produce heads, (cabbage, cauliflower and lettuce, etc.), better heads are produced and stalk appearance is avoided. In beets, root: leaf ratio increases and leaf senescence is delayed and stalk appearance is reduced In strawberry fruiting is favoured at the expense of stolons (Papadakis, 1978).

FUTURE LINE OF ACTION

The existing barrier to higher grain yield in dwarf wheats may be broken by increasing y-biol while maintaining high HI (Kulshrestha and Jain, 1978). In soybeans, with increase in y-biol, seed yield increased proportionately while HI remained constant (Johnson and

Major, 1979). Ruckenbauer (1984) suggested that the optimum value of HI might have been reached already in wheat genotypes carrying dwarfing genes. Thus further increase in grain yield is only possible by increasing y-biol, while maintaining the already attained high HI. This can be exemplified by the data of Goldsworthy (1970) on sorghum (Table 7.10). The local Nigerian sorghum variety Farafara has a very high y-biol but a very low HI, while the high yielding American sorghum hybrid, NK300, has a very low y-biol but a very high HI. But the Samaru hybrid has intermediate values of both y-biol and HI, and the grain yield is higher than the high yielding NK300 (4.54 as against 4 t/ha). Apparently, there is further scope of improving the Samaru hybrid, as its HI is only 0.24 as against 0.41 in NK300.

Table 7.10: A comparison of yield components and HI of a local sorghum variety and hybrids grown in Nigeria (Goldsworthy, 1970).

Variety/Hybrid	Biological yield (t/ha)	Grain yield (t/ha)	HI
Farafara (local)	24.0	1.95	0.08
NK300 (American hybrid)	9.8	4.00	0.41
Samaru hybrid	19.8	4.54	0.24

Thus the present breeding strategy should be to use the high HI parents and then select for high HI in the segregating population. Then select further for high HI at high plant population and fertility levels. Further, the yield barrier of the highest yielding varieties with the highest HI can be broken by increasing their y-biol while maintaining high HI.

REFERENCES

Allan, R.E. Harvest index of backcross-derived wheat lines differing in culm height. *Crop Sci.* 23: 1029–32, 1983.

Angus, J.F., R. Jones and J.H. Wilson. A comparison of barley cultivars with different leaf inclinations. *Aust. J. agric. Res.* 23: 945–57, 1972.

Austin, R.B. Physiological limitations to cereal yields and ways of reducing them by breeding. In: *Opportunities for Increasing Crop Yields*. London, U.K., Pitman Publishing. pp. 3–19, 1980.

Austin, R.B., J. Bingham, R.D. Blackwell, L.T. Evans, M.A. Ford, C.L. Morgan and M. Taylor. Genetic improvements in winter wheat yields since 1900 and associated physiological changes. *J. agric. Sci.*, U. K. 94: 675–89, 1980.

Bajaj, R.K. and P.S. Phul. Inheritance of harvest index and span of maturity in pearl millet. *Indian J. agric. Sci.* 52: 285–88, 1982.

Bajaj, R.K., T.S. Sandhu and S.S. Sra. Inheritance of harvest index and span of maturity in chickpea. *Crop Improvement* 10: 58–60, 1983.

Baker, D.A., G.P. Champan, M. Standish and M. Baily. Assimilate partitioning in a determinate variety of field bean. In: *Temperate Legumes : Physiolgy, Genetics and Nodulation*. D.G. Jones and D.R. Davis, (eds), Pitman, London. pp. 191–99, 1983.

Balkema-Boomstra, A.G. and H.D. Mastebroek. Effect of selection for harvest index in kernel yield in spring barley (*Hordium vulgare* L.). In: *Efficiency in Plant Breeding*. Proc. 10th Congr. Eur. Assoc. Res. Pl. Breeding EUCARPIA, Wageningen, Netherlands, Pudoc, p. 29 (Abstr.), 1984.

Barley, K.P. and N.A. Naidu. The performance of three Australian wheat varieties at high levels of nitrogen supply. *Aust. J. exp. Agric. Anim. Husb.* 4: 39–48, 1964.

Bauer, N.E., J.W. Pendleton, J.E. Beuerlein and S.R. Ghorashy. Influence of terminal bud removal on the growth and seed yield of soybeans. *Agron.* J. 68: 709–11, 1976.

Beuerlien, J. E., J.W. Pendleton, M.E. Bauer and S.R. Ghorashy. Effect of branch removal and plant population at equidistant spacings on yield and light use efficiency of soybean canopies. *Agron.* J. 63: 317–19, 1971.

Bhatt, G.M. Variation of harvest index in several wheat croses. *Euphytica* 25: 41–50, 1976.

Binnie, R.C. and P.E. Clifford. Flower and pod production in *Phaseolus vulgaris. J. agric. Sci.* U.K. 97: 399–402, 1981.

Birecka, H. Influence of 2-chloroethyl trimethyl ammonium chloride (CCC) on photosynthetic activity and assimilate distribution in wheat. In: *Isotopes in plant nutrition and physiology.* Vienna, Int. Atomic Energy Agency. pp. 189–99, 1967.

Bokhari, U.G. and V.B. Younger. Effect of CCC on the growth of wheat plants and their untreated progeny. *Agron.* J. 63: 809–11, 1971a.

Bokhari, U.G. and V.B. Younger. Effect of CCC on tillering and flowering of uniculm barley. *Crop Sci.* 11: 711–18, 1971b.

Buzzell, R.I. and B.R. Buttery. Soybean harvest index in hill plots. *Crop Sci.* 17: 988–90, 1977.

Catchpole, A.H. and J.Hillman. Effect of ethylene on tuber initiation in *Solanum tuberosum* L. *Nature* 223: 1387, 1969.

Chaudhary, B.D.; O.P. Luthra and V.P. Singh. Studies on harvest index and related characters in wheat. *Z. Pflanzehtg.* 79: 336–39, 1977.

Chandler, R.F. Jr. Plant morphology and the stand geometry in relation to nitrogen. In: *Physiological Aspects of Crop Yield.* J.D. Eastin; F.A. Haskins, C.Y. Sullivan, C.H.M. van bavel and R.C. Dinauer (eds), Am. Soc. Agron., Madison, U.S.A. pp. 265–85, 1969.

Chhoukar, V.S., P.V. Rao and C.R. Gupta. Variability in harvest index (HI) and the association of HI and economic sink with biological yield and components of noneconomic sink in chillies (*Capsicum annuum* L.). *Haryana J. Hort. Sci.* 10: 107–10, 1981.

Choulwar, S.B. and S.T. Borikar. Path analysis for harvest index in cowpea. J. *Maharastra agric. Univ.* 10: 356–57, 1985.

Cleves Vargas, B. (Colombia, Instituto Colombiano Agropecuario). Management report. 92, 1980.

Cook, J.H. and S. Yoshida. Accumulation of ^{14}C-labelled carbohydrate before flowering and the subsequent redistribution and respiraion in the rice plant. Proc. Crop Sci. Soc., Japan 41: 226–34, 1972.

Dadson, R.B., J. Joshi, P. Wells and L. Murphy. Harvest index of selected soybean germ-plasm. *Soybean Genetics Newsletter.* 10: 50–54, 1983.

Dayde, J. and R. Ecochard. (Dry matter production in soybean. I. Comparison of determinate and indeterminate types). *Agronomie* 5: 127–33, 1985.

Desai, N.D. and S.N. Goyal. Integrating breeding objectives and agricultural practices for Indian conditions. FAO Plant Production and Protection Paper. 29: 118–20, 1981.

Deshmukh, H.Y. and G.D. Mitkari. Some observations on the practice of topping in gram (*C. arietinum* L.). *Nagpur agric. Coll. Mag.* 33: 10–12, 1959.

De Vos, N.M. Shortcomings of modern wheat experiments. Euphytica 17(Suppl): 267–70, 1968.

Dhinsa, G.S. and K.S. Bains. Genetics of harvest index in bread wheat. *Indian J. agric.* Sci. 57: 529–34, 1987.

Donald C. M. and J. Hamblin. The biological yield and harvest index of cereals as agronomic and plant breeding criteria. *Adv. Agron.* 28: 361–405, 1976.

Duc, G. and J. Picard. Variability in fertility components among different genotypes of faba bean *FABIS Newsletter.* No. 4: 31–32, 1982.

Engledow, F.L. and S.M. Wadham. Investigations on yield in cereals. I. J. agric. Sci., U.K. 13: 390–439, 1923.

Fernandez, G.C.J. and J.C. Millr, Jr. Yield component analysis in five cowpea cultivars. J. Am. Soc. hort. Sci. 110: 553–59, 1985.

Fischer, R.A. and Z. Kertesz. Harvest index in spaced populations and grain weight of microplots as indicators of yielding ability in spring wheat. *Crop Sci.* 16: 55–59, 1976.

Garcia-Torres, L. and C. Gomez-Gampo. Increased tuberization in potatoes by etherel (2-chloethyl phosphonic acid). *Potato Res.* 15: 76–78, 1972.

Gent, M.P.N. and R.K. Kiyomoto. Comparison of canopy and flag leaf net carbondioxide exchange of 1920 and 1977 New York winter wheats. *Crop Sci.* 25: 81–86, 1985.

Gerakis, P.S. and C.Z. Tsangarakis. Response of sorghum, sesame, and groundnuts to plant population density in the central Sudan. *Agron.* J. 61: 872–75, 1969.

Gifford, R.M. and J. Moorby. The effect of CCC on the initiation of potato tubers. Eur. Potato J. 10: 235–38, 1967.

Goldsworthy, P.R. The growth and yield of tall and short sorghums in Nigeria. J. agric. Sci. U.K. 75: 109–22, 1970.

Greer, H.A.L. and I.C. Anderson. Response of soybean to tri-iodobenzoic acid under field conditions. Crop Sci. 5: 229–32, 1965.

Gupta, U.S. Production potential of grain legumes. In: Crop Physiology, U.S. Gupta, (ed), Oxford & IBH Pub. Co., New Delhi. pp. 374–407, 1978.

Gupta, U.S. 90. Crop improvement for assimilate partitioning. This volume, pp. 132–47.

Hagemann, R.H., E.R. Leng and J.W. Dudley. A biochemical approach to corn breeding. *Adv. Agron.* 19: 45–86, 1967.

Hanson, W.D. Selection for differential productivity among juvenile maize plants : associated net photosynthetic rate and leaf area changes. *Crop Sci* 11: 334–639, 1971.

Hardman, L.L. The influence of some production and pod set in soybean (*Glycine max* (L.) Merril) var. Hark. *Diss. Abstr. Int. B.* 31: 2401–08, 1970.

Harper, J.E. Nitrogen and dry matter partitioning between seed and stover of soybean. In: *Agron. Abstr.*, Madison, Wis., U.S.A. p.102, 1979.

Hayashi, K. and H. Ito. Studies on the form of plant in rice varieties with particular reference to the efficiency in utilizing sunlight. *Proc. Crop Sci. Soc.*, Japan. 30: 329–34, 1962.

Hayes, J.D. In: Annu. Rep. Welsh Pl. Breed. Sm. for 1970. pp. 32–33, 1971.

Hedley, C.L. and M.J. Ambrose. Designing 'leafless' plants for improving yields of the dried pea crop. *Adv. Agron.* 34: 225–77, 1981.

Howley, A.E. Genetic and environmental responses of harvest index and other agronomic traits in winter wheat (Triticum aestivum L.) and spring oats (*Avena sativa* L.) *Dis. Abstr. Int. B.* 43(3): 584 B, 1982.

Humpheries, E.C. and P.W. Dyson. Effect of a growth inhibitor, N-dimethyl aminosuccinic acid (B_9), on potato plants in the field. *Eur. Potato* J. 10: 118–28, 1967.

Ishizuka, Y. Engineering for higher yields. In: *Physiological Aspects of Crop Yield.* J.D. Eastin, F.A. Haskins, C.Y. Sullivan and C.H.M. Van Bavel (eds), Am. Soc. Agron., Crop Sci. Soc. Am., Madison, Wis., U.S.A. pp. 15–26, 1969.

Johnson, D.R. and D.J. Major. Harvest index of soybeans as effected by planting date and maturity rating. *Agron.* J. 71: 538–41, 1979.

Kaul, J.N.; K.B. Singh and H.S. Sekhon. The amount of flower shedding in some Kherif pulses. J. agric. Sci. U.K. 86: 219, 1976.

Kertesz, Z. Improvement for harvest index. In: *Efficiency in Plant Breeding.* Proc. 10th Congr. Eur. Assoc. Res. Pl. Breed, EUCARPIA, Wageningen, Netherlands, Pudoc, 1984. pp. 93–104, 1984.

Khalifa, M.A. and Y. A. Al-Saheal, Inheritance of harvest index in wheat. *Cereal Res. Commu.* 12: 159–66. 1984.

Khapre, P.R. and Y.S. Nerkar. Genetic parameters and character association for yield, total dry matter and harvest index in pigeonpea under different cropping systems. *J. Maharastra Agric. Univ.* 11: 310–11, 1986.

Khurana, S.R. and T.P. Yadav. Harvest index in soybean as influenced by varying fertility environments. *Legume Research.* 9: 97–102, 1986.

Kirby, T.A. Fixation of ^{14}C in cotton canopies as influenced by leaf type. *Diss. Abstr. Int.* B. 37: 5476-B–5477-B, 1977.

Kirby. T.A., D.R. Buxtion and K. Matsuda. Carbon source-sink relationships within narrow row cotton canopies. *Crop Sci.* 20: 208–13, 1980.

Kotvics, G. Selection for higher yield in early maturing mutants of soybean. *Mutation Breeding Newsl.* 18: 12–13, 1981.

Kramer, T. Harvest index and grain yield as selection criteria in plant selection. In: *Efficiency in Plant Breeding.* W. Lange; A.C. Zeven and H.G. Hogenboom (eds). EUCARPIA, Pudoc, Wageningen, Netherlands, pp. 109–12, 1984.

Kulshrestha, V.P. Yield trends in wheat in India and future prospects. *Genetica Agraria.* 39: 259–68, 1985.

Kulshrestha, V.P. and H.K. Jain. Wider significance of Norin dwarfing genes. Proc. 5th Int. Wheat Genet. Symp. 2: 988–94, 1978.

Kumbhar, M.B., A.S. Larik and H.M.I. Hafiz. Biometrical association of yield and yield components in durum and bread wheat. *Wheat Information Service.* No. 54: 35–38, 1982.

Lal, S. Relationship between grain and biological yields in chickpea (*Cicer arietinum* L.). *Tropical Grain Legume Bull.* 6: 29–31, 1976.

Lawes, D.A. Field beans : Improving yield and reliability. *Span* 17: 21–23, 1974.

Lawes, D.A. Yield improvement in spring oats. *J. agric. Sci.* U.K. 89: 751–57, 1977.

Lawrence, P.K. Introgression of exotic germ-plasm into oat breeding populations. Ph.D. Thesis, Library, Lowa State Univ., Ames, Iowa. 1974.

Li. L. (Variation in harvest index and its implications in the breeding of sweet potatoes (*Ipomoea batas* (L.). Lam) *J. agric. Assoc.* China. No. 136: 25–36, 1986.

Luftensteiner, H.W. (Harvest index of several soybean varieties in Austria conditions). In: Jahrbuch. Bundesanstalt fur Pflanzebau und Samenprufung in Wien, Vienna, Austria. pp. 213–36, 1981.

Mahon, J.D. Field evaluation of growth and nitrogen fixation in peas selected for high and low photosynthetic CO_2 exchange. *Can. J. Pl. Sci.* 62: 5–17, 1982.

Matsushima, S. and G. Wada. Relationships of grain yield and ripening of rice plants. 2. *Nogyo-oyobiengei* 34: 303–06, 1959.

McNeal, F.H., M.A. Berg, P.L. Brown and C.F.C. Guire. Productivity and quality response of five spring wheat genotypes *Triticum aestivum* L. to nitrogen fertilizer. *Agron. J.* 63: 908–10, 1971.

Meadley, J.T. and G.M. Milboura. The growth of vining peas. II. The effect of density of planting. *J. agric. Sci.,* U.K. 74: 273–78, 1970.

Merritt, R.G. An examination of the yield potential and inheritance of the uniculm character in wheat and barley *Diss. Abstr. Int.B.* 43: 1323B–1324B, 1982.

Misra, M.D. and P.M. Sangal. Effect of topping in gram (T.87) under rainfed conditions at Kanpur. *Kanpur Agric. Coll.* J. 17: 48–52, 1958.

Monsi, M. and Y. Murata. Development of photosynthetic system as influenced by distribution of matter. In: *Prediction and Measurement of Photosynthetic Productivity.* I. Setlik (ed), Centre for Agric. Publ. and Docum., Wageningen. pp. 115–29, 1970.

Morrow, C.E. and T.F. Hunt. III, *Agr. exp. Sta. Bull.* 13, 1891

Murata, Y. Physiological responses to nitrogen in plants. In: *Physiological Aspects of Crop Yield.* J.D. Eastin, F.A. Haskins, C.Y. Sullivan and C.H.M. van Bavel. (eds). Am. Soc. Agron., Crop Sci. Soc. Am. Madison, Wis., U.S.A. pp. 235–59, 1969.

Mukherjee, B.K., R.D. Singh and V.P. Ahuja. Role of Yogoslav germ-plasm in the improvement of relative partitioning of dry matter production in tropical maize cultivars, *Z. Pflanzenzuchtung*. 84: 49–56, 1980.

Narayanan, A. and A.R. Sheldrake. Pulse physiology – Ann. Report 1974–75. Part I. Pigeon pea physiology. ICRISAT, Hyderabad, India. 1976.

Nass, H.G. Harvest index as a selection criterion for grain yield in two spring wheat crosses grown at two population densities. *Can. J. Pl. Sci*. 60: 1141–46, 1980.

Niehans, W.S. Effectiveness of harvest index and the honeycomb design in early generation yield evaluation in durum wheat. *Diss. Abstr. Int*. B. 42: 868B, 1981.

Nichiporovich, A.A. Photosynthesis and the theory of obtaining high crop yields. *Field Crop Abstr*. 13: 169-75, 1960.

Ojehomon, O.O. Effect of continuous removal of open flowers on the seed yield of two varieties of cowpea, *Vigna unguiculata* (L.) Walp. *J. agric. Sci*., U.K. 74: 375–81, 1970.

Ojoma, O.A. The current out look on cowpea improvement at University of Ife. In: Proc. IITA, collaborators meeting on grain legume improvement, Plant improvement. R.A. Luse and K.O. Rachie, (eds) Ibadan, Nigeria. 1976.

Okelana, M.A.O. and N.O. Adedipe. Effects of gibberellic acid, benzyleadenine and 2-chloroethyl phosphonic acid (CEPA) on growth and fruit abscission in cowpea (*Vigna unguiculata* L.). *Ann. Bot*. 49: 485–91, 1982.

Osman, M.B. Selection for harvest index in spring barley (*Hordeum vulgare* L.). *Diss. Abstr. Int. B*. 46: 357B–358B, 1985.

Oyedokum, J.B. Screening for harvest index in upland rice in south-eastern Nigeria. *Int. Rice Res. Newsletter*. 4(3): 9, 1979.

Papadakis, J. Root toxins and crop growth : Allelopathy. In: *Crop Physiology*. (ed) U.S. Gupta, Oxford & IBH Pub. Co., New Delhi. pp. 202–37, 1978.

Popovic, V. (The relationship between harvest index and nutrient uptake in winter wheat and spring barley). *Rostlinna Vyroba*. 27: 313–21, 1981.

Puckridge, D.W. and C.M. Donald. Competition among plants sown at a wide range of densities. *Aust. J. Agr. Res*. 18: 193–211, 1967.

Rosielle, A.A. and K.J. Frey. Inheritance of harvest index and related traits in oats. Crop Sci. 17: 23–28, 1977.

Roy, I. and A.P. Smetanin. (Variation in harvest index among rice hybrids). *Tr. Kuban. S.-kh. in-t*. No. 6241/269: 3–7, 1984.

Ruckenbauer, P. Stability of harvest index. In: *Efficiency in Plant Breeding*. Proc. 10th Congr. Eur. Assoc. Res. on Plant Breeding. EUCARPIA, Wageningen, Netherlands, 19–24 June, 1983. W. Lange, A.C. Zeven and N.G. Hagenboom (eds), Wageningen, Netherlands, Pudoc, 1984.

Sagar, P., R.L. Kapoor and D.S. Jatasra. Harvest index and its stability in pearl millet. *Genetica Agraria* 38: 161–68, 1984.

Sakamoto, C.M. and R.H. Shaw. Apparent photosynthesis in field soybean communities. *Agron. J*. 59: 73–75, 1967.

Sakharova, O.V. (Features of phosphorylation in the chloroplasts during ontogeny in different wheat species in relation to their yields). *Trudy po Prikladnoi Botanike, Genetike i Selekteii*. 67: 135–46, 1980.

Sandhu, T.S., J. S. Barar and R.S. Malhotra. Combining ability for harvest index and grain yield in lentil. *Crop Improvement*, 8: 120–23, 1981.

Sanguineti, M.C. and R. Tuberosa. Genetic control of the main culm harvest index in barley (*Hordium vulgare* L.). *Genetica Agraria* (Abstr.). 38: 345, 1984.

Saxena, N.P. and A.R. Sheldrake. Pulse physiology. Ann. Rep., 1974–75. Part II. Chickpea physiology. ICRISAT, Hyderabad, p. 51, 1976.

Sharma, K. Genetic analysis in bread wheat (*Triticum aestivum* L.) with particular reference to harvest index. Thesis Abstr. H.A.U., Hissar, India 7: 335–36, 1981.

Sharma, R.C. Harvest index in winter wheat : breeding and genetic studies. *Diss. Abstr. Int.* B. 47: 863 B, 1986.

Sharma, R.C. and E.L. Smith. Selection for high and low harvest index in three winter wheat populations. *Crop Sci.* 26: 1147–50, 1986.

Sharma, S.K. and R.K. Singh. Path analysis of harvest index and its related characters in wheat. *Haryana Agric. Univ. J. Res.* 13: 218–26, 1983.

Shibles, R.M. and C.R. Weber. Interception of solar radiation and dry matter productivity by various soybean planting patterns. *Crop Sci.* 6: 55–59, 1966.

Shims, H.J. Changes in hay production and the harvest index of Australian oat varieties. *Aust. J. exp. Agric. Anim Husb.* 3: 198–202, 1963.

Singh, S. Revolution in cotton yield with CCC. *Indian Fmg.* 20: 5–6, 1970.

Singh, T.P. Harvest index in lentil (*Lens culinaris* Medik). *Euphytica.* 25: 833–39, 1977.

Singh, C.B. and M.P. Shrivastava. Variation in harvest index and its utilization in breeding of *Cajanus cajan. In*: Int. Workshop on Pigeonpeas. Vol. 2., ICISAT, Hyderabad, India. pp. 137–42, 1980.

Singh, I.D. and N.C. Stoskopf. Harvest index in cereals. *Agron. J.* 63: 224–26, 1971.

Sinha, S.K. and M.C. Ghildyal. Increase in yield of Bengal gram (*Cicer arietinum* L.) by 2, 3, 5-tri-idobenzoic acid. *Crop Sci.* 13: 283, 1973.

Snyder, F.W. and G.E. Carlson. Selecting for partitioning of photosynthetic products in crops. *Adv. Agron.* 37: 47–72, 1984.

Soper, M.H.R. A study of the principal factors affecting the establishment and development of the field bean (*Vicia faba*). *J. agric. Sci.*, U.K. 42: 335–46, 1952.

Spaeth, S.C., H.C. Randall, T.R. Sinclai and J.S. Vendeland. Stability of soybean harvest index. *Agron. J.* 76: 482–86, 1984.

Subhadrabandhu, S. F.G. Dennis and M.W. Adams. Abscission of flowers and fruits in *Phaseolus vulgaris* L. II. The relationship between pod abscission and endogenous abscisic, phaseic and dihydrophaseic acids in pedicels and pods. *J. Am. Soc. hort. Sci.* 103: 565–67, 1978.

Swamy, L. and M. Hashim. An induced nonbranching mutant in Guar. *J. Cytol. Genet.* 15: 105–06, 1980.

Syme, J.R. A High yielding mexican semidwarf wheat and the relationship of yield to harvest index and other varietal characteristics. *Aust. J. exp. Agr. Anim. Husb.* 10: 350–53, 1970.

Syme, J.R. Single plant characters as a measure of field plot performance of wheat cultivars. *Aust. J. Agric. Res.* 23: 753–60, 1972.

Tanaka, A., K. Kawano and J. Yamaguchi. Photosynthesis, respiration and plant type of the tropical rice plant. *Int. Rice Res. Inst. Bull.* 7, 1966.

Tanner, J.W. Discussion of paper Loomis and Williams, In: *Physiological Aspects of Crop Yield* (eds) J.D. Eastin et al., Am. Soc. Agron., Madison, Wis. U.S.A. pp. 50–51, 1969.

Tanner, J.W., 1969 C.J. Gardener, N.C. Stoskopf and E. Reinbergs. Some observations on upright-leaf-type small grains. *Can. J. Pl. Sci.* 46: 690, 1966.

Tanno, H. and K. Gotoh. (Range of variation of harvest index in spring wheat varieties). Memoirs of the Fac. Agric., Hokkado Univ. 14: 56–63, 1983.

Tanno, H., Y. Komaki and K. Gotoh. (The effectiveness of selection based on harvest index in spring wheat). Memoirs of the Fac. Agric., Hokkaido Univ., Japan. 14(4): 352–56, 1985.

Thakural, S.K., O.P. Luthra and R.K. Singh. Genetics of harvest index *vis-a-vis* biological and grain yield in wheat. *Cereal Res. Commu.* 7: 153–659, 1979.

Tikhonov, V.E. (A model for appraising the significance of the harvest index and the rate of vegetative weight accumulation in spring wheat under drought conditions). *Selektsiya i Semeno-vodstvo*, USSR. No. 3: 22–23, 1985.

Van Dobben, W.H. Influence of temperature and light conditions on the dry matter distribution, development rate and yield in arable crops. *Neth. J. agric. Sci.* 10: 377–89, 1962.

Van Schaik, P.H. and A.H. Probst. The inheritance of inflorescence type, peduncle length, flower per node, and per cent flower shedding in soybeans. *Agron. J.* 50: 98–102, 1958.

Vogel, O.A., R.E. Allan and C.J. Peterson. Plant performance characteristics of semidwarf winter wheats producing most efficiently in eastern Washington. *Agron. J.* 55: 397–98, 1963.

Walker, A.K. and R.J. Fioritto. Effect of cultivar and planting pattern on yield and apparent harvest index in soybean. *Crop Sci.* 24: 154–55, 1984.

Wallace, D.H., M.M. Peet and J.L. Oxbun. Studies of CO_2 metabolism in *Phaseolus vulgaris* L. and applications in breeding. In: *CO_2 Metabolism and Plant Productivity.* R.H. Burris and C.C Black, (eds) Univ. Park Press. pp. 43–58, 1976.

Watson, D.J. and K.J. Witts. The net assimilation rates of wild and cultivated beets. *Ann. Bot. N.S.* 23: 431–39, 1959.

Webster, B.D. M.E. Craig and C.L. Tucker. Effects of ethiphon on abscission of vegetative and reproductive structures of *Phaseolus vulgaris* L. *HortScience* 10: 154–56, 1975.

Weber, C.R. Effect of defoliation and topping simulating hail injury to soybeans. *Agron. J.* 47: 262–66, 1955.

Whan, B.R., R.Knight and A.J. Rathjen. Response to selection for grain yield and harvest index in F_2, F_3 and F_4 derived lines of two wheat crosses. *Euphytica.* 31: 139–50, 1982.

Wien, H.C. The use of ethiphon (2-chloroethyl phosphonic acid) to screen for abscission resistance among cowpea (*Vigna unguiculata* (L) Walp) cultivars. HortScience. 14: 422, 1979.

Wittwer, H.S. Growth retardants in agriculture. *Outlook in Agriculture.* 6: 295–17, 1971.

Yadav, I.S., R.P.S. Tomer and D. Kumar. Phenotypic stability for harvest index in black gram. *Madras Agric. J.* 70: 433–35, 1983.

Yoshida, S. Physiological aspects of grain yield. *Annu. Rev. Pl. Physiol.* 23: 437–64, 1972.

CHAPTER 8

ROOT TYPE

INTRODUCTION

Much has been researched and written about plant ideotype (shoot type only); architecture, determinate versus indeterminate habit, dwarfness, earliness, leaf angle, green area retention, etc., but very little attention has been paid to similar aspects of root growth on which whole plant architecture depends, mainly because of difficulties in study due to its underground position. Improvements in different aspects of shoot growth have resulted in yield increases; similar improvements in roots are bound to result in further yield increases, and under critical stress conditions these may even determine production or failure of the crop. There is very little study on root variability, selection, heritability, heterosis and breeding, and root: (i) architecture — deep penetrating roots for subsoil water and nutrient absorption, root branching behaviour and ramification, diameter (surface-volume ratio), regeneration capacity, etc.; (ii) efficiency in nutrient uptake; and (iii) stress tolerance, for example, the stresses of drought, toxic ions, soil acidity, salinity and sodicity (alkalinity), waterlogging and anoxia, and cold. The little work done is also diffused and not easily accessible.

Vedrov (1974) presented a discussion on the importance of root type in breeding new forms of plants. In breeding wheat for resistance to drought, selection for the vigour of the primary (seminal) root was considered important. In their monograph, "Root systems and yield in crop plants" Ustimenko et al. (1975) have discussed the anatomy, morphology and biology of the root, taking account of genetic, specific and varietal features in the development of root system and differences in hybrids in this respect. Attention has been drawn to the inheritance of growth and activity of the root system and the value of selection for this character. Correlations are indicated between the development of the root system and the various economic characters in such crops as wheat, barley, maize, *Panicum* millet, sunflower, peas and perennial herbage.

Zykin (1976) reports that in bread wheats the number of secondary roots is positively correlated with such characters as 1000-grain weight, grain weight per plant, and productive tillering. Performance of spring wheat varieties and their F_1 and F_2 progenies was studied in relation to the initial rate of primordial root production (Blaha and Toman, 1980). The ratio of dry matter of primordial roots to that of shoots, five days after planting served as a prediction index for performance. High positive correlations between the index and the actual performance of progeny were found.

Phaseolus beans selections with tap root morphology outyielded others possessing fibrous roots (Guadamuz, 1980). In soybean, cultivars were selected with fast tap root elongation rate and slow tap root elongation rate. The cultivars with fast elongation rate had deeper maximum rooting depths and a greater number of roots at 100, 150 and 200 cm depths than those with slow elongation rate (Kaspar, 1982). The cultivars with fast elongation rates obtained a larger percentage of their water from below 120 cm while others with slow elongation rate absorbed from above this zone.

Several of the recently bred varieties, for example AN 402 as against Tashkent 1, develop roots more rapidly, and the main and lateral roots are more vigorous and have greater final root length than Tashkent 1 (Nazirov and Satipov, 1980). Also variety AN 402 requires less water up to the flowering stage. Hayami (1983) reports that the improved rice varieties have a bigger proportion of lower roots, higher respiration rate in the roots and better uptake and distribution of nitrogen and carbohydrates in the leaves. Both the number and length of seminal roots are independent of endosperm or grain size, rather they are controlled genetically (Sandhu and Bhaduri, 1984). The rate of N uptake by these roots is also governed genetically. The N uptake rate by the excised root tips of *japonica* rice cv. Taichung 65 is faster than the *indica* rice cv. Taichung Native 1 (Hou and Lai, 1983). Uptake then became very slow in *japonica* cv. but increased as length and weight of root increased in the *indica* cv. Also the activities of nitrate reductase and glutamate dehydrogenase were higher in the roots of *indica* cv. than those in *japonica* cv.

There is some relation between the number of leaves on the main stem and root growth in rice (Kawashima, 1983). Root elongation ceased seven days before heading in cv. Reiho (22 leaves on the main stem), at heading in Toyonishiki (16 leaves) and 13 days after heading in Ishikari (10 leaves). Secondary and tertiary roots of Ishikari were longer than those of Toyonishiki and Reiho (Kawashima, 1988). Density of tertiary roots was highest in the upper primary roots of Toyonishiki and Reiho. Percentage of total root formation attained at a specific time before heading was highest in Reiho, intermediate in Toyonishiki and lowest in Ishikari; the cultivars completed their root system development in the similar order.

Passioura (1972) emphasized the effect of root geometry on the yield of wheat growing on stored water. Tikhonov (1973) involving 40 varieties of spring bread wheat determined the number of embryonic roots; varieties having the greatest number of roots at germination were most productive in both rain-fed and irrigated conditions. The number of primary roots was positively correlated with the number of grains per ear and with 1000-grain weight (Alekseeva and Malina, 1976). In a study, four drought resistant sorghum lines showed deeper root penetration and produced heavier and more numerous primary and secondary roots than the four susceptible lines (Bhan et al., 1974). In barley, the degree of development of root system is positively correlated with the productive tillering and grain weight per plant (Oleinik, 1977). The correlation between yield and the number and length of embryonic roots was high, while that between yield and the number of crown roots was low (Oleinik, 1977; Gorny and Patyna, 1981).

Wheat varieties with the highest number of primary roots give the highest grain yield, being 8 to 43 per cent greater in bread wheat and 11 to 48 per cent greater in *durum* wheat when plants have five such roots and the others have three (Tyslenko, 1981). The number of crown roots is also correlated with grain yield. Total root volume has also shown a positive correlation with grain yield (Bruns and Croy, 1985). Generally, the roots of semidwarf wheat varieties penetrate deeper into the soil and are better developed than those of the tall varieties (Netis et al., 1985). In addition to total root weight, yield is also related with the depth of root penetration into the soil. In rice, fresh root weight has a positive direct effect on grain yield (Sasmal, 1987).

VARIABILITY

Variability has been observed in root number, length and rate of growth in different crop varieties and the plants have been noted to possess differences in their yield potential, tolerance to drought, water-logging, salinity, cold, fertilizer responsiveness and earliness of the crop. Some of the relevant work will be discussed here taking care that the information reported by O'Toole and Bland (1987) is not repeated.

Observations on six-day-old seedlings of four standard winter wheat varieties and natural and induced mutants of San Pastore, multiplied in the $M_{10} - M_{12}$, were compared (Foltyn, 1972). Positive correlations apparent within varieties or lines and grain size, number of primary roots, length of roots and seedling height did not hold good. A study of 350 winter wheats including 11 amphidiploids revealed intervarietal and interspecific differences in the number of primary roots per germinating seed (Kiryan, 1974). The number of roots is a varietal characteristic, though subject to modification under environmental influences.

In a study of spring wheat varieties including Poltava-Albidum 43, Lutescens 758, Saratov 38 and Albidum 1616, intervarietal differences were found in the number of embryonic roots, length of the main and other embryonic roots, and the extent of their branching (Pavlova et al., 1973). The rapid rate of formation, growth and branching of the embryonic roots in Saratov 38 and Albidum 1616 resulted in a better development of the top growth.

Among the 40 varieties of winter wheat, the mean number of primary roots in 14-day-old plants ranged from 3.3 in Centurk to 5.7 in Vigorka (Tomasovic, 1978). Most of a group of 15 branched–eared and four–rowed lines had over 4.5 primary roots but differed little in this respect among themselves. The lines derived from crosses between branched and normal forms had more primary roots than the respective normal plants. Selection according to the number of primary roots is recommended as a rapid way of eliminating less productive forms early in the breeding process. Further, in a greenhouse experiment with 10 wheat varieties from three different geographical regions, O'Brien (1979) recorded differences between varieties in the number of second-order lateral roots, maximum depth of root penetration and the angle between the seminal root axes of four-week-old plants. Varieties also differed in the number of first-order lateral roots of plants which were between two and four weeks old. To study variations under different conditions, 15 hybrids from crosses between winter and spring wheat varieties were grown under favourable, early and drought conditions (Petukhovskii et al., 1984). Under these three conditions (modificatory variation) for the number of seminal roots was 17.3 to 20.4 per cent; genotypic variation was 5 to 5.9 per cent and phenotypic variation was 18.2 to 21.1 per cent. The number of seminal roots varied from 3.05 to 4.97. Under drought conditions most hybrids were inclined toward the better parent, and Petukhovskii et al. suggested that selection for high seminal root number would be effective in early generations.

Hirota and Watanabe (1981) studied the mechanism of varietal differences in the curvature of seminal roots of maize in water culture. The seminal roots of sweet corn varieties grew strongly curved and those of dent varieties mainly straight, while those of flint varieties were intermediate. They suggested that ethylene produces departures from straightness in seminal roots and that auxins promote its production from methionine. The differences in auxin activity would then account for intervarietal differences in root curvature.

At a given stage of development, the number of primary roots differ in spring wheat varieties (Rudenko and Kiryan, 1975). At the phase of coleoptile rupture, 48 per cent of the plants had four to five roots, whereas in Kharkov 46, the figure was 81 to 94 per cent. The ratio of roots to top growth differed among the varieties even at the early stages of development; in Red River 68 it was 0.25 to 0.26 whereas in Kharkov 46 it was 0.37 to 0.38. But Robertson and Waines (1977) found greater differences in root number and total root length within wheat lines than between lines in seedlings grown from 30 grains taken from each of 50 accessions of diploid and tetraploid types.

Engledow and Wardlaw (1923) studied two barley cultivars and found distinct cultivarial differences in the pattern of root, one being deeper rooted than the other. Apel et al. (1981) studied 957 forms of barley for number of roots per seedling which ranged from 3.70 to 8.47. The number of roots was correlated positively with 1000-grain weight, height and lysine content, and negatively with protein content. The number of roots per plant and maximum root length were measured at 10, 20, 30 and 40 days of age in five early cultivars and 27 F_5 early barley lines (Omara and Hussein, 1987). Both the root traits showed significant genetic variation. Both the root traits at 40 days age showed significant association with grain yield under drought condition. Omara and Hussein suggested that selection for root number per plant may be helpful in increasing drought resistance.

Out of 10 sorghum lines, Jordan and Miller (1977) found three to have shorter and dense root systems than the other lines having greater lateral root growth, higher root densities in the top 60 cm of soil profile and fewer roots below 120 cm. Milyutkin et al. (1974) found a relatively rapid initial root growth in the sorghum varieties Kafrskoe Rannee (Kaffir Early), Durra 2-plodnaya (Durra 2-fruited), Zheltozernoe 10 (Yellow-grained 10), Gaolyan rannii (Kaoliang Early) and Khegari rannee (Early Hegari). Out of 28 sorghum cultivars representing seven species, Yakushevskii and Kochetova (1974) showed that root systems were less developed in *Sorghum technicum* and *S. nervosum* than in forms from warmer habitats, but among the latter certain individuals were found which combined good root systems with low heat requirements. The most vigorous root systems among the seven species were found in *S. guineense*. In pearl millet, Shriniwas and Subbaiah (1976) noted that the hybrids Hyb K559, HB3 and HB4 were deeper rooting than HB1 and D356. HB3 and HB1 had wider root spread than the others.

Raper and Barber (1970) isolated two soybean cultivars with differing roots systems. One cultivar, Harosoy 63, had about twice the root surface of the other, Aoda, though their top growths were similar. Significant differences in the number of primary and secondary roots of nine soybean cultivars were observed (Tian, 1984). Variability in the number of secondary roots was greater than that for primary. Tian suggested that soybean cultivars can be divided in three groups : (a) with well-developed primary roots, (b) with well-developed secondary roots, or (c) intermediate. The last group has wider adaptability. Varietal differences in root system become evident after 35 to 40 days of seed germination when the secondary roots grow from the hypocotyl.

Ali-Khan and Snoad (1977) noted considerable genetic variation in several root characters of 30 *Pisum* genotypes, with length of the main root showing the greatest variability. Estimates of heritability exceeding 50 per cent were noted for the number of laterals, length of shoot, weight of root and weight of shoot. Seed weight was significantly correlated with the length of the longest lateral and weight of root. Stoffella (1980) studied the morphology of two dwarf cultivars (Redkote and Redkloud), two semiviny cultivars (Aurora and BTS) and one viny cultivar (UI 111) of *Phaseolus* beans. Generally, the dwarf and semiviny cultivars had a deeper root system than the viny cultivar.

Significant differences were observed in root length, volume and dry weight, and number of secondary roots among 22 accessions of groundnut (Pandey and Pendleton, 1986). Root length ranged from 91.3 cm in accession 199 to 56.1 cm in 717; root volume from 55.9 cm^3 in 840 to 31.1 cm^3 in 331; root dry weight from 5.3 g in 228 to 2.5 g in 331; and mean number of secondary roots from 32.6 in 330 to 17.4 in 199. Three Virginia and three Spanish groundnut genotypes were studied for tap root growth rate, number of roots and root length density (root length/cm^3 of soil) (Huang and Ketring, 1987). Virginia types had longer tap roots and faster root growth rates than Spanish types, and a lower root : shoot ratio.

Lateral root development and vascular bundle arrangements were compared in seven-day-old seedlings of 120 exotic *Gossypium hirsutum* strains (Mc Michael et al., 1987). Significant differences in lateral root development were observed between the strains. Partitioning of total root length into lateral roots was significantly higher for T 25 and T 256 than for T 269 and T 265. The proportion of plants with high branching intensity (number of lateral roots per cm of tap root) increased as the number of vascular bundles of the strain increased from four to six or more.

Roots in relation to drought tolerance: Derera et al. (1969) studied the genetic variability in root development in relation to drought tolerance of spring wheats. Varietal differences in number of roots, their total and effective surface area and the weight of the aerial plant parts were determined by Danilchuk (1970) in varieties of winter bread and *durum* wheats, barley and rye. Winter wheats with vigorous, well-developed root systems were most productive. Roots of cultivar Odessa 16 and Odessa 26 penetrated to a greater depth, enabling them to resist drought but preventing effective utilization of applied nutrients in the surface layer. In high-yielding varieties (Bezostaya 1), roots are vigorous, but located mainly in the upper layers of the soil, and thus do not confer drought resistance. Nefedov (1970) also observed differences among different forms of wheat in a number of embryonic roots at germination, length of the basal internode, number of days from seedling emergence to the formation of nodal roots, length and thickness of roots, dates of commencement of root branching, extent of branching, density of root hairs, absorbing capacity and cessation of vital activity. Plants raised from seeds which at germination had many roots, gave the highest yields. By selecting twice for root vigour, a yield increment of 19 per cent was achieved in four *durum* varieties. Under favourable conditions, drought resistant varieties had a greater proportion of aerial parts to roots at the tillering stage.

In a study of different varieties, Baitulin (1972) found a greater depth of penetration of the root system in the more drought resistant varieties of wheat and barley than in the less resistant varieties. In five varieties of barley, Pasela (1975) recorded the root mass in successive 5 cm soil layers to a depth of 1 m at various stages of development. The greater root development below 75 cm found in var. Piast at the milk-ripe stage was considered to be a factor contributing to its drought resistance. Under rain-fed conditions, the rice varieties N 22 and Bala developed the greatest spread of lateral roots, whereas the rooting of IR 24 and Jaya improved only under flooded conditions.

The more drought resistant the variety, the better it maintained the vigour of its embryonic roots in drought conditions and more crown roots it developed. Tewari et al. (1974) studied the rooting pattern as a selection parameter of wheat varieties under moisture stress. Of 16 varieties, NP 404, C 306, PKD 14 and Hy 633 had the greatest number of seminal roots; PKD 14, C 306, Hy 11, Hy 633, S 307 and Hy 65, the greatest horizontal spread of seminal roots; HD 1625, Hy 633, C 306, DA 491-5 and Hy 65 the earliest crown root initiation; NP 404, HD 1615, C 306 and PKD 14 the greatest number of functional crown roots; and C306, Hy 11, Hy 633, Hy65, HD 1460 and HD 1625 had the greatest yield. Short crown-node length was directly correlated to crown root number. Yield was directly correlated with number and length of seminal and crown roots. They concluded that C 306, Hy 633, Hy 11, PKD 14, HD 1625 and Hy 65 are most suitable for rain-fed cultivation.

Oleinik (1978) found a high correlation between the number of embryonic roots and yield in 200 forms of barley studied under drought conditions. Under hydroponics, Kazywanski et al. (1976) determined the ratio of the shoot mass to mass of three barley varieties which was greatest in var. Lubuski at tillering, earing and milk stages, but the variety Skrzeszowicki had the highest proportion of active root surface from tillering to milk ripeness.

Nour and Weibel (1978) have reported that the most drought resistant varieties had greater root weight, greater root volume, longer roots and higher root : shoot ratios than in the less drought resistant ones. The large grained barleys have been reported to possess fewer primary and secondary roots but a more vigorously developed root system in the lower horizons of the soil (Borodai and Savchenko, 1977). In a study of four cotton varieties, 159F had the least vigorous and S 4727 the most vigorous root system at flower bud formation, but by the end of growth, 159F had the most vigorous root system and S4727 the shortest main root (Sattarov, 1978). S4727 had the largest percentage of bolls surviving and opening. Thus it is likely that in early selection based only on the number of embryonic roots, these materials might have been lost.

A comparison of the Mexican semidwarf wheat variety Kalyan Sona and the tall Indian var. C 306 showed that the latter is more adaptable to variation in soil moisture owing to greater lateral root spread and higher concentration of roots at lower depths (Singh et al., 1973). Further, Welbank (1974) reports that early spring semidwarf wheat varieties have more roots below 30 cm than the taller varieties, suggesting a possible advantage in dry seasons. On silty clayey loam, wheat var. Sonalika and triticale var. Branco 90 had deep root systems, with over 18 per cent of the roots at 36 to 60 cm depth as compared with wheat varieties UP 302 and Hira which had less than 12 per cent at this depth, and Kalyan Sona which was intermediate in this respect (Sharma et al., 1974).

Differences in grain sorghum yield between F_1 hybrids E57 and Tx 671 under dryland conditions with sufficient rain are associated with differences in extraction of deep soil water (Wright and Smith, 1983). E57 uses less water before anthesis than Tx 671, but uses more water after anthesis. The ability of E 57 to maintain higher rates of water uptake during drought than Tx 673 is associated with the higher . root mass and longer roots of E 57. The rye variety Permontra having an extensive root system is fairly tolerant of heat and drought (Gordon-Werner and Dörfeling, 1988).

Roots in relation to water-logging: Of the two dwarf, high yielding Mexican wheat varieties, Inia and Pato, Pato tolerates flooded soil better than Inia (Yu et al., 1969). Syu (1973) investigated the root growth and resistance to water-logging in 24 wheat and barley varieties at the tillering and flowering stages. The resistant varieties had thicker roots at both stages and longer and heavier adventitious roots on the main stem at flowering. The root systems of maize composites A54 and Ganga 5 are relatively shallow and compact, that of Vijay deep and compact, and those of Hunis, Deccan and Ganga Safed 2, deep and spreading (Kamath et al., 1974). Under water-logged conditions, Deccan and Vijay developed shallow and compact root systems.

Reduced oxygen concentration under water-logged conditions could be another factor responsible for reduction in root growth. Thus the development of rice seedlings in water at controlled oxygen levels was studied by Turner et al. (1981). Inhibition of root elongation occurred at oxygen concentration below 5 per cent in varieties Labelle and Bellemont, but at a concentration below 21 per cent in M101 and Calrose. Though roots of the latter pair of varieties were longer than those of the former pair at the higher oxygen concentration, the results suggest that Labelle and Bellemont were more tolerant to decreases in oxygen concentration from 21 to 5 per cent. When oxygen concentration was reduced from 10.5 per cent, root length decreased.

Root length in relation to salinity: Development of a vigorous root system is associated with salt tolerance (Vekulenko, 1976). Of the several spring wheat varieties tested, Milturum 553 is quite tolerant of saline soils and has a vigorous root system. In an experiment, there was a greater reduction in the number of embryonic roots and the number of root hairs in the barley var. Dokuchaev 1 than in the more salt tolerant var. Krasnodar 35 (Semushina, 1979). Also, Dokuchaev 1 had less rapid accumulation of dry matter at heading under saline conditions than Krasnodar 35. The number of embryonic roots was correlated with both spikelet number and grain number.

Roots in relation to cold tolerance: The deep rooted winter wheat variety Ultuna gives a better performance in field tests for resistance to frost damage than the shallow rooted variety Sv 58315 (Anon., 1974). Tymchuk (1975) demonstrated the effectiveness of selection for vigorous development of the root system in breeding high yielding and cold tolerant winter wheat varieties. Further, with a legume, Gromova (1975) reports that most of the cold hardy soybean varieties have a more vigorous root system in the early stages of growth at optimum temperatures, which can serve as an indirect method of selecting for cold hardiness.

Roots in relation to fertilizer response: Among the 36 wheat varieties and lines and 21 barley varieties, the nitrogen responsive ones showed greater increase in seminal root growth of 20-day-old seedlings after N-application (Szymanska, 1980). But the seedlings of those varieties responding least to N-application, produced a greater mass of seminal roots in the absence of N-treatment than with it. Root distribution in triticale (70-2), and *Triticum durum* HD 4502 and *T. aestivum* Kalyan Sona and Moti increased both vertically and horizontally by N-fertilizer. Maximum N absorption occurred in *T. durum* (Meena and Seth, 1974).

Roots in relation to earliness: The corn hybrid VIR 42, in the medium maturity class, has a root system that covers a larger soil volume than that of the midearly Dneprovsk 98, and absorbs greater amount of water from the soil (Zolotov and Fevralev, 1973). Gorbunov and Chernov (1973) report that in early stages of growth the root system of the early spring wheat varieties penetrate deeper into the soil than that of the late varieties. In the later stages, however, the midseason and late varieties show a higher growth rate of the embryonic roots.

SELECTION

Sathyanarayanaiah and Townley-Smith (1981) evaluated 16 wheat cultivars and breeding lines for their root development at four stages after sowing and found that the optimum time for screening for root development is 30 days after sowing. Hurd (1964) also noted that some wheat cultivars produce more root at the seedling stage than others, and also maintain higher root weight at maturity. However, Kalashnik et al. (1980) stress that selection for increased length of the primary root is more effective at higher rates of fertilization.

Vedrov (1982) recorded a close correlation between the number of embryonic roots and grain yield in spring wheat. Vedrov emphasized the importance of a number of embryonic roots as an important selection criterion in breeding for drought resistance. In order to obtain forms with many seminal (embryonic) roots and improve drought tolerance, the maternal parent should have more seminal roots (Dorofeev and Tyslenko, 1982). Selection for many seminal roots is highly effective from the F_3. Hybrid populations from forms with five seminal roots yielded 31.1 to 43.1 per cent more than those from forms with three such roots. However, Hettinger and Engels (1986) stress that when three drought tolerant and seven susceptible barley genotypes were grown in plexiglass containers, after 16 days the tolerant genotypes had the longest seminal roots, total length of seminal roots and highest total root weight. They emphasized that the length of the longest seminal root appeared to be the most suitable marker in selecting for drought resistance. With rice, however, Loresto et al. (1983) showed that only root diameter had a significant correlation with drought resistance during the vegetative phase, but during the reproductive phase, long roots, thick roots, few roots and a high root : shoot weight ratios showed high correlations. They concluded that root system data at the age of 50 days are reliable for screening rice breeding materials for drought resistance. Selection for desirable root characteristics associated with drought resistance has been practiced in rice. Selection for root type based on individual plant performance in early segregating generations has been recommended (Ekanayake et al., 1985b). Ekanayake reports that root characters are significantly correlated with visual field drought resistance scores and with leaf water potential.

With increasing plant age, root characters become more closely correlated with yield (Kuzmin and Shumeiko, 1980). At the tillering stage, the main role is played by the volume of the root system rather than by the number of roots. They further observed a close positive correlation between the number of roots per plant and the number of productive tillers at the stage of wax ripeness. This enabled forms with a large number of roots plus a high yield to be selected among F_1 hybrids. However, with cotton, Eissa et al. (1983) suggested that selection will be more successful if delayed up to F_3 to allow genetic recombination of additive and additive × additive epistatic genes to occur. Recurrent selection should be a useful method of breeding.

Experimental results show that selection for vigour of development of the root system can be an effective method of obtaining initial material for breeding widely adapted varieties. A number of the new varieties bred have deep penetrating and well-branched roots (Garkavyi et al., 1970). The pea varieties with vigorous root system were found to flower and mature two to three days earlier than the others (Yakimenko and Boitsov, 1971). Sachanski and Naneva (1975) divided the plants of pea varieties differing in their morphological characters into groups according to the length of their radicles, 10 to 12 days after germination. The plants with the longest radicles had the best values for morphological characters and the highest seed yield, which was inherited by the progeny. Further, Dekov et al. (1978) selected seedlings of the French bean Dobrudzha 2 and Ruse 3 for root length on the sixth day after germination. Those with roots 6 to 8 cm long gave the highest yields. Also the lupin plants selected for root vigour gave higher yields of seed and green matter in the F_2 and F_3 than the progeny of unselected plants. In green gram, both seedling root length and the number are positively and significantly correlated with yield; Islam et al. (1987) suggested that both these root traits might be valuable as early selection criteria.

Intravarietal mass selection for germination energy, initial growth vigour and development of the root system in winter wheat varieties Bezostaya 1 and Odessa 16, increased productivity of the plants and the quality of the seed. Stavnuchuk (1972) attributed the considerable variation in germination energy, initial growth vigour and development of the root system within these otherwise relatively constant varieties, largely to the fact that selection for these characters has so far received little attention. Danilchuk (1974) asserted that the features exercising the most beneficial effect on yield are greenness of the leaves and the ability of the roots to proliferate in the soil horizons which constitute the best sources of moisture and nutrients in particular conditions. Danilchuk made use of these criteria in selection of wheat and barley breeding materials. Tymchuk (1975) also demonstrated the effectiveness of selection for vigorous development of the root system in breeding high yielding and cold tolerant winter wheat varieties.

As a result of many years work, Kirichenko et al. (1974) described a method for selecting plants for the vigour of their root development. In this way, the grain yield of maize was increased by 3 to 6 c/ha, grain yield of wheat and barley by 2 to 3 c/ha and the root yield of sugar beet by 10 to 15 per cent and its sugar content by 0.5 to 1 per cent. Zelenskii and Neboka (1971) reported that selection of the best winter wheat plants for yield resulted in simultaneous selection for vigour of the root system, the improvements in root vigour amounting to 14.2 to 73.5 per cent and in grain yield to 10.2 to 39.2 per cent.

Selection of cucumber varieties Nerosimyi 40 and Izyashchnyi by Lebedeva (1971) for root vigour resulted in increase of dry matter accumulation per unit area of leaf. Selection for root vigour was most effective during the cotyledonary and five to six leaves-stages. Four families of Izyashchnyi and five families of Nerosimyi 40 were selected for their improved dry matter accumulation. Selection for root system vigour resulted in higher seed content per fruit, higher absolute seed weight, more female flowers, and higher resistance to cold. After a three-fold selection for vigorous root system, 1000-seed weight rose by 15 to 16 per cent in Izyashchnyi and Nerosimyi 40 (Yurina and Lebedeva, 1974). A selection intensity of 10 per cent proved better than of one 50 per cent. Seed yield compared with the control was increased by 29 per cent and fruit yield in the progeny by 23 per cent. Alpatev and Agapov (1970) also report that cucumber plants selected at the cotyledon stage for vigorous root system developed better and had more shoots and leaves; yields were 20 to 32 per cent higher and the fruits were larger. The increase in yield was also transmitted to the progeny.

Selection Techniques

Selection of genotypes for desirable root traits *in situ* is laborious and time-consuming, and therefore, some simple and short-cut methods had to be developed. Several scientists have tried different methods. For instance, direct selection was practiced mainly for grain yield in a semiarid environment with ample, deep, stored soil moisture (Hurd et al., 1972b; Hurd, 1974; Hurd and Spratt, 1975). However, as we depend on selection for yield, this method cannot be used to select for rooting among diverse, nonadapted genotypes.

Ekanayake et al. (1985b) used a hydroponic technique to observe the root characters of elite rice genotypes, and established a significant relationship between root length, root thickness, and the thick-root numbers with field drought tolerance. Unfortunately this method is expensive and tedious, particularly with regard to the selection for superior cultivars in an improvement programme where one has to deal with large numbers.

It has been shown with rice and maize cultivars that those possessing high "root pulling resistance" (RPR) are more drought tolerant than others possessing low RPR (O'Toole and Soemartono, 1981; Ekanayake et al., 1985a; Fincher et al., 1985). Fincher et al. (1985) showed that RPR is also related to lodging resistance in maize and can be used as a screening technique for this trait. Plants having high RPR also possess the ability to maintain higher leaf water potential under severe drought stress (Ekanayake et al., 1985a). For measuring RPR, Ekanayake et al. (1986a,b) used a modified spring balance (capacity 25 kg) attached to the base of each seedling to be pulled. It is important to improve our understanding of the determinants of pulling resistance, for example, root volume and the number of seminal (primary) and secondary roots, amount of fibrous roots, relative size and symmetry of root system, number of nodes with roots and secondary root development. Nass and Zuber (1971) observed a relationship between seedling root traits such as total root weight, volume, and nodal root weight and RPR of mature plants. Based on path coefficient and correlation analyses of RPR in rice, O'Toole and Soemartono (1981) concluded that root branching, weight, length, and the number of thick roots are important contributors to high pulling resistance and hence the degree of drought resistance.

When the plants are pulled, some roots always get broken. Ekanayake et al. (1986a) addressed this important question of unpulled roots. They studied the characteristics of pulled and unpulled root components in five rice genotypes varying from drought sensitive to drought tolerant. Genotypes with high RPR were characterised by larger, thicker, and denser root systems. Pulling resistance was significantly correlated with root volume, dry weight, thickness, and the number of thick roots in the pulled root component. Several traits of the pulled root component showed highly significant positive correlations with "root length density" (RLD) at 0.3 to 0.6 m soil depth. However, RPR was most highly correlated with the RLD of the unpulled component, particularly the RLD from 0.45 to 0.6 m depth. Their results suggest that high RPR is strongly correlated to the ability of the plant to rapidly develop greater root penetration into deeper soil layers, and confirm the utility of RPR as a practical technique in selection for superior root systems. Time course analysis of RPR among cultivars showed that genotypic differences increased with plant age (Ekanayake et al., 1986a).

In addition to the deep root system with high RPR which exploit soil moisture more completely, especially at depth, and increase the total amount of water available to plants for producing greater dry matter and thus grain yield (Hurd, 1974; Jordan and Miller, 1980), Passioura (1972) argues that plants having slow early season water use, preserve more moisture stored deep in the soil for the critical grain filling period and can tolerate late season drought better. In fact, Richards and Passioura (1981 a,b) are developing wheat genotypes with slow early water use by selecting smaller metaxylem vessel diameters in the seminal roots. This is expected to increase resistance to water flow, and indirectly reduce water use early in the season. Unfortunately, this screening method may not work in dicotyledons where secondary growth in the root increases the amount of xylem tissue. One can select dicot genotypes with slow early water use by selecting slow early leaf area development which is not very reliable. Robertson et al. (1985) developed a method of herbicide banding. They banded the root zone of *Vigna unguiculata* with the herbicide metribuzin at the time of sowing at specific depths and lateral distances from the seed rows. By screening for the appearance of herbicide symptoms, it was possible to follow the progress of root growth. Significant genotypic differences in mean number of days to first herbicide symptoms were detected among five genotypes; California Black Eye 5 and Grant developed symptoms early and 8006 and PI 302457 late, PI 293579 being intermediate. Robertson et al. concluded that the technique is suitable for screening a large number of genotypes in the field for their rate of RLD.

Root pulling resistance is a genetically controlled heritable character. Analysis of data from three F_1 rice hybrids and the three varieties from which they were derived revealed that the F_1s showed significant and positive heterosis for RPR (Ekanayake et al., 1986a). The high RPR of the F_1 hybrids indicated substantially superior root growth. Sahai and Chaudhary (1986) compared the RPR of six parents, six of their F_1 hybrids and two control rice varieties. The F_1's were significantly superior in RPR and showed a positive heterosis. Ekanayake et al. (1985b) observed transgressive segregation for RPR in three crosses (high × high, low × high and intermediate × intermediate). Both additive and dominant genes controlled the variation. F_1 heterosis was positive in all three crosses, and F_2 distribution curves indicated that plants highly resistant to root pulling could be obtained from all the three types of crosses. Heritability estimates for the character ranged from 39 to 47 per cent.

GENETICS

In 20 cultivars of spring wheat of similar growth durations, Kazemi et al. (1976, 1979) found a significant variation in root mass measured after three, six and nine weeks of growth. The root mass of 12 of these cultivars was measured at six weeks, nine of which were used as female and three as male parents with their 27 F_1 hybrids. In F_2 over 32 per cent of the total phenotypic variability for root : short ratio was conditioned by additive gene effects. Although the additive portion of the total variance was not especially high, it suggests that a significant progress by direct selection in early generations should be possible. The inheritance of first leaf length, root length and the number of seminal roots was studied in S413 × C773 (tall × tall) and C273 × K68 (tall × semidwarf) (Virk et al., 1978). Root number and root length in S413 × C273 and leaf length in K 68 × Sharbati Sonora 64 indicated dispersed genes in the parents. Further, five crosses were made between tall spring barley varieties and short mutants and the variability of the strength of the root system was analyzed (Zenisceva, 1973). Results indicated that the root strength is controlled by a large number of dominant factors varying in their effects; additive effects being negligible. The heritability coefficient was low and the level of discrimination high for all the crosses. Zenisceva concluded that recurrent selection in later generations is necessary for improving root strength. The analysis of progeny from a diallel cross involving five spring varieties showed several root characters to be influenced by additive genes (Surma et al., 1978). The significance of specific combining ability effects indicated that dominant and epistatic genes also influence these traits. The heritability of root systems in 14 cultivars of wheat were studied by Roma (1962) who noted that the parental symptoms were passed on in crosses. In a study of seven wheat varieties and 10 F_1 intervarietal hybrids obtained from top crosses in which Saratovskaya 29 and Atlas 66 were the testers used with five maternal varieties, quantitative characters of the primary root system were controlled by additive and dominance genes (Gudinova et al., 1983). The highest general combining ability effects for root length were shown by Atlas 66 and Omskaya 9, and for the weight of the primary root system by Atlas 66, Omskaya 9 and Tselinnaya 20. Combining ability and the inheritance of tillering node length, epicotyl length and a number of primary and secondary roots were studied (Nakhaeva, 1985). Overdominance was noted for the characters, with additive genes making the greatest contribution. The variety Zhigulevskaya was recommended as a donor for improving length of the tillering node and the number of primary roots, while Partizanka and Kadett were recommended for improving length and the number of secondary roots. Hybridization using barleys with many rootlets improved drought resistance (Loshak and Logachev, 1970). Hybrid analysis indicated monogenic inheritance of the embryonic root number, with low number recessive (Danilchuk and Grabovaya, 1981).

The rain-fed rice varieties from Africa generally have deep and dense root systems (Ahmadi, 1983). Additive genetic effects predominate in the control of root characters, but dominance and epistatic effects were also observed. The mean value of a line for a character was correlated with its general combining ability for that character. Ahmadi suggested that different genetic systems control root development in *indica* and *japonica* types. Long roots and high root number are mainly controlled by dominant alleles (Armenta-Soto et al., 1983). Control of thick root tips and high root : shoot ratios is mainly by dominant alleles in some parents and recessive alleles in others. Heritability estimates were high for all four characters. Root length and root tip thickness were positively correlated with plant height. Root tip thickness was positively correlated with plant height. Root tip thickness was negatively correlated with root number and tiller number. Ekanayake et al. (1985b) studied inheritance of root characters in $F_1 - F_3$ of a cross between IR 20 (drought susceptible) and MGL 2 (drought resistant), in hydroponic culture. F_1 plants had thick, deep roots with a higher lateral and vertical distribution than that found in IR 20. Heterosis over the mean parental value was significant for most traits. Additive and dominance genetic effects were equally important. Heritability estimates from progeny parent regressions in the F_2 and F_3 were high for root thickness (0.61 and 0.80, respectively), root dry weight (0.56 and 0.92) and root length density (0.44 and 0.77).

The early cold resistance in maize hybrids having Vukovarski Zuban as a parent is expressed in the form of extensive root development soon after germination (Vekic, 1972). In the F_2 crosses of maize variety Navajo, which lacks seminal roots with the marker Mangelsdorf Tester, absence of seminal roots segregated as a single dominant gene or a pair of complementary dominant genes, one of which is linked with *Su* and is designated Asr 1. Analysis of data from a line × tester cross in *Pennisetum americanum* involving four lines and two testers revealed the importance of both additive and nonadditive gene action for dry root weight and the importance of nonadditive gene action for root length and root number (Manga and Saxena, 1986). Heterosis over the parental value was high for dry root weight and root length.

The F_2 material from a full 6 × 6 diallel set of crosses involving two wild forms, two primitive and two European varieties of peas, seed weight, seedling shoot and root length and the number of lateral roots at six to nine days after germination, and of fresh weight of shoot and root at nine days were under polygenic control, which was principally additive for seed weight, but dominance was important for seedling characters (Snoad and Arthur, 1974). Large differences were found between primitive types and the varieties with respect to fresh weight of shoot and root, and the number of lateral roots, which have resulted from the unconscious selection by plant breeders. Nine characters were studied in 10-day-old seedlings of 27 *Vigna radiata* cultivars (Rangasamy and Shanmugam, 1984). Phenotypic and genotypic coefficients of variation were high for fresh root weight. Fresh and dry weights of roots showed high heritability combined with moderate predicted genetic advance, suggesting additive gene effects. At the genotypic level, the number of rootlets, length of rooting zone, fresh root weight, dry root weight, and 100-seed weight were positively correlated with seed yield.

A diallel cross between seven genetically distinct homozygous lines of cotton showed that root growth exhibits overdominance and is controlled by at least three effective factors. Heritability is low for this character (Abdalla and Bird, 1973). From 150 lines of cotton, four lines with long roots and four lines with dense roots were selected (Eissa, 1982). Two lines with long roots, two with dense roots and the commercial cultivar ST 213 were crossed in a diallel and an analysis of the F_1s indicated that the genetic variation for root length and tap root weight were primarily additive. Eissa et al. (1983) further reported that large amounts of additive, dominance and additive × additive epistatic effects were present for the two primary root traits (roots longer and higher in relative root weight – mg/mm length). But in jute the number of adventitious roots is controlled by nonadditive gene action (Ghosh et al., 1979).

Ploidy Relations

Philipchuk (1971) reports that in general, the tetraploid forms of rye developed a stronger root system and had greater growth above ground. This is particularly noticeable in Vyatka moskovskaya poliploidnaya compared with Vyatka moskovskaya. The size of the root system was most often inherited through the maternal form, but in the F_2 segregation a considerable degree of variability was observed. All the tetraploid forms of rye were more resistant to lodging (better root system) than the diploid forms (Gladysheva et al., 1977).

Among the four species of wheat and *Aegilops* studied (Fritsch, 1977), the average number of seminal roots per plant was three to five and it was a species and variety specific character, being lower in hexaploid than in tetraploid wheats. Modern bread wheats have 5.5 seminal roots per plant. *Triticum sphaerococcum* have a seminal root number below that of wild diploid species. *A. mutica* has the smallest number among the *Aegilops* species. Robertson et al. (1979) observed genetic variability in seedling root number in wild and domesticated wheats. In 143 diploid, tetraploid and hexaploid accessions, different mean seedling root numbers were found within each ploidy level, and usually within each species or variety. The highest seedling root number was 6.45 in a *T. durum* accession and the lowest was 2.50 in a *T. araraticum* accession. Seedling root number was generally stable from one generation to the next and across environments. There was a positive correlation between seedling root number and grain weight. Mac Key (1979) presented data showing that: (a) an increase in grain weight in the diploid, tetraploid and hexaploid series of *Aegilops* and *Triticum* species is directly correlated with an increase in seminal root number, and (b) grain weight and seminal root number are dependent on the position in the ear at which grain develops. Richards and Passioura (1981b) studied the extent of genetic variation in maximum vessel diameter, number of seminal axes and frequency of multiple metaxylem vessels in over 1000 accessions of both modern and primitive wheats with different ploidy levels and in populations derived from them. The values in the tetraploids were similar to those of the hexaploids but the diploids had lower values for all the root characters. Parent-offspring regression established that the heritability of vessel diameter was considerably higher than that of the number of axes. There were significant responses to selection for increased or decreased vessel diameter in six different populations.

BREEDING ACHIEVEMENTS

Breeding for morphological or physiological traits depends on first establishing their significance and their inheritance characteristics. So far, there is very little established relationship of this kind and even where correlations have been found there is a cause-effect problem to solve. A few crosses have been made to study the inheritance of rooting characters and a much smaller number still has been made with the objective of transferring a particular root character from one parent into a new cultivar. Troughton and Whittington (1969) in a review of genetic variation in root systems refer to several root characters that have been shown to be controlled by single genes. The number of root characters is very large covering many aspects of growth, form and metabolism. Root weight has been most frequently studied but volume, length, diameter, depth of penetration, distribution in soil, degree of branching and number of root hairs are all aspects of form (Hurd and Spratt, 1975).

An extensive root system is positively associated with drought resistance. Early rapid growth builds up a reserve that will carry the plant through severe drought and contribute to yield even when that plant is not particularly resistant to desiccation (Hurd, 1974). For example, during the high stress days, the wheat variety Pitic 62 continues to photosynthesize and produce more root. It maintains a higher assimilation rate and utilizes the available soil moisture more efficiently than other varieties tested. The genetics of characters associated with drought are complex. To breed for a combination of such attributes requires the use of (a) a few carefully studied parents, (b) large population to permit the combination of many favourable genes, and (c) yield testing from early generation (F_3 if possible) onward to homogeneity.

Hurd (1969, 1971) outlined a successful method of breeding for yield in a droughty environment. Because genetic balance is so important in plant environment interaction, methods of breeding must be designed to allow and encourage the superior types to be present in segregating populations and to be identified. Hurd emphasized on careful study of potential parents and judicious selection of a few three-way or double crosses. Once a parent with a superior character for avoiding or tolerating drought is identified, complementary well-adapted parents are chosen to combine with it.

Pohjakallio (1945) crossed two oat cultivars — a deep rooted one (Klock III) with another possessing physiological drought resistance (Golden Rain); and promising high-yielding drought tolerant selections were obtained. Development of a creeping rooted alfalfa is one of a few examples of breeding for a root characteristic (Heinrichs, 1963).

Hurd et al. (1972a,b, 1973) transferred the Pelissier root pattern into Wascana and Wakooma, which are high quality cultivars with a 15 to 20 per cent increase in yield over cultivars previously grown in the fairly arid durum growing area of Saskatchewan. Pelissier has been grown for many years in spite of being downgraded and not recommended. The reason of its persistence is that Pelissier under average drought has always given an economic return. Waskana and Wakooma withstand average drought, as well, as Pelissier. Pelissier has a massive root system at lower depths, especially at the 60 to 120 cm depths relative to the other wheat cultivars studied (Hurd, 1964). Pelissier was crossed and backcrossed to Lakota, a good quality disease resistant durum cultivar. About 1500 lines from the backcross were yield tested in F_4 and again in F_6 generations (Hurd et al., 1972 a, b). Selection was made for yield on the dry prairie. They concluded that root patterns are heriditary; extensive root systems are an advantage to plants grown under moisture stress, and selection for yield in large populations grown under droughty conditions will result in choice of lines that possess an extensive root system.

Kalashnik et al. (1980) reported that the spring wheat variety Pyrothrix 28 is the best donor for improving length and weight of primary roots. Using high yielding forms in which 81 to 84 per cent of seedlings produced five seminal roots as maternal parents and forms with six seminal roots as paternal parents, lines were obtained which had six times as many seedlings producing six seminal roots as the better parental form (Sidorov, 1984). As the number of seminal roots has been shown to be associated with drought resistance, Sidorov suggested that by using this trait, varieties can be produced to improve yields and increase yield stability in areas of low rainfall. The best hybrids were obtained when the maternal parent had a very well-developed root system, particularly in terms of primary root number (Kuzmin and Shumeiko, 1985). Selection for this trait in cv. Zhnitsa raised yields by 20 per cent.

Heterosis Observed

Dorovskaya et al. (1971) compared the growth of root systems in interline hybrids of maize with their parental inbred lines and with an open pollinated variety. The interline hybrids had a faster and stronger root growth from the early stages of growth onwards. On a three-year average, root weight of the hybrids was 140.5 per cent of that of the initial lines. Early evaluation by root system of interline hybrids appears possible. Baligar and Barber (1979) grew seedlings of two maize hybrids and their parents in nutrient solution for 15 days. All hybrids exhibited heterosis for dry weight and root length. Root length per gram of shoot was higher in hybrids than in their parents. Further, three grain sorghum hybrids and their parents were compared until 40 days after emergence (Blum et al., 1977). Heterosis was significant in all hybrids for length of the seminal roots, growth rate of adventitious roots and root volume (Table 8.1, Fig. 8.1). Heterosis for root volume was probably a result of increased number and improved growth of lateral root branches. Consistent heterosis, however, was neither apparent for the number of adventitious roots per plant, nor for seminal or adventitious root dry matter. However, Damodar et al. (1978) observed heterosis for root activity in the sorghum hybrid CSH 1 but at the knee-high stage only, occurring at 10 cm depth. In another hybrid CSH 2 heterosis occurred at 12 to 22 cm depth.

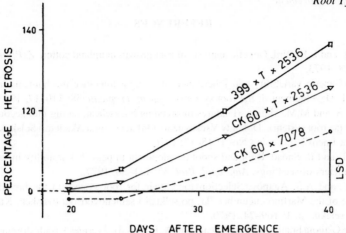

Figure 8.1. Heterosis in root volume (expressed in each hybrid as per cent of the best parent at several stages of growth) (After Blum et al., 1977)

In intervarietal crosses of bread wheat, Zykin (1976) recorded the highest heterosis for number of secondary roots and the lowest for number of primary roots. Bacon and Beyrouty (1987) grew seedlings of wheat hybrid HW 3015 and its parental lines for 37 days. The hybrid significantly exceeded the parental mean for root fresh weight by 51.8 per cent. Also in rice, heterosis was highest for root weight, followed by root number per plant and root length (Sasmal and Banerjee, 1986). Raj and Siddiq (1986) also reported that roots of all rice hybrids showed appreciable heterosis over the midparental, better parent and control values. One hybrid showed the highest heterosis for root density at 0 to 15 and 15 to 30 cm depths.

In a greenhouse experiment, Aycock and Mc Kee (1978), weighed the roots of seven cultivars and 21 F_1 hybrids of tobacco, 20, 35 and 52 days after transplantation. Heterosis estimates of 21.4, 13.6 and 15.5 per cent were obtained at the respective sampling dates. A significant general combining ability was found at each sampling date and a significant specific combining ability at 35 and 52 days after transplanting. The value of selection for large roots in order to reduce field lodging, is suggested. Papova and Mikhailan (1976) also reported heterosis for root length in red pepper.

Table 8.1: Root attributes and calculated per cent heterosis in three sorghum hybrids measured between 34 and 40 days after emergence (After Blum et al., 1977).

Hybrids	Adventitious roots per plant		Seminal root length rate		Root growth of adventitious roots		Total length of adventitious roots		Root volume		Dry weight of adventitious roots	
	No.	%H*	cm	%H	cm/ day	%H	cm	%H	cm³	%H	g	%H
CK 60 × Tx 7078	11.0	132.0	78	127.9	2.9	136.6	392	138.5	11.9	121.4	0.53	155.9
CK 60 × Tx 2536	8.7	72.5	79	129.5	3.0	156.4	353	114.2	15.1	131.3	0.65	102.7
Wheatland × Tx 2536	12.3	92.5	95	158.3	2.9	124.2	460	124.0	16.3	141.7	0.65	102.7

* Heterosis calculated for each hybrid as per cent of the best parental line.

REFERENCES

Abd-alla, S.A. and L.B. Bird. Genetic analysis of root growth in upland cotton. *Z. Pflanzenzuchtung* 69: 123–28, 1973.

Ahmadi, N. (Genetic variability and inheritance of drought tolerance mechanisms in rice, *Oryza sativa* L. 1. Development of the root system). *Agron. Tropicale* 38: 110–17, 1983.

Alekseeva, E.S. and M.M. Malina. (Features of breeding buckwheat, taking into account the development of the root system). Doklady Vsesoyuznoi Ordena Lenina Akademii Selskokhozyaistvennykh Nauk Imeni V.I. Lenina No. 1: 15–17, 1976.

Ali-Khan, S.T. and B. Snoad. Root and shoot development in peas. I. Variability in seven root and shoot characters of seedlings. *Ann. appl. Biol.* 85: 131–36, 1977.

Alpatev, A.V. and A.S. Agapov. Selection for vigour for root system in the breeding and seed production of the Marfino cucumber. Tr. po selktsii i semenovodstvu ovoshch. Kultur. Gribov. Ovoshch. selekts. st. 1: 169–74, 1970.

Anon. *In:* The Ostgota branch station 60 years old. A. Borg (ed), Sveriges Utsadesforenings Tidskrift. 84: 177–86, 1974.

Apel, P., A. Bergmann, H.W. Jank and C.O. Lehmann. (Variability in the numbers of seedling roots in barley). *Kulturpflanze* 29. 403–15, 1981.

Armenta-Soto, J., T.T. Chang, G.C. Loresto and J.C.O. Toole. Genetic analysis of root characters in rice. *SABRAO Journal.* 15: 103–16, 1983.

Aycock, M.K. Jr. and C.G. Mc Kee. Heterosis and combining ability estimates for root size in Maryland tobacco (Abstr.). *In: Agron. Abstr.*, p. 46, Madison, Am. Soc. Agron., U.S.A. 1978.

Bacon, R.K. and C.A. Beyrouty. Seedling root and shoot growth of a commercial wheat hybrid and its parents. *Cereal Res. Communications.* 15: 237–40, 1987.

Baitulin, I. O. Root system of spring wheat, barley and oats in the dry conditions of the foothill plain of the Trans-Ili Alatan. *In:* Biol. Nauki, 4. Alma-Ata, Kazakh SSR : 41–61, 1972.

Baligar, V.C. and S.A. Barber. Genotypic differences of corn for ion uptake. *Agron. J.* 71: 870–73, 1979.

Bhan, S., H.G. Singh and A. Singh. Note on root development as an index of drought resistance in sorghum (*Sorghum bicolor* (L.) Moench). *Indian J. agric. Sci.* 43: 828–30, 1974.

Blaha, L. and K. Toman. (Possibilities of primordial roots in breeding wheat for grain yield). *Genetika a Slechteni.* 16: 187–92, 1980.

Blum, A., W.R. Jordan and G.F. Arkin. Sorghum root morphogenesis and growth. II. Manifestation of heterosis. *Crop Sci.* 17: 153–57, 1977.

Borodai, Yu.G. and M.P. Savchenko. Development of the root system in barley in the south of western Siberia and in northern Kazakhstan. *Nauch. tr. Omsk. s-kh. in-t.* 161: 29–31, 1977.

Bruns, H.A. and L.I. Croy. Root volume and root dry weight measuring system for wheat cultivars. *Cereal Research Commu.* 13 (2/3): 177–83, 1985.

Collins, F.C., D.A. Hinkle and R.E. Baser. Tolerance among wheat varieties to low soil pH. *Arkansas Farm Research* 30 (4): 15, 1981.

Damodar, R., I.V. Subba Rao and N.G.P. Rao. Heterosis for root activity in grain sorghums. *Indian J. Genet. Pl. Breed.* 38: 431–36, 1978.

Danilchuk, P.V. The development of the roots and above ground parts of cereals in relation to their productiveness and drought resistance. *Sb. nauch. tr. Vseb. selekts-genet. in-t.* No. 9: 163–71, 1970.

Danilchuk, P.V. Morphological and physiological peculiarities of the development of the above ground mass and of the root system in wheat and barley plants, considered as yield factors. *In:* L.Natr (ed). The Seminar 'Physiological and Biochemical Foundations of Breeding for Yield in Cereals.' Vestnik Ceskoslovenske Akad. Zemed. pp. 343–44, 1974.

Danilchuk, P.V. and L.I. Grabovaya. (Inheritance of a number of embryonic roots in plants of spring barley and the relation of this character to yield). Nauchnotekhnicheskii Byulleten Vsesoyuznogo Selektsionnogeneticheskogo Instituta No. 2: 63–65, 1981.

Dekov, D., P. Radkov and S. Pavlova. Study of the efficacy of selection for root length in the early developmental stages of French bean in relation to breeding. *Rastenievdni Nauki* 15(6): 43–47, 1978.

Dorofeev, V.F. and A.M. Tyslenko. (Number of seminal roots in spring wheat in the course of selecting pairs for hybridization). *Vestnik Selskokhozyaistvennoi Nauki*. No. 8: 50–56, 1982.

Dorovskaya, I.F., V.F. Kovalenko and F.Ya. Konovalov.Comparison of interline hybrids of maize with their parental inbred lines and with an open pollinated variety in respect of the growth of the root system. *Tr. Gorsk. s-kh. inta.* 31: 123–31, 1971.

Derera, N.F., D.R. Marshall and L.N. Balam. Genetic variability in root development in relation to drought tolerance in spring wheats. *Exp. Agric.* 5: 327–37, 1969.

Eissa, A.G.M. Studies of root length and density in 150 primitive lines of cotton and several commercial cultivars. I. Classification. II. Inheritance of root length and density. *Diss. Abstr. Int. B.* 42: 3526 B, 1982.

Eissa, A.G.M, J.N. Jenkins and C.E. Vaughan. Inheritance of root length and relative root weight in cotton. *Crop Sci.* 23: 1107–11, 1983.

Ekanayake, I.J., D.P. Garrity, T.M. Masajo and J.C.O'Toole. Root pulling resistance in rice : inheritance and association with drought tolerance. *Euphytica* 34: 905–13, 1985a.

Ekanayake, I.J., J.C.O'Toole, D.P. Garrity and T.M. Masajo. Inheritance of root characters and their relations to drought resistance in rice. *Crop Sci.* 25: 927–33, 1985b.

Ekanayake, I.J., D.P. Garrity and J.C.O'Toole. Influence of deep root density on root pulling resistances in rice. *Crop Sci.* 26: 1181–86, 1986a.

Ekanayake, I.J., D.P. Garrity and S.S. Virmani. Heterosis for root pulling resistance in F$_1$ rice hybrids. *Int. Rice Res. Newsl.* 11(3): 6, 1986b.

Engledow, F.L. and S.Wardlaw. Investigation on yield in cereals. *J. agric. Sci.* 13: 390–439, 1923.

Fincher, R.R., L.L. Darrah and M.S. Zuber. Root development in maize as measured by vertical pulling resistance. *Maydica* 30: 383–94, 1985.

Foltyn, J. Varietal differences in the number of primary roots in winter wheats. Vedecke Prace Vyzkumnych Ustavu Rostlinne Vyroby v Praze Ruzyni. 17: 251–55, 1972.

Fritsch, R. On morphological root characters in *Triticum* L. and *Aegilops* L. (Graminae). *Kulturpflanze* 25: 45–70, 1977.

Garkavyi, P.F., P.V. Danilchuk and A.A. Linchevskii. Breeding malting varieties of spring barley on the basis of the vigour of development of the root system. *In*: Sb. nauchn. tr. Vses. Selekts-genet. in-t No. 9: 53–65, 1970.

Ghosh, S.N. P. Paria and S.L. Basak. Combining ability analysis in jute (*Corchorus capsularis* L.) *Bangladesh. J. Bot.* 8: 91–97, 1979.

Gladysheva, N.M., N.S. Rostova and V.G. Smirnov. Analysis of the system of correlation between morphological characters and lodging resistance in diploid and tetraploid forms of winter rye. *Tsitologiya in Genetika* 11: 145–50, 1977.

Gorbunov, V.V. and V.K. Chernov. Depth of penetration of the roots and their daily growth in spring wheat varieties. Sb. nauch. tekhn. inform. NII s.-kh. Yugo-Vost. No. 8: 20–21, 1973.

Gordon-Werner, E. and K. Dorfeling. Morphological and physiological studies concerning the drought tolerance of the *Scale cereale* × *S. montanum* cross Permontra. *J. Agron. & Crop Sci.* 160: 277–85, 1988.

Gorny, A.G. and H. Patyna. Genetic variation of the seedling shoot and root system and its relationship with adult plant characters in spring barley (*Hordium vulgare* L.). *Genetica Polonina.* 22: 419–28, 1981.

Gromova, A.I. Varietal specificity of soybean in relation to the temperature regime during seed germination and emergence. *In:* Vopr. restenievodstva v Priamure. Blagoveshchensk, USSR. pp. 32–37, 1975.

Guadamuz, M.E. (Relationship between root morphology and yield components in French bean, *Phaseolus vulgaris* L.). Thesis, Universidad de Costa Rica, San Jose, Costarica. 1980.

Gudinova, L.G., N.A. Kalashnik, O.I. Gamzikova, G.P. Antipova and O.T. Kachur. (Variation in and combining ability for indices of the primary root system of spring bread wheat in Siberia). *Selskokhozyaistvennaya Biologiya* No. 6: 51–54, 1983.

Hayami, K. (Studies on the physiological and ecological characteristics of high yielding rice varieties with good fertilizer response. 3. Characteristics relating to root distribution, respiratory activity and nutrient uptake ability). *Bull. Tohoku National Agric. Exp. Sta.* No. 68: 45–68, 1983.

Heinriches, D.H. Creeping alfalfas. *Adv. Agron.* 15: 317–37, 1963.

Hettinger, B. and J.M.M. Engels. Screening methods for drought resistance in indigenous Ethiopian barley. II. The use of plexiglass containers. *PGRC/E-ILCA Germ-plasm Newsl.* No. 13: 26–30, 1986.

Hirota, H. and Watanabe. Endogenous factors affecting the varietal differences in the curvature of seminal roots of *Zea mays* L. seedlings in water culture. *Jap. J. Crop Sci.* 50: 148–56, 1981.

Hou, C.R. and K.L. Lai. Root physiology of *japonica* and *indica* rices (*Oryza sativa* L.). 2. Nitrogen uptake and the enzyme activities of nitrogen metabolism. *J. Agric. Assoc.* China. No. 124: 10–18, 1983.

Huang, M.T. and D.L. Ketring. Root growth characteristics of peanut genotypes. *J. agric. Res.* China 36: 41–52, 1987.

Hurd, E.A. Root study of three wheat varieties and their resistance to drought and damage by soil cracking. *Can. J. Pl. Sci.* 44: 148–56, 1964.

Hurd, E.A. A method of breeding for yield of wheat in semiarid climates. *Euphytica* 18: 217–26, 1969.

Hurd, E.A. Can we breed for drought resistance. *In*: Drought Injury and Resistance in Crops. K.L. Larson (ed), *Crop Sci. Soc. Publ.* No. 2: 77–88, 1971.

Hurd, E.A. Phenotype and drought tolerance in wheat. *Agric. Meteorol.* 14: 39–55, 1974.

Hurd, E.A. and E.D. Spratt. Root patterns in crops as related to water and nutrient uptake. *In*: Physiological Aspects of Dryland Farming. U.S. Gupta (ed), Oxford & IBH Pub. Co., New Delhi. pp. 167–235, 1975.

Hurd, E.A., T.F. Townley-Smith, L.A. Patterson and C.H. Owen. Wascana, a new durum wheat. *Can. J. Pl. Sci.* 52: 687–88, 1972a.

Hurd, E.A., T.F. Townley-Smith, L.Á. Patterson and C.H. Owen. Techniques used in producing Wascana wheat. *Can. J. Pl. Sci.* 52: 689–91, 1972b.

Hurd, E.A., T.F. Townley-Smith, D.Mallough and L.A. Patterson. Wakooma durum wheat. *Can. J. Pl. Sci.* 53: 261–62, 1973.

Islam, M.O., S. Sen and T. Dasgupta. Seedling root length and seedling root number in mung bean. *TVIS News*, Taiwan. 2 (2): 17–18, 1987.

Jordan, W.R. and F.R. Miller. Genotypic variations in root systems of sorghum. *Agron. Anstr.* Madison, Am. Soc. Agron., U.S.A. p. 87, 1977.

Jordan, W.R. and F.R. Miller. Genetic variability in sorghum root systems: Implications for drought tolerance. *In*: Adaptation of Plants to Water and Higher Temperature Stress. N.C. Turner and P.J. Kramer (eds). John Wiley and Sons, New York. pp. 383–99, 1980.

Kalashnik, N.A., O.I. Gamzikova and I.R. Kolmakova. Comparative assessment of the combining ability of spring wheat varieties for root system vigour under different fertilizer conditions. *In*: *Selektsiya i semenovod, Zern. Kultur.* Novosibirsk, USSR. pp. 12–16, 1980.

Kamath, M.B., N.N. Goswami, A.M. Oza, M.S. Dravid and B.Sen. Rooting behaviour of maize (*Zea mays* L.) under normal and adverse soil conditions. *In*: Symp. on use of radiation and radioisotopes in studies of plant productivity (Abstr.), Food & Agric. Comm., Dept. Atomic Energy, Govt. of India. p. 154 and p. 112, 1974.

Kaspar, T.C. Evaluation of the tap root elongation rates of soybean cultivars. *Diss. Abstr. Int. B.* 43: 1323 B, 1982.

Kawashima, C. (Differences in the finishing time of crown root elongation among rice cultivars with different numbers of leaves on the main stem). *Jap. J. Crop Sci.* 52: 475–83, 1983.

Kawashima, C. (Root system formation in rice. IV. Differences among cultivars with different numbers of leaves on the main stem). *Jap. J. Crop Sci.* 57: 37–47, 1988.

Kazemi, H., S.R. Chapman and H.F. Mc Neal. Inheritance of root mass in 12 spring wheat cultivars. *Agron. Abstr.* Madison, Am. Soc. Agron., U.S.A. p. 54, 1976.

Kazemi, H., S.R. Chapman and F.H.Mc Neal. Components of genetic variance for root/shoot ratio in spring wheat. *In*: Proc. 5th Int. Wheat Genetics Symp. Vol. 2, Session IV, Genetics of Quantitative Variation. Feb. 23–28, S. Ramanujam (ed). Indian Soc. Genet. Pl. Breed., New Delhi. pp. 597–605, 1979.

Kazywanski, Z, D.Wojcik-Wojtkowiak, M.W. Borys, A. Stroinski and J.Wojciechowski. Inheritance of the root system in barley. II. Indices of the dimensions of the root systems of the varieties Gryf, Lubuski and Skrzes zowicki in hydroponic culture. *Prace Komisji Nauk Rolniczych i Komisji Nauk Lesnych.* 41: 231–38, 1976.

Kirichenko, F.G., P.V. Danilchuk and A.I. Kostenko. Method of selecting plants for vigour of root development for breeding and seed production. *In*: Sort i udobrenie Irkutsk, USSR, pp. 93–104, 1974.

Kiryan, M.V. The discovery of wheat forms valuable for their root system. *Selektsiya i Semenovodstvo.* No. 2: 37–39, 1974.

Kuzmin, N.A. and A.F. Shumeiko. A study of characters of the root system in spring bread wheat in relation to breeding for yield. Nauchn. tr. NII s.kh. Tsentrchernozemn. polosy. 17(3): 24–34, 1980.

Kuzmin, N.A. and A.F. Shumeiko. (The root system of spring wheat and the possibility of improving varieties by strengthening its development). *Selektsiya i Semenovodstvo,* USSR. No. 3: 14–16, 1985.

Lebedeva, A.T. The effect of selection for vigour of the root system in cucumber on the photosynthetic activity of the leaves. Tr. VNII Selektsii i semenovodstva ovoshch. Kultur No. 4: 26–32, 1971.

ebedeva, A.T. and O.V. Yurina. Effect of selection for root system vigour in cucumber on seed content of the fruits and their absolute weight under the conditions of plastic greenhouses. Tr. VNII selektsii i semenovodstva ovoshch. Kultur. No. 3: 50–54, 1970.

Loresto, G.C., W.X. Zhang, D. Chaudhary and T.T. Chang. Aeroponic technique for screening the drought avoidance mechanism of rice genotypes by the root characters. *Garcia de Orta, Estudos Agronomicos.* 10: 77–82, 1983.

Loshak, I.F. and N.D. Logachev. Breeding spring barley for resistance to drought in North Kazakhstan. *In*: Poryshenie zasukhoustoich. Zern. Kultur, Moscow, USSR, Kolos. pp. 172–76, 1970.

Mac Key, J. Wheat domestication as a shoot: root interaction process. *In*: S. Ramanujam (ed), Proc. 5th Int. Wheat Genetics Symp., Feb., 1978. Vol. 2. Indian Soc. Genet. and Pl. Breed., New Delhi. pp. 875–90, 1979.

Mc Michael, B.L., J.E. Quisenberry and D.R. Upchurch. Lateral root development in exotic cottons. *Environ. Exptl. Bot.* 27: 499–502, 1987.

Manga, V.K. and M.B.L. Saxena. Combining ability and heterosis for root and related traits in pearl millet. *Indian J. agric. Sci.* 56: 164–67, 1986.

Maurya, P.R. and B.P. Ghildyal. Root distribution pattern of rice varieties evaluated under upland and flooded field conditions. *Il Riso* 24: 239–44, 1975.

Meena, N.L. and J. Seth. Root distribution pattern and nitrogen uptake of some wheat and triticale germ-plasms in relation to rates and methods of nitrogen application. *In*: Symp. on use of radiation and radioisotopes in studies of plant productivity (Abstr.). Food & Agric. Committee of the Dept. of Atomic Energy, Govt. of India. p. 154 and p. 107, 1974.

Milyutkin, A.F., V.N. Ogurtsov and L.M. Dorshina. Some characteristics of the growth and development of the root system in sorghum on chernozem soils in the central Volga region. *In*: *Selektsiya i Kuibyshev,* USSR : 46–50, 1974.

Mirauda, L.T. de, Inheritance of linkage of root characteristics from Pueblo maize. *Maize Genetics Cooperation Newsl.* No. 54: 18–19, 1980.

Nagorskaya, M.D. and T.P. Polkanova. Efficacy of selecting plants of yellow lupins for vigour of the root system. *In*: Pate povysheniya urozhainosti polevykh Kultur, 7. Minsk, Belorussian SSR, Uradzhai. pp. 70–75, 1976.

Nakhaeva, V.I. (Genetics of root system characters of spring bread wheat). Nauchno-teckhnicheskii Byulleten SO VASKh NIL. No. 8: 3–4, 1985.

Nass, H.G. and M.S. Zuber. Correlation of corn roots early in development to mature root development. *Crop Sci.* 11: 655–58, 1971.

Nazirov, N.N. and G.S. Satipov. (Root system and yield in the cotton varieties AN402 and Tashkent 1). Dok. Vses. Ordena Lenina i Ordena Trudovogo Krasnogo Znameni Akad. Selskokhzyaistvennykh Nauk Imeni V.I. Lenina. No. 11: 5–7, 1980.

Nefedov, A.V. The role of the root system in improving the drought resistance of spring wheat. *In*: Povyshenie Zasukhovstoich, Zern. Kultur, Moscow, USSR, Kolos. pp. 104–08, 1970.

Netis, I.T., P.P. Borovik, R.V. Morozov and S.A. Zaets. (Features of the development of the root system of semidwarf and tall varieties of winter wheat). *Oroshaemoe Zemledelie*. No. 30: 28–31, 1985.

Nour, A.E.M. and D.E. Weibel. Evaluation of root characteristics in grain sorghum. *Agron. J.* 70: 217–18, 1978.

O'Brien, L. Genetic variability of root growth in wheat (*Triticum aestivum* L.). *Aust. J. agric. Res.* 30: 579–95, 1979.

Oleinik, A.A. Efficacy of selecting plants of spring barley for number of embryonic roots. *Selskokhozyaistvennaya Biologia.* 12: 42–44, 1977.

Oleinik, A.A. Evaluation of spring barley varieties by the root system. *Selektsiya i Semenovodstvo,* No. 3: 28, 1978.

Omara, M.K. and M.Y. Hussein. Genetic variability in root characteristics in relation to yield under drought in early maturing barley. *Assiut. J. Agric. Sci.* 18: 319–32, 1987.

O'Toole, J.C. and W.L. Bland. Genetic variation in crop plant root system. *Adv. Agron.* 41: 91–145, 1987.

O'Toole, J.C. and Soemartono. Evaluation of a simple technique for characterizing rice root systems in relation to drought resistance. *Euphytica* 30: 283–90, 1981.

Pandey, R.K. and J.W. Pendleton. Genetic variation in root and shoot growth of peanut in hydroponics. *Philippine J. Crop Sci.* 11: 189–93, 1986.

Papova, D. and L. Mikhailan. Study of heterosis in red pepper (*Capsicum annuum* L.). Vestsi A.N. BSSR. Ser. Biyaln. No. 5: 55–57; 139, 1976.

Pasela, E. The formation of the root mass of spring barley in individual layers of the soil profile in relation to soil moisture and some meteorological factors. Zeszyty Nauk Akad. Rolniczej v Krakowie, Rozprawy. 35(103): p. 86, 1975.

Passioura, J.B. The effect of root geometry on the yield of wheat growing on stored water. *Aust. J. agric.Res.* 23: 745–52, 1972.

Pavlova, S.S., V.K. Chernov and V.F. Kravchenko. Characteristics of formation of embryonic roots in various varieties of spring wheat. Sb. nauch. rabot. Saratov. s-kh. in-t. No. 26: 52–56, 1973.

Petukhovskii, S.L., V.P. Shamanin, and S.A. Polikarpov. (Variation and inheritance of seminal root number in wheat hybrids). *In*: Selektsiya i semenovodstvo yarov. pshenitsy v Zap. Sib. Omsk, USSR (1984): 21–24, 1984.

Philipchuk, B.Z. A comparative examination of the root system of diploid and tetraploid forms of rye. Dep. 3521–71; 8, pp., 1971.

Pohjakallio, O. The question of the resistance of plants to drought periods in Finland. *Nord. Jordbr. Forskn.* 5–6: 206–26, 1945.

Raj, K.G. and E.A. Siddiq. Hybrid vigour in rice with reference to morphophysiological components of yield and root density. *SABRAO Journal*. 18: 1–7, 1986.

Rangasamy, S.R. and A.S. Shanmugam. Variability and correlation analysis in seedling characters in green gram. *Indian J. agric. Sci.* 54: 6–9, 1984.

Raper, C.D. and S.A. Barber. Rooting systems of soybeans. I. Differences in root morphoplogy among varieties. *Agron. J.* 62: 581–84, 1970.

Richards, R.A. and J.B. Passioura. Seminal root morphology and water use of wheat. I. Environmental effects. *Crop Sci.* 21: 249–52, 1981a.

Richards, R.A. and J.B. Passioura. Seminal root morphology and water use of wheat. II. Genetic variation. *Crop Sci.* 21: 253–55, 1981b.

Robertson, B.M. and J.G. Waines. Genetic variation in seminal root numbers of wheat. *Agron. Abstr.* Madison, Am. Soc. Agron., U.S.A. p. 68, 1977.

Robertson, B.M., J.G. Waines and B.S. Gill. Genetic variability for seedling root numbers in wild and domesticated wheats. *Crop Sci.* 19: 843–47, 1979.

Robertson, B.M., A.E. Hall and K.W. Foster. A field technique for screening for genotypic differences in root growth. *Crop Sci.* 25: 1084–90, 1985.

Roma, A. Heritability in root system characters of wheat in correlation with lodging resistance. *Lucrari Stuntifice* (Clij). 18: 81–90, 1962.

Rudenko, M.I. and M.V. Kiryan. Features of development of the root system in short-strawed spring wheats of the U.S.A. Byulleten Vsesoyuznogo Ordena Lenina i Ordena Druzhby Naratov Instituta Rastenievodstva Imeni N.I. Vavilova. No. 51: 10–14, 1975.

Sachanski, S. and D. Naneva. The effectiveness of selection for length of root system in the early stage of development in pea. *In*: Selektsionnogenetichni i agrotekhnicheski Prouchvaniya na Z'rnenobobovite Kulturi. Sofia, Bulgaria; Izdatelstvo no BAN. pp. 141–50, 1975.

Sahai, V.N. and R.C. Chaudhary. Root systems in hybrid rice. *Int. Rice Res. Newsl.* 11(5): 13–14, 1986.

Sandhu, D. and P.N. Bhaduri. Variable traits of root and shoot of wheat. II. Under different cultural conditions and ages. *Z. Acker-und Pflanzenbau* 153: 216–24, 1984.

Sasmal, B. Relationship of root and shoot characters in parent, F_1 and F_2 populations of rice. *J. Agron. & Crop Sci.* 159(4): 260–63, 1987.

Sasmal, B. and J. Banerjee. Heterosis and inbreeding depression estimation of root and shoot characters in rice (*Oryza sativa* L.). *J. Agron. & Crop Sci.* 156: 117–22, 1986.

Sathyanarayanaiah, K. and T.F. Townley-Smith. Optimum time to screen for wheat (*Triticum durum*) root development. (Abstr.). *Can. J. Pl. Sci.* 61: 492, 1981.

Sattarov, D. Role of the root system in varietal specificity of reaction to fertilizer treatment in cotton. *Uzbekistan Biologija Zurnali* No. 1: 23–26, 1978.

Semushina, L.A. Correlation of yield with some morphological and physiological characters in barley under saline and non-saline conditions. *Trudy po Prikladnoi Botanike, Genetike i Selektsii.* 64: 101–09, 1979.

Sharma, R.B., B.P. Ghildyal and D. Sharma. Root distribution studies of some wheat and triticale varieties. *In*: Symp. on use of radiations and radioisotopes in studies of plant productivity (Abstr.). Food & Agric. Comm., Dept. Atomic Energy, Govt. of India. p. 110, 1974.

Shriniwas, and B.V. Subbaiah. Root distribution studies of some *bajra* hybrids (*Pennisetum typhoides* Stapf). *J. Nucl. Agric. Biol.* 5: 15–16, 1976.

Sidorov, A.V. (The possibility on increasing the number of seminal roots in spring wheat). *Selektsiya i Semenovodstvo*, USSR. No. 11: 12–13, 1984.

Singh, K., R.S. Narang and S.M. Virmani. Rooting behaviour of Mexican and indigenous wheats under different soil moisture regimes. *J. Res.*, P.A.U. 10: 296–300, 1973.

Snoad, B. and A.E. Aurthur. Genetic studies of quantitative characters in peas. 3. Seed and seedling characters in the F_2 of a six parent diallel set of crosses. *Euphytica* 23: 105–13, 1974.

Stavnuchuk, V.G. Formation of the principal yield components and seed quality of winter wheat progeny in relation to intravarietal selection. *In*: Nauch. tr. po s-kh. biol. Odessa, Ukrainian SSR : pp. 62–64, 1972.

Stoffella, P.J. Morphological studies of dry beans (*Phaseolus vulgaris* L.). *Diss. Abstr. Int. B*. 41: 751 B, 1980.

Surma, M., M. Borys, Z. Kaczmarek, Z. Krzywanski and D.Wojcik-Wojkowiak. An attempt to determine genetic basis of the root system morphological characters in spring barley (*Hordium vulgare* L.). *Genetica Polonica* 19: 437–45, 1978.

Syu, H.S. Studies on the resistance of cereals to water-logging. II. The relation between the resistance of cereals to water-logging and the root growth. *Korean J. Breeding* 5: 91–97, 1973.

Szymanska, L. Early assessment of the intensive character in winter wheat and spring barley, on the basis of the response to N of their seminal roots. Hodowla Roslin, Aklimatyzacja i Nasiennictwo. 24: 203–10, 1980.

Tian, P.Z. (Studies of root system ecotypes in soybean cultivars). *Acta Agronomica Sinica*. 10: 173–78, 1984.

Tikhonov, V.E. The role of the number of embryonic roots in spring bread wheats in the semidesert conditions of northern Priarale. Byulleten Vsesoyuznogo Ordena Lenina Instituta Rastenievodstva Imeni N.I. Vavilova. No. 33: 3–7, 1973.

Tewari, D.K., D.P. Nema and J.P. Tiwari. Rooting pattern as a selection parameter of wheat varieties under moisture stress. *Madras agric. J.* 61: 334–39, 1974.

Tomasovic, S. The number of primary rootlets in various genotypes of winter wheat (*Triticum aestivum* Subsp. *vulgare*) and their importance in breeding. *Archiv. za Poljoprivredno Nauka*. 31(114): 127–35, 1978.

Troughton, A. and W.J. Whittington. The significance of genetic variation in root systems. *In*: Root Growth. W.J. Whittington (ed), Butterworths, London. pp. 296–314, 1969.

Turner, F.T., C.C. Chen and G.N. McCauley. Morphological development of rice seedlings in water at controlled oxygen levels. *Agron. J.* 73: 566–70, 1981.

Tymchuk, M.Ya. Complex selections in breeding winter wheat of the intensive type in the eastern forest steppe of the Ukrainian SSR. Tr. Khartov. s-kh. in-t. 204: 70–75, 1975.

Tyslenko, A.M. (Effect of root number on yield in spring wheat). Byulleten Vsesoyuznogo Ordena Lenina i Ordena Druzhby Narodov Instituta Rastenievodstva Imeni N.I. Vavilova. No. 114: 15–17, 1981.

Ustimenko, A.S., P.P. Danialchuk and A.T.Grozdikovskaya. "Root Systems and Yield in Crop Plants" Kiev, Ukrainian SSR, Urozhai. p. 368, 1975.

Vakulenko, G.M. Solonets tolerance in cereal crops and factors affecting it. *Nauk. tr. Onsk. s-kh. in-t*., 148: 92–96, 1976.

Vedrov, N.G. Root system of creal crops as an objective in breeding: *In*: Sort i udobrenie. Irkutsk, USSR. 139–45, 1974.

Vedrov, N.G. (Nature of root system development in spring wheat in the drought zone of eastern Siberia). *Selskhozyaistvennaya Biologiya* 17: 196–98, 1982.

Vekic, N. Investigation of germination and the early development of the root system in the cold chamber in simple interline maize hybrids. Zbornik Radova Poljoprivrednog Instituta, Osijek 2: 141–76, 1972.

Virk, D.S., G.S. Bhuller, K.S. Gill and G.S. Poonia. Inheritance of juvenile leaf and root characters in three bread wheat crosses. *Cereal Res. Commu.* 6: 75–84, 1978.

Welbank, P.J. Root growth of different species and varieties of cereals (Abstr.). *J.Sci. Food & Agric.* 25: 231–32, 1974.

Wright, G.C. and R.C.G. Smith. Differences between two grain sorghum genotypes in adaptation to drought stress. II. Root water uptake and water use. *Aust. J. Agric. Res.* 34: 627–36, 1983.

Yakushevskii, E.S. and E.A. Kochetova. Some characteristic features in the growth of the root system in cultivated species of sorghum (*Sorghum* Moench subsp. *sorghum*) in the initial period of plant development. Trudy po Prikladnoi Botanike, Genetike i Selektsii. 51: 127–44, 1974.

Yakimenko, N.P. and I.I.Boitsov. The selection of pea plants for vigour of development of the root systems at the first stage of growth. *Nauch. tr. VNII Zernobob. Kultur.* 3: 227–34, 1971.

Yu, T.; L.H. Stolzy and J. Letey. Survival of plants under prolonged flooded conditions. *Agron. J.* 61: 844–47, 1969.

Yurina, O.V. and A.T. Lebedeva. The effectiveness of selection for vigour of the root system in the breeding and seed production of cucumber. *In*: Selektsiya i Semonovodstvo ovoskch. Kulture, 1, Moscow, USSR. pp. 66–74, 1974.

Zelenskii, M.O. and V.S. Neboka. Heterogeneity of winter wheat varieties and its expression in suitable conditions. *Nauk. pratsi Ukr. silskogospod. akad.* No. 32: 13–16, 1971.

Zenisceva, L. A genetical analysis of the strength of the root system in spring barley. *Genetika a Slechtni.* 9: 107–12, 1973.

Zolotov, V.I. and V.S. Fevralev. Development of the root system in hybrids differing in earliness. *Kukuruza* No. 1: 31, 1973.

Zykin, V.A. The root system of wheat and the possibility of improving it by breeding. *Vsetnik Selskokhozyaistvennoi Nauki* No. 11: 43–48, 1976.

CHAPTER 9

IMPROVEMENT OF GRAIN LEGUMES FOR NODULATION AND NITROGEN FIXATION

INTRODUCTION

The most needed plant nutrient is nitrogen. Although submerged in an ocean of nitrogen (ca. 78 per cent of the atmosphere) higher plants cannot make use of this gaseous nitrogen but some bacteria, fungi, algae and ferns possess the requisite enzyme system to use N_2. Among all these, *Rhizobium* spp. in association with legumes are best known. They enter through the root hair, form nodules and N_2 is fixed into a compound form, which is either translocated upward into the shoot and used directly or enters into the soil after the nodules decay. Thus the N requirement of leguminous plants is met but is not enough for maximum production. When N-fertilizers are used as a supplemental dose, nodulation and N_2 fixation are suppressed (Wynne et al. 1979), and therefore, a full dose of N-fertilizer, as required by non-legumes, has to be supplied. Since the symbiont drains the plant carbon source, and the fixed nitrogen may not be enough, anodulating varieties were developed (Singh et al., 1974; Gorbet and Burton, 1979; Peterson and Barnes, 1981; Nigam et al., 1980, 1982). But these are suitable only for affluent countries where N-fertilizers are available easily and cheaply. In developing countries where fertilizers are costly and scarce, and for marginal lands anodulating varieties are not desirable. However, *Rhizobium* inoculation has been tried successfully (Bajpai et al., 1974; Fernandez, 1986; Greder et al., 1986). The data of Greder et al. (loc. cit.) clearly show that inoculation of soybean with *Brdyrhizobium japonicum* increased nodulation and grain yield significantly (Table 9.1). The technology of preparing rhizobium culture, its easy and cheap availability and unknown competition with other soil organisms make it impracticable in underdeveloped countries. Their is a lot of variability in the effectiveness and competitiveness of different *Rhizobium* strains. There screening and culture demands expertise, lacking so much in such countries.

Table 9.1: Yield of soybean (kg/ha) with different nodule mass ratings as affected by inoculation with *Bradyrhizobium japonicum* (After Greder et al., 1986).

Treatments	Nodule mass ratings [x]				
	1	2	3	4	5
Inoculated	2270[*]	2380[*]	2440[*]	2440[*]	2530[*]
Uninoculated	2197	2300	2350	2360	2350

[x] 1 = low nodule mass, 5 = high nodule mass

[*] Means within a column are significantly different at the 0.05 level, based on the F-test on inoculation mean squares.

Gelin and Blixt (1964) reported that plant genes affect N_2 fixation in soybeans, peas and clover. With the same inoculation, T_1, T_2, T_3, K_4 and K_5 varieties of chickpea showed increases in grain of 38, 25, 18, 36 and 14 per cent, respectively (Bajpai et al., 1974). The genetics of the host crop and rhizobium, and their interaction in N_2 fixation have been reviewed several times (Caldwell and Vest, 1977; Schwinghamer, 1977; Postgate, 1979; Gibson, 1980; Ham, 1980; Brill, 1986).

In studies on four types of cowpea with high N_2 fixing capacities and one with a low capacity, fixation was shown to increase exponentially from the third to the sixth week of growth (full flowering), and thereafter decreased rapidly (Zary and Miller, 1978). Thus for finding the maximum genotypic differences in N_2 fixation, sampling should be done at the full flowering stage. After this stage nodules decay, but the plants' N-need for better grain filling increases. Late N-fertilizer application at this stage can be made with no danger of decrease in symbiotic N_2 fixation (Fernandez, 1986).

Soybeans infected with both the vesicular-arbuscular mycorrhizal (VAM) fungus and *Bradyrhizobium* were similar in total dry weight, leaf area, and development to plants that received 1.0 or 2.0 mMN. The presence of the VAM fungus can decrease nutrient stress in environments limited in P, Zn and Cu, elements essential in N_2 fixation.

Since considerable genetic variability has been recorded in the host genotypes in their ability of developing effective nodules and fixing N_2 (Tarkashvili et al., 1974; Kumar Rao and Vishvanatha, 1974; Kuzmin, 1977; Bello et al., 1980; Wynne et al., 1980; Smith et al.,

1982; Arrendell et al., 1985; Askin et al., 1985), there is a possibility of breeding superior genotypes which can make the best use of the symbiotic fixation and have good agronomic characters and yield potential. Further, nitrate tolerance can be built up in these genotypes, so that the benefits of supplemental N can be achieved. Recent developments on these aspects will be discussed in the following pages.

Management of N_2 fixing bacteria or P-scavenging endomycorrhiza may lead to decreased fertilizer use on extensively cropped lands. Leguminous crop plants form mutualistic associations with the bacterium *Bradyrhizobium* and with VAM fungi. In soils marginal in N and P, these microsymbionts play a crucial role in legume development and yield since enhanced nodulation and N_2 fixation result from improved P nutrition. Under conditions where P is plentiful but N is not, improved growth and nutrition of nodulated legumes depends on the effectiveness (Nutman, 1959) of symbiotic N_2 fixation.

GENETIC VARIABILITY

Host Genotypes

The first report on variability in nodulation of a host to strains of *Rhizobium* was by Vorhees (1915). He observed that soybean cultivar 'Haberlandt' did not nodulate when planted in plots containing *R. japonicum*. Five other soybean cultivars planted in the same plots were well nodulated. Interplanting of Haberlandt seed and seeds of nodulating cultivars did not result in nodulation of Haberlandt. Therefore, Vorhees concluded that different cultivars of the same legume species have different powers of resistance to association with the symbiotic bacteria. Wilson et al. (1937) also concluded that host genetics were involved in the control of N_2 fixation in the nodules of sweet clover. These conclusions were based on their studies of four species of sweet clover and several strains of *R. meliloti*. They observed that some sweet clover species inoculated with certain strains of *R. meliloti* always nodulated well and fixed N_2. However, some strains fixed high amounts of N_2 with one cultivar but not with another.

Studies with 10 varieties of soybean showed significant differences in dry weight (Rao and Vishvanatha, 1974). Intervarietal differences were also found in nodulation (Kuzmin, 1977); fewer nodules were formed in short duration varieties. The determinate variety, Williams, fixed more N_2 after pod filling than the determinate variety, Elf, which was constantly low in N_2 fixation ability throughout its development (Bello et al., 1980).

A study on nodulation of 48 groundnut types revealed that Virginia types such as Florigiant, Va72r, NC5 and NC6 nodulated more than the Spanish or Valentia types (Wynne et al., 1980). In the families from a cross between Virginia × Spanish types, broad sense heritability estimates for nodule number, nodule weight, nitrogenase activity, short weight and fruit weight were moderate to high, indicating that superior genotypes within this population could be readily identified (Arrendell et al., 1985). Nodule number and weight were positively correlated with each other and with nitrogenase activity, suggesting that only nitrogenase activity was required to identify superior families. They concluded that selection for families with increased N_2 fixing activity should be possible and should result in direct selection for yield.

In a green house study on 100 lines of cowpea, Zary et al. (1978) reported that nitrogenase activity measured as μ moles C_2H_2 produced per plant per hour ranged from 0.6 to 43.3, nodule mass from 0.1 to 3 g per plant, nodule number from 10 to 146 per plant and dry weight per plant top from 2.5 to 8.4 g. Variations in number and weight of nodules per plant and in μ moles C_2H_2 per g nodules per hour in seven cultivars of cowpea studied at IITA, Ibadan, Nigeria, are given in Table 9.2.

Table 9.2: Nodulation and efficiency of seven uninoculated cowpea cultivars (Ayanaba and Nangju, 1977).

Cultivars	Number of nodules per plant	Dry weight of nodules per plant (mg)	μ moles C_2H_2 per g nodules per hour
TVu37 (Pale green)	44	283.4	89.3
TVu76 (Prima)	40	198.5	53.5
TVu3616 (K2809)	52	92.8	53.2
TVu6198	71	202.3	46.5
TVu1190 (V.u.5)	39	263.0	94.7
TVu3629 (Ife Brown)	54	324.6	90.0
TVu29-17	40	175.5	53.3

A study on over 100 varieties of French bean inoculated with *Rhizobium* showed marked differences in their susceptibility as evidenced by the number and size of nodules produced (Tarkashvili et al., 1974). The greatest susceptibility was shown by Shchedraya and Rachulitsiteli and the least by the dwarf forms such as Abkhazcura. Among 85 lines of *Pisum sativum* screened, mean N_2 fixation rates ranged from 1.5 to 12.5 μ mole C_2H_2 per hour per plant and from 10 to 49 μ mole C_2H_2 per hour per g nodulated root (Hobbs and Mahon, 1982).

Eight *Vicia faba* genotypes exposed to a single strain of *Rhizobium* differed in their response to the bacterium in dry matter yield per plant (Welsh Pl. Breed. Sta., UK., 1976). El Sherbeeny et al. (1977b) also noted genetic variability among different varieties of *Vicia faba* in dry matter production and N_2 fixation when inoculated with a single strain of *R. leguminosarum*. A study on eight cultivars of peas (*Pisum sativum*) revealed considerable varietal variability in seasonal profiles of N_2 fixation (Askin et al, 1985). The cultivars Partridge and Whero reached peak N_2 fixation before the onset of flowering, while others reached it shortly after flowering has started. Earlier, Smith et al. (1982) had noted significant differences among 14 crimson clover lines for acetylene reduction. Top dry matter, leaf number and visual scores of nodulation, root growth and top growth were significantly correlated with the acetylene reduction assay.

Both plant and rhizobium populations exhibit genetically determined variations in nodulation characteristics and in the quantity of N_2 fixed (Burton and Wilson, 1939; Nutman, 1946; El-Sherbeeny et al., 1977a, b). The genetics of the host has been reviewed (Nutman, 1969; Holl and LaRue, 1976; Caldwell and Vest, 1977; Beringer and Johnson, 1978; Gibson, 1980). Much attention has been given to the quantitative differences and the effects of major genes of the host (Gibson, 1980).

Quantitative host genetic variability for N_2 fixation has been reported in Spanish clover (Pinchbeck et al., 1980), cowpea (Zary et al., 1978) and in alfalfa, (Duhigg et al., 1978; Hoffman and Melton, 1981). In alfalfa, there is some evidence of intracultivar variability (Hoffman and Melton, 1981). The genetic control on N_2 fixation has been investigated in a few instances; major genes in pea affected both nodulation and N_2 fixation (Holl, 1975) and quantitative genetic variation in overall fixation was due solely to general combining ability in Spanish clover (Pinchbeck et al., 1980). Furthermore, plant selection has actually resulted in some improvement in fixation by alfalfa (Duhigg et al., 1978). The known major genes in both plants and rhizobia have been reviewed a number of times (Beringer and Johnston, 1978; Gibson, 1980; Nutman, 1981) but in general, no immediate application for the use of major plant gene variants has been possible.

Of 75 varieties of *Cicer arietinum* assessed for nodulation characters, several proved superior to the standard C235 (Maherchandani and Rana, 1977). Path and Moniz (1974) reported that out of the five varieties of chickpea, N_3, was most heavily nodulated by a *Rhizobium* isolate from *C. arietinum* var. N59, and gave the highest recovery of nitrogen. Out of the 17 varieties of *Cymopsis tetragonoloba*, FSR 77 was the best in N_2 fixation and nitrogenase activity and Jodhpur local was the best among 16 varieties of *Vigna aconitifolia* (Rao et al., 1984).

Rhizobium Strains

Highly effective strains of rhizobia occur naturally and there is no difficulty in selecting and multiplying such individuals. The problem is that they invariably occur at low frequencies in natural populations and thus appear to be at a selective disadvantage in nature (Jones, 1964; El-Sherbeeny et al., 1977a.) Even when such strains are multiplied in the laboratory and reintroduced into the soil in large numbers, the frequency with which they nodulate plants is not always improved (Brockwell, 1981). The average effectiveness of *R. leguminosarum* populations from various sites in Britain differed by up to 30 per cent. It seems likely that the potential exists for improving fixation, and through it increasing yield.

Out of the 14 *R. japonicum* strains associated with soybean; five from Thailand, two from Taiwan, five from the U.S.A. and two from Japan; strain 110 from the U.S.A. was most effective (Tansiri et al., 1974). Valdes Ramirez et al. (1984) used three serologically distinct strains of *R. japonicum* and the commercial inoculum Nitragin, to inoculate seeds of soybean cr. Jupiter, just before sowing. The effectiveness of each strain was assessed according to the number and weight of nodules per plant at flowering and seed yield. ENCB516 was the most effective strain in all respects, producing a greater effect on seed yield than urea applied as a N-source at 80 kg/ha. It was also the dominant strain when a mixed inoculum was used. Alwi (1985) showed variability in the rhizobial strains in their ability to infect groundnut roots and fix N_2. Plant response to the effective strains was correlated with pod and seed characteristics.

GENETIC STUDIES

Host Legumes

Host determined genetic control of N_2 fixation in the *Pisum - Rhizobium* symbiosis was studied by Holl (1973, 1975). In a cross between nodulating and non-nodulating genotypes, nodulation was controlled by a single dominant factor; F_2 segregates were obtained which formed nodules but did not fix N_2 as determined by acetylene reduction assay (Holl, 1973). Both nodulation and N_2 fixation were reported to be genetically determined by the host plant. Holl (1975) examined over 200 lines of *Pisum* for resistance to nodulation or inability to fix N_2. Data indicate that nodulation and N_2 fixation are controlled by two independent dominant genes. The genes were designated Sym 2 (for nodulation) and Sym 3 (for N_2 fixation). The data from a diallel of six lines of *Pisum* indicate heterosis for N_2 fixation per plant, partly caused by non-allelic interaction (Hobbs and Mahon, 1982). For N_2 fixation per g nodulated root, non-additive effects were small. Heterosis occurred for total shoot nitrogen and root weight, and was attributed in the latter to non-allelic interaction and overdominance. Total dominance associated with decreased expression was evident for percentage shoot nitrogen. High narrow sense heritability estimates were found for N_2 fixation. Significant positive correlations were found between N_2 fixation per plant and total shoot nitrogen and between both of these characters and shoot weight. Significant negative correlations were found between percentage N and both N_2 fixation per plant and shoot weight.

Intracultivar variability and heritability of N_2 fixation ability in two green gram populations were determined (Fernandez and Miller, 1983). Significant differences were observed in nodule number and nodule weight and genetically controlled intracultivar variation was suggested. In an 8×8 diallel excluding reciprocals, where the parents and the F_1 s were grown under controlled conditions with two *Rhizobium* strains, both additive and non-additive effects were detected for nodule number, nodule volume and nodule dry weight per plant with both strains; additive effects were more important (Singh et al., 1985). The parental lines T44 and K851 had high general combining ability estimates for the three nodulation characters. Heritability estimates (narrow sense) were moderate to high. Correlation coefficient between the three characters on one hand and plant N content on the other were significant and high, and were highest for the nodule number. They suggested that this character could be used in screening for N content.

Thomas et al. (1986) studied the inheritance and expression of the three genes controlling root nodule formation in chickpea. Genetic studies showed that recessive alleles at three different loci were responsible for the *Nod* (non-nodulating) phenotypes in PM 233, PM 655 and PM679. They proposed that the symbols rn1, rn2 and rn3 be assigned to the loci producing *Nod* phenotype in the above mutants, respectively.

A study on effectively and ineffectively nodulating cultivars of *Phaseolus vulgaris* indicated the presence of a single dominant gene (In) in ineffectively nodulating cultivars determining the low nitrogenase activity in nodules (Asokan, 1981). Positive and significant correlations between nitrogenase activity and both nodule weight and nodule number were noted in a diallel cross of five varieties of cowpea using a mixed strain of *Rhizobium* inoculum (Miller et al., 1986). Analysis of reciprocal F_1, F_2, BC_1 and BC_2 generations of a cross between H-Brown Crowder and L-Bush Purple Hill with a single strain inoculum showed that additive gene action was more important than dominance and inter allelic gene action for nodule number and nitrogenase activity, while the reverse was true for nodule weight and top dry weight. Narrow sense heritability estimates were 0.55 for nodule number and 0.62 for nitrogenase activity, but low for the other two traits.

Host plant genetics of N_2 - fixation efficiency in crimson clover - *R. trifoli* symbiosis was investigated using a diallel mating design and mass selection (Smith et al., 1982). Crosses differed significantly in acetylene reduction (AR) rate. In a combining ability analysis, both GCA and SCA were significant sources of variation. Only one of the six parental lines showed a significant, positive GCA effect. This line uniformly transmitted its high AR ability to all but one of its progeny. In terms of variance components, SCA was more important in determining the AR rate than GCA for all lines tested. After one cycle of mass selection, AR rate was increased by 8 per cent. The realized heritability of this trait was 63.1 per cent. Heritability estimates for a nodule mass of soybean on an entry mean basis over locations ranged from 54 to 67 per cent (Greder et al., 1986). This result suggests progress in selection for increased nodule mass with native *Bradyrhizobia* should be possible. A positive relationship of nodule mass to seed yield was observed indicating that further selection for increased nodulation with the native *Bradyrhizobia* may be warranted.

Rhizobium Symbionts

The molecular genetics of symbiotic N_2 fixation by *Rhizobium* when in symbiotic association with legumes has been briefly reviewed (Ansubel, 1982). The genetic mapping and cloning of the symbiotic genes of rhizobium have been described and the clustering of symbiotic genes including *nif* genes on large plasmids has been reported. The similarity of plasmids of *Rhizobium* and *Azotobacterium* has also been discussed. By using cloned *Rhizobium meliloti* nodulation (nod) genes and nitrogen fixation (*nif*) genes, Prakash and Atherly (1984) reported that the genes for both nodulation and N_2 fixation were on a plasmid present in fast growing *Rhizobium japonicum* strains.

IMPROVEMENT

Grain Legumes

Nitrate Susceptibility — Anodulating Types: A full understanding of the genetic control of a character is necessary before it can be successfully improved upon in a breeding programme. Although this improvement may be achieved by genetic modification of either the plant or bacterial symbiont, investigations have concentrated mainly on the bacteria (Schwinghamer, 1977). The information available concerning the potential for increasing the N_2 fixing ability of the host is, therefore, sparce.

We know that the present level of N_2 fixed through rhizobium is not enough for achieving maximum production; especially because of after flowering when N-requirement of the developing grains increases tremendously, the nodules start degenerating as a result of diminished carbon source availability to the symbiont, as the plant can hardly spare any amount at this period of maximum requirement. The N-deficit can be made good by supplementing the balance amount through N-fertilization. Unfortunately, the supplemental N-fertilization reduces nodulation (Latimore et al., 1977; Wynne et al., 1979). This suggests that moderate N-fertilization can limit productivity of the heavily nodulated cultivars rather than increasing it. However, the cultivarial differences in NO_3- tolerance suggests that a breeding programme can be launched for increasing NO_3- tolerance of the nodular system. The other possibilities are developing anodulating cultivars and fertilizing them like cereals, or giving a supplemental urea spray to the nodulating cultivars at the later developmental stages.

Varietal differences in N-translocation from vegetative tissue to the developing grain have been observed (Perez et al., 1973; Rhodes and Jenkins, 1976; Johnson and Mattern, 1975). Harvest index (HI) and harvest nitrogen index (HNI) are positively correlated with seed yield which suggest that certain genotypes are efficient (E) in mobilizing both N and dry matter to the developing seed (Jeppson et al., 1978). Difference in HNI at maturity result from differential N mobilization from leaflets; petioles, and stems of efficient (E) (high HNI) as compared to inefficient (I) (low HNI) genotypes. Thus the anodulating varieties have to be further developed for their translocation characteristics. A comparison of two nodulating and two non-nodulating, and of three efficient and three inefficient soybean genotypes are given in Tables 9.3 and 9.4. Even with a heavy dose of N

Table 9.3: Plant characteristics of nodulating and non-nodulating soybean genotypes (Jeppson et al., 1978).

Geno-types	DM 30 days after flowering (g/m²)	Maturity date (days after 1st Sept.)	Grain yield (kg/ha)	Total N content (kg/ha)	N in stover (%)	Seed protein (%)	Seed oil (%)	HI	HNI
Naro-soy, *nod*	632	4	3.356	213	0.77	42.6	20.7	60	93
nonnod	578	4	2.633	156	0.67	38.3	22.7	52	91
Clark, *nod*	974	26	2.733	184	0.89	38.3	22.7	51	89
nonnod	773	27	2.528	157	0.79	38.5	21.8	46	87

Table 9.4: HI and HNI of three efficient (E) and three inefficient (I) soybean genotypes measured after 150 days of emergence (Jeppson et al., 1978).

Genotype	HI	HNI
21(E)	44	70
27(E)	53	79
30(E)	51	76
3(I)	42	68
24(I)	43	71
31(I)	41	67

(200kg/ha) to the non-nodulating soybean line, the yield of the near isogenic nodulating line of the variety Clark 63 was always higher (Singh et al., 1974). Thus unless the N-translocation efficiency of the non-nodulating genotypes is improved, the scarce and costly fertilizer application, which developing countries can hardly afford, is also not helpful. Such anodulating genotypes have been developed in crops like groundnut (Gorbet and Burton, 1979; Nigam et al., 1980, 1982), alfalfa (Peterson and Barnes, 1981); soybean (Caldwell, 1966; Weber, 1966a,b; Williams and Lynch, 1954), chickpea (Dahiya and Khurana, 1981; Davis et al., 1985; 1986), beans (Davis et al., 1988), and peas (Kneen and La Rue, 1984). Nigam et al. (1980, 1982) suggested that non-nodulation in groundnut is governed by two duplicate recessive genes. In alfalfa, the *nonnod* trait is conditioned by two tetrasomically inherited recessive genes (Peterson and Barnes, 1981).

Supplemental Urea Spray: Maximum yield of pulses may not be achieved on natural nodulation (Smartt, 1976). The relationship between nodulation, N-fertilization and maximum yield is somewhat problematic. The problem is how to supplement the N_2 fixed by the root nodule bacteria without impairing their activity unduly, as it is well known that the available nitrates in the soil depress nodulation (Latimore et al., 1977; Wynne et al., 1979). Further, Chen and Phillips (1977) exhibited decrease in nitrogenase activity followed by a decline in leghaemoglobin content in nodules supplied with KNO_3. The inhibition of nodulation observed by N-application in soil will not take place if N is applied foliarly (N-effect on nodulation is local - Hinson, 1975), and the plants will be able to derive simultaneous benefits of both elemental and combined nitrogen. Thus foliar fertilization during the grain filling period (when nodules are already decayed) promises to increase yield (Garcia and Hanway, 1976) by avoiding undue depletion in the leaves and the resultant reduction in photosynthetic rate. This should be a very practical method of increasing grain yield in legumes (Garcia and Hanway, 1976).

Nodulation and Nitrogen Fixation: Many developing countries lack the facilities to produce and distribute high quality rhizobia inoculants. If genotypes are available that can nodulate effectively with the locally present rhizobia, farmers can grow a successful crop without inoculation or N-fertilization. Rewari et al. (1974) using four soybean genotypes and two *Rhizobium* strains concluded that host specificity controls the efficiency of *Rhizobium*. Several soybean genotypes have been identified which nodulate effectively with many strains of the cowpea inoculation group, which is ubiquitous in tropical soils of Africa (Kueneman et al., 1984; Pulver et al., 1985). Of the 400 soybean germ-plasm accessions screened for such promiscuous nodulation with indigenous rhizobia at five locations in Nigeria, only 10 established effective symbiosis at all sites. Further, five soybean lines differing in N_2 fixation were crossed and representative F_2 derived F_4 families were studied at three sites for shoot dry weight and seed yield (Neuhansen, 1987). Expected gain estimates from selection for dry weight were from 0 to 10 per cent and for yield from 4 to 27 per cent.

From the M_4 population obtained after γ-ray treatment of cowpea var. C15-2, 10 lines were identified that proved superior in nodulation to the original variety (Maherchandani and Rana, 1977). From progeny row studies in M5, 35 lines with improved nodulation were selected (Maherchandani, 1979).

Studies on host factors on N_2 fixation in *Phaseolus vulgaris* showed a close association between high N_2 fixation and the late maturing, indeterminate character (McFerson et al., 1980). A donor parent with good N_2 fixation properties and five commercial cultivars differing in N_2 fixation, yield potential, plant type and maturity were used to generate populations of C 70 near homozygous recombinant families (McFerson and Bliss, 1982). In spite of low broad sense heritability for N_2 fixation, lines were identified which consistently exceeded the recurrent parents for both N_2 fixation and yield. Some lines were capable of competitive yields even under conditions of extreme soil nitrogen deficiency.

The F_1 generations of a diallel cross of 10 South American groundnut cultivars were evaluated for six N_2 - fixation traits (Wynne et al., 1980). Hybrid progenies were significantly different for all traits. GCA was significant and greater than SCA for nodulation, N_2 fixed, plant weight, N content and total N. Field tests of 30 F_4 lines from a Virginia × Spanish cross showed that heritability estimates range from 0.45 for number of nodules to 0.80 for N_2 fixed. Berestetskii and Tikhonovich (1985) selected some lines of *Pisum sativum* on the basis of nitrogenase activity. After two to three cycles of selection, some lines showed increased activity over their initial forms and higher N_2 fixing ability and efficiency.

White clover plants selected for high total nodule mass were crossed and further selections made after inoculation with a known effective and efficient strain of *R. trifolii* (Jones and Burrows, 1968). Selection of these plants was evidently successful in that both the F_1 and F_2 progeny of the crosses had significantly higher nodule scores than did the parental population. Nodule mass increased significantly in F_1 but there was only a slight increase from F_1 to F_2. They further observed that efficiency of N_2 fixation on a nodule basis decreased as nodule size increased. As selection increased the total amount of nodule tissue, there was also a significant increase in nodule number in the F_1. Nutman (1980) showed that selected lines of red clover generally maintain their symbiotic superiority when tested against a number of environmental variables including a range of different

Rhizobium strains. The F_1 progeny from high N_2 fixation lines showed a 31 per cent increase in total N accumulation and a 20 per cent increase in dry matter yield over the base population (Mytton and Hughes, 1985). When grown with non-limiting applications of nitrate, the same material gave 17 per cent and 6 per cent, respectively, more than the base population.

Nitrate Tolerance: *Pisum sativum* was inoculated with 38 strains of *R. leguminosarum* in the presence of NH_4NO_3 for four weeks with weekly additions of 2mM NH_4NO_3 (Nelson, 1983). Acetylene reduction was inhibited relative to the N-free control in all isolates, but the inhibition varied from 60 to 100 per cent. Nelson concluded that N_2 fixation enhancement is possible through selection of rhizobia with maximum effectiveness at low levels of NO_3 - N. Gibson and Harper (1985) also indicated that selection and breeding enhanced NO_3 - tolerance of nodulation between soybean and *Bradyrhizobium japonicum* is possible. In fact, following screening soybean mutants in a high NO_3 - culture (5mm KNO_3), 15 independent NO_3 - tolerant symbiotic mutants were obtained by Carroll et al. (1985) from 2500 M_2 families derived from seed treatment of Bragg with ethylmethane-sulphonate.

Nodulation of a mutant of Rondo and wild type Rondo (*Pisum sativum*) was also investigated in the presence and absence of 15mM KNO_3 in a medium containing *R. leguminosarum* (Jacobsen and Feenstra, 1984). They concluded that the Rondo mutant contains a single, recessive, nodulating gene other than nod_1 and nod_2 designated as nod_3.

Rhizobium

Effectiveness: Different available races of rhizobium vary in their effectiveness in fixing N_2. The most effective strains which are different for different crop species and varieties can be selected. The effectiveness of selected non-selected/strains can further be increased by mutation.

Inoculation of soybean plants with two strains of *R. japonicum* (Nanking and USDAb 136) showed that Nanking strain was more effective in increasing shoot weight, grain yield, protein content and N-uptake, eight weeks after inoculation (Jain and Rewari, 1973). Further, Jain and Rewari (1975) using six varieties of soybean reported that in all except improved Pedican, the Nanking race of *Rhizobium* resulted in a higher grain yield than the USDA race. The increase in yield of inoculated varieties compared with controls ranged from twice in Punjab 1 to 10 times in Masterpiece. The mixture of *Rhizobium* races was found beneficial only in the variety Lee. The Nanking race gave the best N_2 fixation in Bragg, Clark 63 and Masterpiece, and the USDA race in Improved Pedican. The varietal specificity for rhizobial serotypes in relation to nodulation and crop yield was further investigated by Dadarwal and Sen (1974). Nodulation and yield with native and introduced rhizobia were examined in five varieties of chickpea, eight varieties of pea, three varieties of soybean and two varieties of black gram. Both nodulation and yield increased after inoculation in four, three, two and two varieties, respectively. They also reported that both nodulation and response to yield were quantitative characters under independent genetic control. In the same year, Chandra (1974) obtained 15 serotypes of the *Rhizobium* strain BU/S_4 by UV radiation and compared with the wild strain for their biochemical properties and efficiency of N_2 fixation on two soybean varieties. The serotypes U/11 and U/49 were most effective in increasing yields.

By X-radiation of efficient rhizobial strains of *Vigna mungo, V. radiata* and *Dolichos biflorus,* Sudersanam and Prasad (1972/73) produced three mutants with increased ability to fix N$_2$. Further, Maier and Brill (1978) treated a strain of *R. japonicum* with N-methyl-N-dinitro-N-nitrosoguanidine, the mutant strains SM31 and SM35 were isolated and were found to nodulate roots earlier, fixed more N$_2$ and caused plants to achieve greater dry weight and contain more N than did the wild strain. A mutant of *R. japonicum* strain USDA110 with approximately 100 per cent greater C$_2$H$_2$ reduction capacity in free living culture than the wild type strain was analysed on soybean (Williams and Phillips, 1983). The mutant strain, C33, which also was deficient in nitrate reductase activity, increased dry weight and N content approximately 40 per cent relative to the wild type USDA110.

Stress Tolerance

Agrochemicals: A variety of agrochemicals (insecticides, fungicides, herbicides, etc.) are added to the soil and they do affect the native rhizobial population and their activity. Gupta and Moolani (1970) reviewed the effects of applied herbicides on soil microorganisms and their activity. Generally the toxicity to the rhizobia is increased with additional halogen substitution and decreased when butyric substituted for acetic as the aliphatic acid portion of the molecule. The inhibiting concentration of some herbicides are given in Table 9.5.

Table 9.5: Concentrations (ppm) of some herbicides that start inhibiting N$_2$ fixation (After Vincent, 1977).

Herbicides	Inhibiting concentration (ppm) after 14 days of inoculation
2,4,5-T (2,4,5-trichlorophenoxy acetic acid)	50
2,4-D (2,4-dichlophenoxy acetic acid)	200
MCPA (Methylchlorophenoxy acetate)	200
MCPB (Methylchlorophenoxy buterate)	500
2,4DB (2,4-dichlorophenoxy buterate)	500

There are clear indications of toxic effects of many fungicides and insecticides on N$_2$ fixation, when used as seed inoculants (Vincent, 1958; Goss and Shipton, 1965). But many fungicides based on thiram (tetramethyl-thiuram disulphate) are generally least toxic to rhizobia (Vincent, 1977).

Salinity: Legumes are known to be very sensitive to salt stress and rhizobia are also very sensitive, thus nodulation, N$_2$ fixation, crop growth and yield are all reduced under salinity stress. While breeders are trying to breed salt tolerant genotypes, similar efforts for increasing salt tolerance of symbionts are lacking. But before that the existing rhizobial strains should be screened. Subba Rao et al (1972) screened 13 strains of *R. meliloti*, of which three strains tolerated up to 3 per cent NaCl solutions while lucerne seeds failed to germinate even in 1.5 per cent solution.

Drought: Drought has been known to affect the development of nodules and N$_2$ fixation activity (Huang et al., 1975a, b; Pankhurst and Sprent, 1975; Patterson et al., 1979; Sprent, 1976), with Huang et al. (1975a, b) concluding that the decline in N$_2$ fixation in water-stressed nodules is due mainly to a deficiency in the photosynthate supply from the host plant. Pankhurst and Sprent (1975) and Sprent (1976) have reported that the inhibitory effect of water stress on N$_2$ fixation is associated with lower nodule respiratory activity.

Sung (1982) observed genotypic differences in nodule growth reduction following water stress in soybean, with genotype Clark having greater reduction (50 per cent) than genotype TK5 (30 per cent) when subjected to severe moisture stress (−16 to −18 bars). The inhibition in nodule growth is mainly attributed to a lowered activity of *Rhizobium* in the root and subsequent nodulation process (Sprent, 1976).

Several hypotheses have been proposed to explain the inverse relationship between N_2 fixation and moisture stress. Huang et al. (1975a, b) have suggested that the inhibition of photosynthesis in the aerial portion of the plant accounted for the decline in nodule N_2 fixation. Pankhurst and Sprent (1975) and Sprent (1976) suggested that the prime effect of drought on nodules is to depress O_2 uptake; this reduces the supply of available metabolites for example ATP, essential for N_2 fixation. Further, Patterson et al. (1979) have indicated that N_2 fixation in moisture stressed nodules is closely related to their energy charge.

Acidity: *Rhizobium* is often more affected by low pH than the host plant, in as much as the host can grow in soils in which these bacteria perish rapidly. Freire (1976) pointed out the need for rhizobial strains that are tolerant to low pH and to high concentrations of Al and Mn for use in areas of low input agriculture. Schreven (1972) noted that the growth of three strains of *R. trifolii* was retarded in cultures at pH 4.5 and was very poor at pH 4.0 and 3.5. After 12 subcultures on media from pH 3.5 to 7.0 the symbiotic effectiveness of the strains was estimated by inoculating white clover seedlings. Most nodules were found on plants inoculated with strain A136 which had been substituted repeatedly on a medium of pH 3.5. With strain A139, most nodules were found on plants inoculated with the strain which had been substituted at pH 3.5 but A142 resulted in fewest nodules when grown at this pH.

LEGUME GENOTYPE × RHIZOBIUM STRAIN INTERACTION

Burton and Wilson (1939) clearly indicated host-symbiont specificity between nine strains of *R. meliliti*, three cultivars of *Medicago sativa* (alfalfa), and four other *Medicago* species. In this way when large numbers of host genotypes and symbiont strains are screened, specific combinations can be found which are exceptional fixers of N_2. Specific genotype × strain combinations that result in enhanced nodulation, increased N_2 fixation and increased yields have been identified (Caldwell and Vest, 1968; Sloger, 1969; Kvien et al., 1981). The success of these genotype-strain combinations in a heterogenous soil environment will depend on the survival and competitive ability of the inoculant strain. Zobel (1980) in evaluating 500 plant introductions and 342 named cultivars and breeding lines, found considerable genetic variability in nodule number and size, with the indigenous *Bradyrhizobium japonicum* strains. He concluded that genetic manipulation of these traits should be possible. Further, Kvien et al. (1981) evaluated 1600 soybean genotypes for ability to nodulate with *B. japonicum* in Minnesota and for their ability to preferentially select an effective inoculant strain, *B. japonicum* strain USDA110, in the presence of a large indigenous *B. japonicum* population.

Dart et al. (1976) have shown an interactive effect between five rhizobium strains and three chickpea genotypes (Table 9.6). The data of Chowdhury (1977) using four soybean genotypes and five rhizobium strains and a mixture of strains further show the interactive effects on grain yield (Table 9.7).

Table 9.6: Interactive effects of five rhizobium strains and three chickpea genotypes (After Dart et al., 1976).

Genotypes	Rhizobium strains				
	CB 1189	27A2	27A9	Ca-1	Ca-2
	Dry weight of nodules per plant (mg)				
Deshi	89	79	84	82	87
Kabuli	98	93	108	78	103
Iranian	100	86	96	104	96
	Nitrogen fixed (mg/plant)				
Deshi	26	21	27	28	28
Kabuli	30	27	35	24	33
Iranian	33	29	32	35	21

Table 9.7: Interactive effects of soybean genotypes and rhizobium strains on the grain yield (After Chowdhury, 1977).

Rhizobium strains	Grain yield (t/ha) of indicated variety				
	IH/192	HLS223	7H/101	3H/49/1	Mean
CB1809	2.05	2.71	2.44	2.28	2.37
M15	2.39	2.82	2.21	2.65	2.52
M17	3.02	3.50	2.30	2.57	2.85
M18	2.80	2.64	2.40	2.26	2.52
M19	2.83	2.86	3.23	3.13	3.01
Mixture	2.92	3.84	3.17	2.29	3.05
Uninoculated	2.13	2.38	2.60	2.42	2.38

The potential of breeding to utilize the aforementioned diversity in nodulation ability was addressed by Lawson (1980). Lawson evaluated the genetic variability for nodule number, nodule weight and per cent recovery of strain USDA110 of 21 diverse soybean genotypes at three locations in Minnesota. Heritability estimates, averaged over locations, were 89 per cent for nodule number per plant, 81 per cent for nodule weight per plant, and 31 per cent for percentage recovery of strain USDA110. Gupta et al. (1984) using 36 F_4 derived lines and several cultivars in a field study at two locations in India, reported estimates of heritability on a plot basis of 15 per cent for nodule number per plant and less than 10 per cent for nodule dry weight per plant.

Zary (1980) reports that in cowpea, N_2 fixation is controlled by several quantitatively inherited genes. The interaction of genotypes of cowpea with the strain of rhizobia used was important in determining the amount of N_2 fixed and the size and number of nodules produced. Broad sense heritability estimates for these characters were high. Nodule mass, nodule number and nodule specific activity were correlated with N_2 fixation. The capacity of genotypes to fix carbon is influenced by the rhizobium strain, and the interaction of genotypes and strain influence the amount of CO_2 fixed (Ganeshaiah, 1984). During the nodule growth more efficient rhizobium strain received less C (in the form of ^{14}C than the less efficient strain. However, while the percentage of ^{14}C translocated to the nodule was not related to N_2 fixation efficiency, total ^{14}C fixed was positively associated with N_2-fixation percentage. Genotype and strains varied and interacted for nodule tissue formation.

Vincent and co-workers (1953, 1954) showed that when strains of *R. trifolii* were mixed and placed on different hosts, the host exercised selection pressure and determined the strains relative success in induction of nodulation. They mixed two strains of *R. trifolii* and inoculated red and subterranean clover. Strain 157 occurred most frequently on red clover while strain 204 was most common on subterranean clover. They found a similar relationship for cultivars within a species.

In an interaction experiment between 15 alfalfa genotypes and 10 rhizobium strains, Tan and Tan (1986) reported that genotypes contributed more than 30 per cent of the total variance for acetylene reduction rate, strains contributed 26 per cent and the genotype × strain interaction more than 36 per cent.

REFERENCES

Alwi, N. Peanut groups and their symbiotic relationship with *Rhizobium* strains. (Abstr.) *Diss. Abstr. Int. B.* 46: 353B, 1985.

Ansubel, F.M. Molecular genetics of symbiotic nitrogen fixation. *Cell* 29: 1–2, 1982.

Arrendell, S.,J.C. Wynne, G.H. Elkan and T.G. Isleib. Variation for nitrogen fixation among progenies of Virginia × Spanish peanut cross. *Crop Sci.* 25: 865–69, 1985.

Askin, D.C., J.G.H. White and P.J. Rhodes. Nitrogen fixation by peas and their effect on soil fertility. *In*: The Pea Crop. P.D. Hebblethwaite; M.C. Heath and T.C.K. Dawkins (eds), London, Butterworths. pp. 421–30, 1985.

Asokan, M.P. Genetical and physiological studies on effective and ineffective nodulation of some bean (*Phaseolus vulgaris* L.) cultivars. *Diss. Abstr. Int. B.* 42: 15 B, 1981.

Ayanaba, A. and D. Nangju. Personal communication, 1977.

Bajpai, P.D., L.K. Lehri and A.N. Pathak. Effect of seed inoculation with *Rhizobium* strains on the yield of leguminous crops. *Proc. Indian Natl. Sci. Acad. B.* 40: 571–75, 1974.

Bello, A.B., W.A. Ceron-Diaz, C.D. Nickell, E.O. El-Sherif and L.C. Davis. Influence of cultivar, between row spacing and plant population on fixation of soybeans. *Crop Sci.* 20: 751–55, 1980.

Berestetskii, O.A. and I.A. Tikhonovich. (Increasing the effectiveness of the biological nitrogen fixation by breeding legumes for symbiosis). Doklady Vsesoguznoi Ordena Lenina i Ordena Trudovogo Krasnogo Znameni Akademii Selskokhozyaistvennykh Nauk Imeni V.I. Lenina. No. 6: 9–11, 1985.

Beringer, J.E. and A.W.B. Johnston. The genetics of the *Rhizobium* legume symbiosis. Isotopes in Biological Dinitrogen Fixation. Proc. Advisory group meeting organized by the joint FAO/IAEA Division of Atomic Energy in Food and Agriculture, Vienna, 1977. pp 107–133, 1978.

Brill, W.J. Agricultural opportunities from nitrogen fixation research. *In*: Biotechnology for Solving Agricultural Problems. P.C. Augustine, H.D. Danforth and M.R. Bakst (eds), Dordrecht, Netherlands, Martinus Nijhoff Publishers. pp. 183–93, 1986.

Brockwell, J. Can inoculant strains ever compete successfully with established soil populations? *In*: Current Perspectives in Nitrogen Fixation. Proc. IV Int. Symp. on Nitrogen Fixation. Canberra, 1980 (eds). A.N. Gibson and W.E. Newton (eds), 277. 1981.

Burton. J.C. and P.W. Wilson. Host plant specificity among the *Medicago* in association with root nodule bacteria. *Soil Sci.* 47: 293–303, 1939.

Caldwell, B.E. Inheritance of a strain specific ineffective nodulation in soybeans. *Crop Sci.* 6: 427–28, 1966.

Caldwell, B.E. and H.G. Vest. Nodulation interaction between soybean genotypes and serogroups of *Rhizobium japonicum. Crop Sci.* 680–82, 1968.

Caldwell, B.E. and H.G. Vest. Genetic aspects of nodulation and nitrogen fixation by legumes: the macrosymbiont. *In:* A Treatise on Dinitrogen Fixation. Section III. Biology (eds) R.W.F. Hardy and W.S. Silver, John Wiley & Sons, N.Y., U.S.A. pp. 557–76, 1977.

Carroll, B.J., D.L. Mc Neil and P.M. Gresshoff. Isolation and properties of soybean [*Glycine max* (L) Merr] mutants that Modulate in the presence of high nitrate concentrations. *Proc. Nat. Acad. Sci.,* U.S.A. 82: 4162–66, 1965.

Chandra, A.K. Induction of mutation in *Rhizobium japonicum* by ultraviolet radiation and selection of agronomically superior serotypes. *Proc. Indian Nat. Sci. Acad. B.* 40: 491–B, 1974.

Chen, P.C and D.A. Phillips. Induction of root nodule senescence by combined nitrogen in *Pisum sativum* L. *Pl. Physiol.* 59: 440–42, 1977.

Chowdhury, M.S. Paper presented at Int. Symp. on Limitations and Potentials of Biological N$_2$ Fixation in the Tropics, July 1977, Brazil. 1977.

Dadarval, K.R. and A.N. Sen. Varietal specificity for rhizobial serotypes in relation to nodulation and crop yield. *Proc. Indian Nat. Sci. Acad. B.* 40: 548–53, 1974.

Dahiya, B.S. and A.L. Khurana. Non-fixing genotype in chickpea. *Int. Chickpea Newsl.* 5: 16, 1981.

Dart, P.J, J.M. Day, R. Islam and J. Döberiner. *In:* Symbiotic Nitrogen Fixation in Plant, P.S. Nutman (ed), Cambridge Univ. Press, Cambridge. 1976.

Davis, J.H.C., K.E. Giller, J.Kipe-Nolt and M. Awah. Non-nodulating mutants in common bean. *Crop Sci.* 28: 859–60, 1988.

Davis, T.M., K.W. Foster and D.A. Phillips. Nodulation mutants in chickpea. *Crop Sci.* 25: 345–48, 1985.

Davis, T.M., K.W. Foster and D.A. Phillips. Inheritance and expression of three genes controlling root nodule formation in chickpea. *Crop Sci.* 26: 719–23, 1986.

Duhigg, P; B. Melton and A. Baltensperger. Selection for acetylene reduction rates in "Mesilla" alfalfa. *Crop Sci.* 18: 813–16, 1978.

El-Sherbeeny, M.H., L.R. Mytton and D.A. Lawes. Symbiotic variability in *Vicia faba* 1. Genetic variation in the *Rhizobium leguminosarum* population. *Euphytica* 26: 149–56, 1977a.

El-Sherbeeny, M.H., D.A. Lawes and L.R. Mytton. Symbiotic variability in *Vicia faba.* 2 Genetic variation in *Vicia faba. Euphytica* 26: 377–83, 1977b.

Fernandez, G.C.J. Interrelationships between N$_2$ fixation and seed yield in cowpea (Abstr.). *Diss. Abstr. Int. B.* 47: 62B, 1986.

Fernandez, G.C.J. and J.C. Miller, Jr. Intracultivar variability and heritability estimation for nitrogen fixation variables in *mung* bean populations. (Abstr.) *HortScience* 18: 171, 1983.

Freire, J.R.S. Inoculation of soybeans. *In:* Exploiting the Legume Rhizobium Symbiosis in Tropical Agriculture. J.M. Vincent; A.S. Whitney and J. Bose (eds), Coll. Trop. Agric. Mis. Pub. 145. Univ. of Hawaii, Honolulu. pp. 335–80, 1976.

Ganeshaiah, K.N. Host-genotypic variation for symbiotic nitrogen fixation in cowpea, *Vigna unguiculata* (L) Walp. (Abstr.) *Mysore J. agric. Sci.* 18: 237, 1984.

Garcia, L.R. and J.J. Hanway. Foliar fertilization of soybean during the seed filling period. *Agron. J.* 68: 653–57, 1976.

Gelin, O. and S. Blixt. Root nodulation in peas. *Agric. hort. genetica.* 22: 149–59, 1964.

Gibson, A.H. Host determinants in nodulation and nitrogen fixation. *In:* Advances in Legume Science. R.J. Summerfield and A.H. Bunting (eds), Kew, Royal Botanical Gardens, U.K. pp. 69–75, 1980.

Gibson, A.H. and J.E. Harper. Nitrate effect on soybean by *Bradyrhizobium japonicum. Crop Sci.* 25: 497–501, 1985.

Gorbet, D.W. and J.C. Burton. A nonnodulating peanut. *Agron. Abstr.* Madison, Wis., U.S.A. p. 62, 1979.

Goss, O.M. and W.A. Shipton. *J. Agric.* (W.A.) 6: 659, 1965.

Greder, R.R., J.H. Orf and J.W. Lambert. Heritabilities and associations of nodule mass and recovery of *Bradyrhizobium japonicum* serogroup USDA110 in soybean. *Crop Sci.* 26: 33–37, 1986.

Gupta, R.P., M.S. Kalra and Y.P.S. Bajaj. Nitrogen fixation in cell cultures of some legumes and non-legumes. *Indian J. exp. Biol.* 22: 560–63, 1984.

Gupta, U.S. and M.K. Moolani. Effects of applied herbicides on soil microorganism and their activity. *Indian J. Weed Sci.* 2: 141–59, 1970.

Ham, G.E. Interactions of *Glycine max* and *Rhizobium japonicum*. *In*: Advances in Legume Science. R.J. Summerfield and A.H. Bunting (eds), Kew, Royal Botanical Gardens, UK. pp. 289–96, 1980.

Hinson, K. Nodulating responses to nitrogen applied half root system. *Agron. J.* 67: 799–804, 1975.

Hobbs, S.L.A. and J.D. Mahon. Heritability of N_2 (C_2H_2) fixation rates and related characters in peas (*Pisum sativum* L). *Can. J. Pl. Sci.* 62: 265–76, 1982.

Holl, F.B. Host-determined genetic control of nitrogen fixation in the *Pisum-Rhizobium* symbiosis. (Abstr.) *Can. J. Genet. Cytol.* 15: 659, 1973.

Holl, F.B. Host plant control of the inheritance of dinitrogen fixation in the *Pisum-Rhizobium* symbiosis. *Euphytica* 24: 767–70, 1975.

Holl, F.B. and T.A. LaRue. Genetics of legume plant hosts. *In*: Proc. 1st Int. Symp. on Nitrogen Fixation. W.E. Newman and C.J. Nymans (eds). Washington State Univ. Press, Pullman. pp. 391–99, 1976.

Hoffman, D. and B. Melton. Variation among alfalfa cultivars for indices of nitrogen fixation. *Crop Sci.* 21: 8–10, 1981.

Huang, C.Y., J.S. Boyer and L.N. Vanderhoef. Acetylene reduction (nitrogen fixation) and metabolic activities of soybean having various leaf and nodule potentials. *Pl. Physiol.* 56: 222–27, 1975a.

Huang, C.Y., J.S. Boyer and L.N. Vanderhoef. Limitation of acetylene reduction (nitrogen fixation) by photosynthesis in soybean having low water potentials. *Pl. Physiol.* 56: 228–32, 1975b.

Jacobsen, E. and W.J. Feenstra. A new pea mutant with efficient nodulation in the presence of nitrate. *Plant Sci. Letters* 33: 337–44, 1984.

Jain, M.K. and R.B. Rewari. Inoculation experiment with different bacterial cultures on soybean. *Curr. Sci.* 42: 749–50, 1973.

Jain, M.K. and R.B. Rewari. Studies of *Azotobacter* and *Rhizobium* inoculation on soyabean. *Agriculture*, Belgium. 23: 37–47, 1975.

Jeppson, R.G., R.R. Johnson and H.H. Hadley. Variation in mobilization of plant nitrogen to the grain in nodulating and non-nodulating soybean genotypes. *Crop Sci.* 18: 1058–62, 1978.

Johnson, V.A. and P.J. Mattern. Report of research findings — improvement of nutritional quality of wheat through increased protein content and improved amino-acid balance. Jan. 1973-March, 1975 (U.S.A.I.D., Washington D.C.). p. 86, 1975.

Jones, D.G. Symbiotic variation of *Rhizobium trifolii* with S100 Nomark white clover (*Trifolium repens* L.) *J. Sci. Food & Agric.* 14: 740–43, 1964.

Jones D.G. and A.C..Burrows. Breeding for increased nodule tissue in white clover (*Trifolium repens* L.) *J. Agric. Sci. U.K.* 71: 73–79, 1968.

Kneen, B.E. and T.A. La Rue. EMS derived mutant of *Pisum sativum* resistant to nodulation. *In*: Advances in Nitrogen Fixation Research, C. Veeger and W.E. Newton, (eds), Martinus Nijhoff/ Dr. Junk W. Pub., Wageningen, Netherlands. pp. 599, 1984.

Kueneman, E.A., W.R. Root, K.E. Dashiell and J. Hobenberg. Breeding soybeans for the tropic capable of nodulating effectively with indigenous *Rhizobium* spp. *Plant & Soil*. 82: 387–96, 1984.

Kumar Rao, J.V.D.K. and S.R. Viswanatha. Nodulation interaction between soybean genotypes and *Rhizobium japonicum*. *Curr. Research*. No. 12: 163–64, 1974.

Kuzmin, M.S. (Features of nodule formation in different soybean varieties). *In*: Selektsiya semeno-vodi agrotekhn soi Novosibirsk, USSR. pp. 44–47, 1977.

Kvien, C.S., G.E. Han and J.W. Lambert. Recovery of introduced *Rhizobium japonicum* strains by soybean genotypes. *Agron. J.* 73: 900–05, 1981.

Latimore, M. Jr., J.G. Giddens and D.A. Ashley. Effect of ammonium and nitrate nitrogen upon photosynthate supply and nitrogen fixation by soybean. *Crop Sci.* 17: 399–404, 1977.

Lawson, R.M. Genetic variability in soybeans for nodule number and weight, and recovery of *Rhizobium japonicum* strain 110. Ph.D. Diss. Univ. of Minnesota. 1980.

Maherchandani, N. Studies on the cowpea mutants for nodulation (Abstr.) *In*: Symp. on the Role of Induced Mutation in Crop Improvement; Hyderabad, Sept. 1979. p.22, 1979.

Maherchandani, N. and O.P.S. Rana. Gamma radiation induced natural variability for nodulation in legumes. *J. nucl. Agric. Biol.* 7: 75–77, 1977.

Maier, R.J. and W.J. Brill. Mutant strains of *Rhizobium japonicum* with increased ability to fix nitrogen for soybean. *Science* 201: 448–50, 1978.

McFerson, J.R. and F.A. Bliss. Selection for enhanced nitrogen fixation in common bean (Abstr.). *HortScience.* 17: 476, 1982.

McFerson, J.R., J.C. Rosas and F.A. Bliss. Host factors affecting nitrogen fixation in common bean. (Abstr.) *HortScience.* 15: 431, 1980.

Miller, J.C. Jr., K.W. Zary and G.C.J. Fernandez. Inheritance of N_2 fixation efficiency in cowpea. *Euphytica* 35: 551–60, 1986.

Mytton, L.R. and D.M. Hughes Breeding for increased N_2 fixation in white clover. *In*: Report, Welsh Plant Breeding Station, 1984. Aberystwyth, U.K. p. 44, 1985.

Nelson, L.M. Variation in ability of *Rhizobium leguminosarum* isolates to fix dinitrogen symbiotically in the presence of ammonium nitrate. *Can. J. Microbiol.* 29: 1626–33, 1983.

Neuhansen, S.L. Screening and selection for enhanced dinitrogen fixation in soybean (*Glycine max* (L.) Merill) of maturity groups 00 and 0. (Abstr.) *Diss. Abstr. Int.B.* 47: 2684B, 1987.

Nigam, S.N., V. Arunachalam, R.W. Gibbsons, A. Bandyopadhyay and P.T.C. Nambiar. Genetics of nonnodulation in groundnut (*Arachis hypogaea* L.). *Oleagineux* 35: 453–55, 1980.

Nigam, S.N., P.T.C. Nambiar, S.L. Dwivedi, R.W. Gibbons and P.J. Dart. Genetics of nonnodulation in groundnut (*Arachis hypogaea* L.). Studies with single and mixed *Rhizobium* strains. *Euphytica* 31: 691–93, 1982.

Nutman, P.S. Genetical factors concerned in the symbiosis of clover and nodule bacteria. *Nature* 157: 463–65, 1946.

Nutman, P.S. *Symp. Soc. exp. Biol.* 13: 42, 1959.

Nutman, P.S. Genetics of symbiosis and nitrogen fixation in legumes. *Proc. Royal Soc. Series B.* 172: 417–37, 1969.

Nutman, P.S. Adaptation: *In*: Nitrogen Fixation. Proc. of the Phytochemical Soc. of Europe Symp. Sussex 1979. W.D.P. Stewart and J.R. Gallon (eds), Academic Press. pp. 335–54, 1980.

Nutman, P.S. Hereditary host factors affecting nodulation and nitrogen fixation. *In*: Current Perspective in Nitrogen Fixation. Proc. IV Int. Symp. on Nitrogen Fixation, Canberra 1980. A.H. Gibson and W.E. Newton (eds). pp. 194–204, 1981.

Pankhurst, C.E. and J.I. Sprent. Effects of water stress on respiratory and nitrogen fixation activity of soybean root noduls. *Jour.expt. Bot.* 26: 287–304, 1975.

Path, B.D. and L. Moniz. Differential response of gram varieties to an efficient isolate of *Rhizobia* from gram (*Cicer arietinum* L.). *Res. J. Mahatma Phule Agric. Univ.* 5: 42–46, 1974.

Patterson, R.P., C.D. Raper Jr. and H.D. Gross. Growth and specific nodule activity of soybean during application and recovery of a leaf moisture stress. *Pl. Physiol.* 64: 551–56, 1979.

Perez, C.M., G.B. Cajampang, B.V. Esmana, R.U. Monserrate and B.O. Juliano. Protein metabolism in leaves and developing grains of rice differing in grain protein content. *Pl. Physiol.* 51: 537–42, 1973.

Peterson, M.A. and D.K. Barnes. Inheritance of ineffective nodulation and non-nodulation traits in alfalfa. *Crop Sci.* 21: 611–16, 1981.

Pinchbeck, B.R., R.T. Hardin, F.D. Cook and I.R. Kennedy. Genetic studies of symbiotic nitrogen fixation in Spanish clover. *Can. J. Pl. Sci.* 60: 509–18, 1980.

Postgate, J. Nitrogen fixation and the future of the world's food supply. *Biologist* 26: 165–66, 1979.

Prakash, R.K. and A.G. Atherly. Reiteration of genes involved in symbiotic nitrogen fixation by fast growing *Rhizobium japonicum. J. Bact.* 160: 785–87, 1984.

Pulver, E.L., E.A. Kueneman and V. Ranga-Rao. Identification of promiscuous nodulating soybean efficient in N_2 fixation. *Crop Sci.* 25: 660–63, 1985.

Rao, A.V., B. Venkateswarlu and A.Henry. Genetic variation in nodulation and nitrogenase activity in *guar* and *moth. Indian J. Genet. Pl. Breed.* 44: 425–28, 1984.

Rewari, R.B., M.K. Jain and R.S. Bhatnagar. Varietal response of soybean to different strains of *Rhizobium japonicum. Indian J. agric. Sci.* 43: 801–04, 1974.

Rhodes, A.P. and G.Jenkins. Nitrogen and dry matter accumulation in high lysine and normal varieties of barley. *J. agric. Sci.* UK 86: 57–64, 1976.

Schreven, D.A. Van. On the resistance of effectiveness of *Rhizobium trifolii* to a low pH. *Plant & Soil* 37: 49–55, 1972.

Schwinghamer, E.A. Genetic aspects of nodulation and dinitrogen fixation by legumes; the microsymbiont. *In:* A treatise on dinitrogen fixation. Section III: Biology. R.W.H. Hardy and W.S. Silver (eds). John Wiley & Sons. N.Y., U.S.A. pp. 577–622, 1977.

Singh, B.D., R.M. Singh, B.K. Murty and R.B. Singh. Combining ability analysis of nodulation characters in *mung* bean (*Vigna radiata (L.)* Wilczek). *Indian J. Genet. Pl. Breed.* 45: 271–77, 1985.

Singh, J.N., S.K. Tripathi and P.S. Negi. Yield and nitrogen content of nodulating soybean grown in association with nodulating lines. *Proc. Indian Natl. Sci. Acad.* B. 40: 507–11, 1974.

Sloger, C. Symbiotic effectiveness and N_2 fixation in nodulated soybean. *Pl. Physiol.* 44: 1666–68, 1969.

Smartt, J. "Tropical Pulses." 348 pp. Tropical Agric. Series, Longman Group Ltd., London, 1976.

Smith G.R., W.E. Knight and H.H. Peterson. Variation among inbred lines of crimson clover for N_2 fixation (C_2H_2) efficiency. *Crop Sci.* 22: 716–19, 1982.

Sprent, J.I. Water deficits and nitrogen-fixing root nodules. *In:* Water Deficits and Plant Growth. (ed) T.T. Kozlowski (ed), Academic Press, In: New York. 1976.

Subba Rao, N.S., M.L. Kumari, C.S. Singh and S.P. Magu. Nodulation of lucerne (*Medicago sativa* L.) under the influence of sodium chloride. *Indian J. agric. Sci.* 42: 384–86, 1972.

Sundarsanam, S. and N.N. Prasad. Studies on the performance of certain X-ray mutants of rhizobial strains isolated from black gram, green gram and horse gram. AUARA. 4/5: 60–74, 1972/73.

Sung, J.M. The effect of water stress on nitrogen fixation efficiency of field grown soybeans. *In:* Food & Fertilizer Technology Centre. *Technical Bulletin.* No. 65: 9–18, 1982.

Tan, G.Y. and W.K. Tan. Interaction between alfalfa cultivars and *Rhizobium* strains for nitrogen fixation. *Theo. & Appl. Genet.* 71: 724–29, 1986.

Tansiri, B., Y. Vasuwat and S. Vangnai. (*Rhizobium japonicum* strains effective on nitrogen fixation with soybean S.J. 2 in Korat soil series.) *Kasertsart J.* 8: 19–22, 1974.

Tarkshvili, D.V., M.Z. Machavarlani, T.A. Patshvili and V.N. Kandelaki. (The problem of the inoculation of different varieties and populations of French bean). *Bull. Acad. Sci.* Georgian SSR 73: 697–700, 1974.

Thomas, M.D.; K.W. Foster and D.A. Phillips. Inheritance and expression of three genes controlling root nodule formation in chickpea. *Crop Sci.* 26: 719–23, 1986.

Valdes Ramirez, M., M. Velazquez del Valle and J.F. Aguirre Medina. (Efficiency in N fixation and nodulation in the soybean, *Glycine max*(L.) Merr.) *Agricultura Teonica en Mexico.* 10: 99–110, 1984.

Vincent, J.M. and L.M. Waters. The influence of host on competition among clover root nodule bacteria. *J. Gen. Microbiol.* 9: 357, 1953.

Vincent, J.M. and L.M. Waters. The root nodule bacteria as factors in clover establishment in the red basaltic soils of the Lismore district, New South Wales. *Aust. J. agric. Res.* 5: 61–76, 1954.

Vincent, J.M. *In*: Nutrition of the Legumes E.G. Hallsworth (ed), Butterworths, London. pp. 108–123, 1958.

Vincent, J.M. *In*: A Treatise on Dinitrogen Fixation. Section III. John Wiley & Sons, New York. pp. 277–366, 1977.

Vorhees, J.H.J. *Am.Soc. Agron.* 7: 139, 1915.

Weber, C.R. Nodulating and non-nodulating soybean isolines. 1. Agronomic attributes. *Agron. J.* 58: 43–46, 1966a. Weber, C.R. Nodulating and non-nodulating soybean isolines. 2. Response to applied nitrogen and modified soil conditions. *Agron. J.* 58: 46–49, 1966b.

Welsh Pl. Breed. Sta., U.K. *Vicia*. Annual Report, 1976.

Williams, L.E. and D.L. Lynch. Inheritance of non-nodulating character in soybean. *Agron. J.* 46: 28–29, 1954.

Williams, L.E. and D.A. Phillips. Increased soybean productivity with a *Rhizobium japonicum* mutant. *Crop Sci* 23: 246–50.

Wilson, P.W., J.C. Burton and V.S. Bond. Effect of species of host plant on nitrogen fixation in melilotus. *J. Agric. Res.* 55: 619–29, 1937.

Wynne, J.C., S.T. Ball, G.H. Elkan and T.J. Schneewels. Cultivar, inoculation and nitrogen effects on nitrogen fixation of peanuts. *Agron. Abstr.* Madison, Wis., USA. p. 82, 1979.

Wynne J.C., H. Elkan and T.J. Schneeweis. Increasing nitrogen fixation of groundnut by strain and host selection. *In*: Proc. Int. Workshop on Groundnuts. ICRISAT, Hyderabad, India. pp. 13–17, 1980.

Zary, K.W. The genetics and heritability of quantitative differences in N_2 fixation in cowpea, *Vigna unguiculata*(L.) Walp. *Diss. Abstr. Int. B.* 41: 2424B, 1980.

Zary, K.W. and J.C. Miller, Jr. The influence of genotype and growth stage on N_2 fixation on cowpea (*Vigna unguiculata* (L.) Walp). *HortScience.* 13: 344, 1978.

Zary, K.W. J.C. Miller Jr., R.W. Weaver and L.W. Barnes. Intraspecific variability for nitrogen fixation in southern pea (*Vigna unguiculata* (L.) Walp). *J. Am. Soc. hort. Sci.* 103: 806–08, 1978.

Zobel, R.W. Rhizogenetics of soybeans. *In*: World Soybean Research Conference — II. F.T. Corbin (ed), Proc. Westview Press, Boulder Co. pp. 73–87, 1980.

CHAPTER 10

IMPROVEMENT OF NON-LEGUMES FOR NITROGEN FIXATION

INTRODUCTION

Unlike legumes, cereals and other non-legumes require large amounts of N-fertilization for optimum production. In addition to the free living N$_2$-fixing microorganisms inhabiting in most soils, some of them have been reported to be actively fixing atmospheric nitrogen in association with cereals, grasses and other diverse crop plants like potato, sweet potato and tomato (Table 10.1). The amounts of N$_2$ fixed by such non-leguminous plants have been recorded by various researchers on per day, per 100 days or per season/year basis (Table 10.2), or, as nmoles of acetylene reduced per g root per hour (Table 10.3). Different organisms active in N$_2$ fixation in association with different non-leguminous plants are given in Table 10.1. From a glance at these Tables, one is convinced that there is substantial N$_2$ fixation taking place, which is improved either by improving the host or the microbial strain. N-fertilizer requirement can be reduced substantially.

The subject has been covered several times to which the more interested readers are

Table 10.1: Organisms active in associative N_2 fixation with different crop species.

Organisms fixing N_2	'Non-legumes in association	Authority
Alcaligens	Rice	You and Qui, 1982.
Azolla-Anabana	Rice	Lamberg, 1979
Azospirillum brasilense	Barley	Qui et al. 1984; Tilak and Murthi, 1981.
	Eleusine coracana	Rai et al., 1984
	Guinea grass	Smith et al., 1976
	Maize	von Bulow and Döbereiner, 1975; Döbereiner, 1977; Albrecht et al. 1981.
	Pearl millet	Smith et al. 1976; Taylor, 1979; Bouton and Brooks, 1982, Bouton et al., 1985; Wani et al., 1985.
	Rice	Kumari et al., 1976
	Sweet potato	Hill et al., 1983
	Wheat	Barber et al., 1976; Rennie and Larson, 1979; Millet et al., 1984; Anderson, 1986.
Azotobacter chroococcum	Potato	Imam and Badawy, 1978
	Wheat	Kavimardan et al., 1978.
Bacillus macerans	Wheat	Barber et al., 1976.
B. Polymyxa	Wheat	Barber et al., 1976
B. species	Wheat	Rovira, 1965; Larson and Neal, 1976.
Beijerinkia	Rice	Yoshida, 1971b; Balandreau, 1975.
	Sugarcane	Döbereiner, 1961.
Clostridium	Rice	Qui et al., 1984.
Enterobacter clocacae	Rice	Yoshida, 1971; Hirota et al., 1978; Barraquio and Watanabe, 1981; Bally et al., 1983.
Klebsiella oxytoca	Rice	Yoo et al., 1986.
Pseudomonas	Rice	Qui et al., 1984
Rhizobia	*Trema cannabina*	Trinick, 1975.

Table 10.2: Varying rates of associative N_2 fixation with some crop and grass species.

Nonlegumes fixing N_2	Rate of fixation	Authority
Rice	By algal component at flowering - 50-200g N_2/ha/day	Watanabe, 1976.
Wheat -	0.5kg N_2/ha/season	Raju et al., 1972
	2 kg N_2/ha/day	Barber et al., 1976.
Maize	2 kg N_2/ha/day	yon Bulow and Döbereiner, 1975.
Panicum varigatum	3.6kg N_2/ha/year	Tjepkema and Burris, 1976.
Pennisatum purpurecum	1kg N_2/ha/day	Day et al., 1975.
Sorghum	2.5kg N_2/ha/100 days	Pedersen et al., 1978
Andropogon girardi	20kg N_2/ha/100 days	
Brachiaria sp	20kg N_2/ha/100 days	
Cynodon dactylon	33kg N_2/ha/100 days	Weaver et al., 1980.
Paspalum urvillei	26kg N_2/ha/100 days	
Agrostis tenuis	3kg N_2/ha/season	Nelson et al., 1976.
Sporobolus heterolepis	2.9kg N_2/ha/year	Tjepkema and Burris, 1976.

Table 10.3: Nitrogen fixatin (nmoles acetylene reduced per g root per hour) associated with the roots of some cereals and grasses (Rothamsted Exp. Sta.)

Plant species	Common name	Activity
Andropogon gayanus	Gambara grass	15-270
Brachiaria mutica	Paragrass	150-750
Cynodon dactylon	Bermuda grass	20-270
Digitaria decumens	Pangola grass	20-400
Oryza sativa	Rice	8-80
Panicum maximum	Guinea grass	75
Paspalum notatum	Bahia grass	2-300
Pennisetum americanum	Pearl millet	3-195
P. clandestinum	Kikuya grass	21-140
P. purpureum	Elephant grass	5-1000
Saccharum officinarum	Sugarcane	5-20
Zea mays	Maize	14-16
	Maize seedlings	1000-3000
Active leguminous plants		1000-3000

referred (Döbereiner, 1974; Neyra and Döbereiner, 1978; Jagnow, 1979; Van Berkum and Bohlool, 1980).

SOME STUDIES ON N_2-FIXATION IN ASSOCIATION WITH NON-LEGUMES

Such associative N_2-fixation occurs with both C_3 and C_4 species. Whether there is any significance of these mechanisms with association and/or N_2 fixation is not clear at this stage.

C_3 Species

Rice: There are several reports on N_2 fixation associated with the rhizosphere of rice plants (Eskew et al., 1981; Hirota et al., 1978; Iyama et al., 1983; Rimaudo et al., 1971; Sano et al., 1981; Watanabe, 1981; Yoshida and Yoneyama, 1980). The occurrence of rhizosphere N_2 fixation and the transport of fixed N from soil to plants was proved by exposing plants to ^{15}N-labelled N_2 gas for a specific time of plant growth (Eskew et al., 1981; Ito et al., 1980; Yoshida and Yoneyama, 1980). Several types of N_2 fixing bacteria including *Azospirillum*, *Clostridium*, *Enterobacter*, *Alcaligens* and *Pseudomonas* have been isolated from the rhizosphere of rice plants (Bally et al., 1983; Barraguio and Watanabe, 1981; Hirota et al., 1978; Kumari et al., 1976; Qui et al., 1981), and each may be a candidate for promoting N_2 fixation in association with rice. However, only a few detailed descriptions of N_2 fixation capacity of these bacteria in association with rice plants have been reported (Qui et al., 1984; Watanabe, 1981).

Strain A15 of *Alcaligens facealis* which is widespread in the soil of rice growing areas in China was isolated by You and Qui (1982). It has a N_2 fixation efficiency of 40mg N_2 assimilated/g malic acid consumed. A15 can fix N_2 in the rice root system without any extra carbon source.

Unusual non-legume symbiotic N_2-fixing associations in between the water tern *Azolla* and its blue green alga symbiont, *Anabana*. These are found in fresh water streams and ponds. This combination is likely to have considerable agronomic impact in South East Asia because it can be grown in dual culture with paddy rice acting as a slow release N-fertilizer (Lamberg, 1979). *Azolla* has been used as a green manure in Vietnam, Thailand and China (Moore, 1969). During the fallow period of rice, a crop of *Azolla* is grown and incorporated. Another crop is grown during the rice growing season — the cover crop (Talley et al., 1977). It is possible to treble the yield of rice grain compared with an unfertilized control by management of *Azolla* in this way.

Yoo et al. (1986) reported N_2 fixation by *Klebsiella oxytoca*, strain NG13, isolated from the rhizosphere in association with Indica type rice strain C5444. Inoculation of C5444 with NG13 resulted in a six per cent increase of total plant plus soil N content. By optimizing such rice plant-bacterial N_2 fixation capacity, the N uptake by rice could be enhanced (Yoo et al., 1986).

Bacterial count indicates that *Beijerinekia* sp. and *Enterobacter cloacae* are most common N_2 - fixing bacteria in the rhizosphere of rice. Most of the N_2 fixation in the rice system has been attributed to rhizosphere soil rather than roots themselves (Yoshida and Ancajas, 1973). Higher number of aerobic than anaerobic N_2-fixing bacteria in the rhizosphere of rice were also found by Dommergues et al. (1973) and Watanabe and Kuk-ki-Lee

(1975). Very high number (up to 3.6×10^7) of N_2 fixing CH_4-oxidizing organisms were also found in the rice rhizosphere (De Bont et al., 1976).

A total of 23 rice crops, in an 11-year experiment at IRRI, were obtained from a non-fertilized field with no apparent decline in N-fertility of the soil. About 45 to 60 kg N/ha per crop was removed through straw and grain (Watanabe and Kuk-ki-Lee, 1975). This represents a substantial amount of N which had to be replaced in order to maintain the fertility level of the soil. Blue-green algae and photosynthetic bacteria account for a large part of the N_2 fixed (Watanabe and Kuk-ki-Lee, 1975).

Bacterial N_2 fixation in intact rice cultures grown in test tubes has been shown by Rinaudo et al. (1971). Excised root assays of field grown rice roots (Yoshida, 1971; Yoshida and Ancajas, 1973) confirmed "rhizosphere N_2 fixation". Results from intact soil plant systems in the field gave about 50 to 200gN/ha /day at the flowering stage, by the non-algal component. The algae were separated by removing the flooding water and assayed separately for N_2 fixation (Watanabe, 1976).

Wheat: Wheat cores assayed in Oregon have been calculated to fix 2kg N/ha/day (Barber et al., 1976). Barber and associates (loc. cit) isolated the N_2 - fixing strains *Entero-bacter cloacae, Bacillus macerans*, and *B. polymyx* from wheat roots. On the other hand, enrichment cultures in semisolid N-free malate medium inoculated with surface sterilized wheat roots obtained from different locations in Brazil yielded almost 100 per cent positive samples of *Azospirillum brasilense*.

Larson and Neal (1976) described a highly specific association of a facultative *Bacillus* sp. with a disomic chromosome substitution line of wheat. *Bacillus* was isolated from soil where wheat had been growing for 30 years without N-fertilizer. Abundant numbers of bacterial cells were found on the root surface, as well as in the intercellular spaces between the cortical root cells. Rovira (1965) also reported establishment of a N_2 - fixing *Bacillus* sp. in wheat roots. The fine structure studies by Foster and Rovira (1976) showed active penetration of wheat cortex cell walls by bacteria, including the *Bacillus* sp.

C_4 Species

Sugarcane: In many parts of the world this crop has been grown in monoculture for more than 100 years without addition of N-fertilizer (Verdade, 1967). Selective stimulation of the N_2-fixing *Beijerinkia* under sugarcane vegetation has been shown. Maximal soil nitrogenase activities were found in rhizosphere soil and between the rows, where the canopy closes (Döbereiner et al., 1972). Sugarcane seedlings exposed to $^{15}N_2$ indicated fixation, incorporation, and translocation of nitrogen to the leaves.

Maize: Reports by Döbereiner (1974) and von Bulow and Döbereiner (1975) in Brazil have focussed attention on the associative symbiosis of N_2-fixing bacteria, such as *Azospirillum brasilense* and the roots of agriculturally important plants like maize and sorghum. Sloger and Owens (1976) also isolated this organism from maize roots. High nitrogenase activities (up to 9000 nmoles C_2H_4/g roots/hour) have been recorded (von Bulow and Döbereiner, 1975; Barber et al., 1976; Okon et al., 1977; Sloger and Owens, 1976). Nitrogenase activity increases in maize seedlings up to about three months in inoculated seedlings, but continues to increase up to a later date in the control plants, though at a lower rate (Table 10.4).

Table 10.4: Nitrogenase activity of excised maize roots during the growing season of both control and inoculated plants (After Albrecht et al., 1981).

Days after planting	Nitrogenase activity* (nmoles C_2H_4 formed/ha/g dry root weight)	
	Control	Inoculation
41	5.0 ± 2.6	14.3 ± 9.2
53	9.1 ± 4.0	69.1 ± 28.4
67	99.1 ± 66.4	185.5 ± 37.1
82	88.1 ± 48.1	340.1 ± 80.0
109	95.0 ± 32.9	150.9 ± 45.9

* Mean of at least 10 replicates ± SE

In Oregon, maize has been reported to fix less than 0.5 kg/ha of N_2 per growing season (Raju et al., 1972). Albrecht et al. (1981) reported that inoculation of maize with *Azospirillum* showed slight increases in plant nitrogen, the mean difference due to inoculation corresponded to 14 to 15kg N/ha.

Pearl millet: Field inoculation with *Azospirillum brasilence,* strain Sp13t, produced significantly higher yields of dry matter than did uninoculated controls in field grown, lightly fertilized pearl millet and guinea grass (Smith et al., 1976). Taylor (1979) also reports beneficial N_2-fixing association between pearl millet and *Azospirillum*. Further to this, Bouton and Brooks (1982) found that individual pearl millet plants differed in their ability to support bacterial acetylene reduction. There were also significant differences among the selfed and hybrid progeny of these plants.

Sorghum : In Nebraska, Pedersen et al. (1978) observed a maximum of 2.5 kg/ha of N_2 fixed per 100 days growing season from the associative dinitrogen fixation.

Some Grass Species : Tjepkema and Burris (1976) estimated the activity of *Panicum varigatum* and *Sporobolus heterolepis* as 3.6 and 2.9 kg N/ha/year in Wisconsin. At one site, the *S. heterolepis* activity extrapolated to 9kg N/ha/year. The rates of C_2H_2 reduction are low compared to those reported for grasses in Brazil. Day et al. (1975) extrapolated C_2H_2 reduction rates for *Pennisetum purpureum* to a maximum N_2 fixation rate of 1kg N/ha/day. Nelson et al. (1976) surveyed grasses in Oregon and discovered that *Agrostis tenuis* has an estimated fixation rate of 3kg N/ha/season. Further, Weaver et al. (1980) surveyed the subtropical region of Texas to determine the rate of biological N_2 fixation naturally occurring in the rhizosphere of native and introduced forage grasses. The most active rhizosphere samples were extrapolated to fix N_2 at the rates of 33, 26, 20 and 20 kg/ha/100 days of *Cynodon dactylon, Paspalum urvillei, Brachiaria* sp. and *Andropogon girardi,* respectively.

These high rates indicate that under the proper environmental conditions a potential exists for agronomically significant rate of biological N_2 fixation to occur in the rhizosphere of forage grasses. Even genotypic variations have been noted and the possibility of improvement in the rates of N_2 fixation exists. For instance, Döbereiner (1977a) and Neyra and Döbereiner (1977) observed that certain grass genotypes achieve better associations between their roots and N_2 fixing microrganisms when compared with other genotypes of the same species.

Non-legume Dicot Species

Response of three genotypes of potato to inoculation with *Azotobacter* was studied (Imam and Badawy, 1978). Tubers of the three potato genotypes were inoculated with four strains of *A. chroococcum* and then planted. The treatment stimulated plant growth of one genotype only, which resulted in an increase of 42.6 per cent in the marketable tuber as compared to the untreated control. When five sweet potato cultivars were grown in pots the following *Azospirillum* species and strains were identified on isolation (Hill et al., 1983): *A. lipoferum* (*brasilense*) from all cultivars except Jewel; *A. brasilense* nir-(TIsp1, TIsp12, TIsp13) from Jewel and Southern Queen; and *A. brasilense* nir (TIsp5, TIsp6, TIsp7, TIsp8, TIsp9, TIsP14) from Centennial, Rojo Blanco and Carver.

Rhizobium in Association with Non-legumes

Tissue cultures of many plant cells can turn N_2 fixing associations with rhizobium and, in particular, a culture of plant cells from a non-legume such as tobacco is as effective in forming an association as a culture from a legume such as soybean (Child, 1975; Scowcroft and Gibson, 1975). An instance is known in which *Trema cannabina* (a non-legume) is the natural host to a rhizobium sp. (Trinick, 1975). The types of rhizobium able to form these 'exotic' associations are now known to be capable of fixing N_2 in the total absence of plant material, they are in fact free living N_2 fixing bacteria of exceptional oxygen sensitivity. The possibility obviously exists of setting up associations between rhizobia or other aerobic N_2 fixing bacteria and new plants. For example, a mutualistic association between carrot callus and a mutant of azotobacter was set up by Carlson and Chaleff (1974).

METHOD OF MEASURING FIXATION EFFICIENCY

In addition to the field evaluation of soil N content in the rhizosphere, precise laboratory methods have been developed (Godse, et al., 1982; Wani et al., 1982). Total number of N_2-fixers from the rhizosphere soil, rhizoplane and root macerate of sorghum plants were estimated by plate counts using N-free sucrose and malate media (Godse et al., 1982). Using these inocula, nitrogenase activity was measured by the most probable number method on sucrose and malate semisolid media. Nitrogenase activity was also measured by the acetylene reduction technique. Value of the spermosphere model was demonstrated with plants grown in tubes on Fahraeus agar medium for selection of the most abundant and host compatible bacteria from the rhizosphere soil, rhizoplane soil and root macerate. Based on colony morphology, various isolates were picked up from plates and purified. Purified isolates were tested for nitrogenase activity on malate and sucrose semisolid media with 20 per cent acetylene incubated for three hours at 33° C. The applicability of analytical profile index tests was demonstrated for quick identification of nitrogen fixers belonging to *Enterobacteriaceae*. Further to this, Wani et al. (1982) developed a test tube culture technique to test the effect of host genotype and bacterial culture on nitrogenase activity. A method of assaying intact plants for nitrogenase activity was developed. By using this method the same plant can be assayed several times during its growth cycle and seed can also be obtained for further multiplication.

In order to verify the lightly variable field responses with regard to acetylene reduction. Schank et al. (1982) developed an *in vitro* screening technique. Axenic systems using diverse sorghum germ-plasm were established by inoculating sorghum plants with *Azospirillum* or other N_2-fixing bacteria. The seedlings were grown in test tubes for 10 days on a Fahraeus, nitrogen and carbon-free medium. In addition, bacterial populations of *Azospirillum* were studied at the end of the growth period using fluorescent antibody labelling. Roots were scoured for root bacteria associations.

GENOTYPIC VARIABILITY IN N_2-FIXATION

C_3 Species

Rice : From a study on variation in N_2-fixing ability of excised roots of 200 rice lines and on reducing activity of 80 lines, Fujii et al. (1977) concluded that symbiosis with bacteria fixing atmospheric nitrogen is genetically controlled. Wide variations in N_2 fixation in the rhizosphere of 50 rice strains were also observed (Hirota et al., 1977 1978). Root weight was positively correlated with the N_2 fixing activity (Hirota et al., 1977). The rice strains with the highest N_2 fixation activity were from India and Thailand (Hirota et al., 1978). The acetylene reduction activity of 10 wild *Oryza* species (76 strains) and two cultivated species (156 strains) was investigated (Ohta et al., 1984). In general, *O. sativa* showed higher acetylene reduction activity than the other species. The Japanese cultivars studied had a lower activity than the tropical cultivars from India and Thailand.

Of the 41 rice varieties tested, JBS 236 showed the highest nitrogen fixation activity (Lee et al., 1977). Maximum activity was reached at heading and thereafter declined. The varietal levels of N_2 fixation were highly correlated with the root dry weight at heading. In the 28 rice varieties studied (Rinaudo, 1977), acetylene reduction activity varied by a factor of up to 5, the highest being in Sefa 319G and lowest in Apura. However, Beunard and Pichot (1981) reported that nitrogenase activity varied by a factor of up to 10, the highest being in IRAT 13 and the lowest in Liberia 208. In a separate study conducted in Senegal (Institute of Tropical Agronomic Research and Food Crops, France, 1978), IRAT 13 again showed the highest N_2 fixation and Lung Shen 1 the lowest. The N_2 fixation rate, dwarf growth habit and high yielding ability appear to be related (Charyulu et al., 1981). Charyulu et al. collected rhizosphere soil samples from 20 rice varieties grown under uniform field conditions and found variations with variety in ability fo fix $^{15}N_2$. Rhizosphere soil samples from UVT3-V14, Surya, UV3-V19, Padma, Kalinga I, Kalinga II and Shakti, all high yielding semidwarf varieties, exhibited high fixation.

Wheat : In a study on the number of *Azotobacter chroococcum* in the rhizosphere of 20 wheat varieties, Kavimardan et al. (1978) showed slight increases in the rhizosphere of HD 2160 and K7435 after 40 days growth and in the rhizosphere of seven other varieties after 65 days of growth. The number of anaerobic N_2 fixing bacteria greatly increased in the rhizosphere of eight varieties. This response was most marked with HD 2009 and HD 2237. In some wheat lines the fixed nitrogen was available to plants (Rennie, 1983). Greenhouse tests of more than 40 varieties showed that about 25 per cent of them were able to benefit from N_2 fixed by *Baccilus* C-11-25. Further, 20 wheat genotypes, comprising lines and cultivars of *Triticum aestivum*, two cv. of *T. turgidum* (*T. durum*) and two derivatives of the cross *T. dicoccoides* × *T. durum* were also studied with and without inoculation at two rates of N-fertilizer application (Millet et al., 1984). Only two semidwarf genotypes (an early heading cv. of *T. aestivum* and a midseason durum wheat cv.) gave significantly increased grain yield per plant after inoculation in comparison with uninoculated controls.

Anderson (1986) reported that the response of wheat plants to inoculation with *A. brasilense* varied with genotype, bacterial strain and nitrogen concentration in the nutrient solution. Root bacterial population differences between genotypes became clear soon after inoculation and early in the vegetative stage of the seedlings. Improved plant growth was correlated with the bacterial population size in the rhizosphere and endorhizosphere. The selective colonization of genetically defined line of wheat by a facultative N_2-fixing species (Larson and Neal, 1976) reveals the importance of plant genotypes for optimal associations and suggests the possibility of improving N_2-fixing associations by plant breeding.

Barley: Association of *Azospirillum* with the roots and stem of hulled and hulless lines/ varieties of barley was studied (Tilak and Murthy, 1981). They measured N_2 fixation in the root isolates of *A. brasilense* and "enrichment cultures" of stems of eight barley varieties. Both measurements showed that N_2 fixation was greater in the two hulless lines than in the six hulled varieties. It may be noted that both the hulless lines had higher protein content.

C_4 Species

Pearl millet: In a test of 142 lines of pearl millet (*Pennisetum americanum*) conducted at ICRISAT, 10 lines were particularly active in stimulating N_2 fixation by bacteria associated with the roots (ICRISAT, 1978). The cross *R americanum* × P. *purpureum* showed particularly higher rate of N uptake and high dry matter yield. Nitrogenase activity associated with the roots, equivalent to 100 mg N_2 fixed per 15 cm diameter soil core per day was recorded for 10 out of 142 lines of *P. americanum* grown in the field (Rao and Dart, 1981). Soil cores without plant roots had mean activities of 0.10mg N per core per day. In a field trial, inoculation with *A. lipoferum* (*brasilense*) increased grain yield of the cv. IP 2787 by 11.7 per cent over the uninoculated control (Wani et al., 1985) but the cultivar ICMS 7819 did not respond to inoculation.

Individual plants from the *P. americanum* "Tift 1 S_1" population differed in ability to support acetylene reduction by *A. brasilense* (Strain J.M 125) when screened in enclosed seedling agar tubes. There were also significant differences among selfed and hybrid progeny of those plants (Bouton and Brooks, 1982; Bouton et al., 1985).

Maize : In the rhizosphere soil of two maize varieties, the number of *Azotobacter chroococcum* increased more than did the number of anaerobic N_2 fixing bacteria (Kavimardan et al., 1978). This response was more marked with the maize variety Ganga 5. A study on excised root assay has revealed a wide range of nitrogenase activities in maize genotypes. Mean nitrogenase activities of some S_1 lines were 10 to 20 times higher than the original cultivar (Von Bulow and Döbereiner, 1975). The crosses between higher fixing versus low fixing maize cultivars showed significant heterosis effects (Von Bulow et al., 1976).

Sorghum : It was observed that in the rhizosphere of three genotypes of sorghum, the number of *A. chroococcum* increased more than the number of anaerobic N_2 fixing bacteria (Kavimardan et al., 1978). Acetylene reduction studies in 51 lines of sorghum revealed a range of 0-1935 nmoles/g dry root/hour. In a separate study on acetylene reduction of 334 lines of sorghum; 55 per cent of the lines stimulated nitrogenase activity in the rhizosphere but 14 per cent of these stimulated a very high activity (100 mg N/core/day – Wani et al., 1982). Further, Berestetskii et al. (1985) inoculated several sweet and grain sorghum cultivars with seven strains of N_2 fixing bacteria. Sweet sorghum was more responsive than the grain sorghum cultivars. Strain 27-35 raised fresh yield in sweet sorghum by 21.5 per cent compared with the uninoculated control. However, in grain sorghum, strain 2r47 raised grain yield by 29.2 per cent.

Sugarcane: Ruschel and Ruschel (1977) recorded nitrogenase activity in the rhizosphere of newly sprouted cuttings and two months old plants of sugarcane and also recorded genotypic variability. The rate of activity increased with time and was high in the varieties NA 56-79 and CB 46-47 but was low in the variety CB 41-76.

IMPROVEMENT STUDIES

The Crop Genetics and Improvement

Crosses between the rice varieties C 5444 and T 65 indicate that the ability to fix N_2 is controlled by more than one gene (Fujii et al., 1979). However, Iyama et al. (1983) reported partial to complete dominance and suggested that the high N_2 fixing activities in the rhizospheres of rice varieties C 6063 and C 5053 are governed by recessive alleles. Balandreau (1982) developed rice mutants 11, 17 and 28 from the rice variety Cesariot by γ-irradiation which were capable of fixing more N_2 in the rhizosphere than was Cesariot. Mutant 17 fixed 50 per cent more N_2 than did Cesariot.

Rennie and Larson (1979) have reported that in the disomic substitution lines of Cadet and Rescue variety of wheat, chromosomes 5B and 5D influence the induction of N_2-fixation in bacteria associated with rhizosphere. Evaluation of the parental, F_1, F_2 and BC generations of crosses between wheat genotypes with high and low responses suggest that plant response to *A. brasilense* is controlled by a small number of genes (Anderson, 1986).

Differences in acetylene reduction activity (ARA) of different selfed lines of pearl millet were recorded (Bouton et al., 1985). They hybridized the high and low ARA lines. The high ARA lines supported higher numbers of bacteria and lost greater quantities of ^{14}C from their roots than did the low ARA lines. Analysis of data from a 5x5 incomplete diallel cross revealed that both additive and non-additive effects were important for non-symbiotic N_2 fixation (Manga et al., 1987). P 631 × 76 K2 showed significant heterosis for nitrogenase activity over the midparental value, while AIB2 × 76K2 and 76K2 × CR28 showed significant heterosis for the trait over the midparental and better parent values.

Strain Improvement

Rai et al. (1984) reported that five drug resistant mutant strains of *A. brasilense* significantly increased grain yield and nitrogenase activity in finger millet (*Eleusine coracana*). The mutant strain STR11 was the best N_2 fixer. By protoplast fusion a hybrid between *Agrobacterium tumefaciens* and *A. chroococcum* was produced (Du and Fan, 1981) which formed crown galls with nitrogenase activity on tomato stems.

The *nif* (N_2 fixing) genes have been transferred to several new bacterial hosts. Most such transfers have made use of *nif* genes from *Klebsiella pheumoniae*, but *nif* from *Rhizobium trifolii* has also been transferred to *K. aerogenes* (Dunican et al., 1976) and preliminary reports suggesting *nif* gene transfers in photosynthetic bacteria have also been published. Incorporation of *nif* may result in N_2 fixing ability.

REFERENCES

Albrecht, S.L.,Y. Okon, J.Lounquist and R.H. Burris, Nitrogen fixation by corn-*Azospirillum* association ina temperate climate,. *Crop Sci.* 21: 301–06, 1981.

Anderson, T.M. Interaction of spring wheat (*Triticum aestivum* L.) genotypes and *Azospirillum brasilense*. (Abstr.) *Diss. Abstr, Int*. B. 47: 8563, 1986.

Balandreau. J. Breeding rise for better N_2-fixation: a step forward. *Mutation Breeding Newsl*. No. 20: 4–5, 1982.

Bally, R., D. Thomas-Bauzon, T. Heulin, J. Balandreau, C. Richard and J. Dely. Determination of the most frequent N_2-fixing bacteria in a rice rhizosphere. *Can. J. Microbiol*. 29: 881–87, 1983.

Barraquio, W.L. and I. Watanabe. Occurrence of aerobic nitrogen fixing bacteria in wet-land and dryland plants. *Soil Sci. Pl. Nutr*. 27: 121–25, 1981.

Barber, L.E., J.D. Tjepkema and H.J. Evans. Environ. Role Nitrogen-Fixing Blue-Green Algac Symbiotic Bacteria, Int. Symp. Uppsala. 1976.

Berestetskii, O.A., B.N. Malinovskii and V.K. Chebotar. (The possibility of using symbiotic N_2 fixers to raise sorghum yields). Doklady Vsesoyuznoi Ordena Lenina i Ordena Trudovogo Krasnogo Znameni Akademii Sel' skokhozyaistvennkykh Nank Imeni V.I. Lenina, No. 12: 6–8, 1985.

Beunard, P. and J. Pichot. (Nitrogen fixation in rhizosphere of six rice varieties. Study in a controlled environment on a tropical feruginous soil from Senegal.) *Agron. Trop*. 36: 212–16, 1981.

Bouton, J.H. and C.O. Brooks. Screening pearl millet for variability in supporting bacterial acetylene reduction activity. *Crop Sci*. 22: 680–82, 1982.

Bouton, J.H., S.L. Albrecht and D.A. Zuberer. Screening and selection of pearl millet for root associated bacterial nitrogen fixation. *Field Crops Res*. 11: 131–40, 1985.

Carlson, P.S. and R. S. Chaleff. Forced association between higher plant and bacterial cells *in vitro*, *Nature* 252: 393–94, 1974.

Child J.J. Nitrogen fixation by a *Rhizobium* sp. in association with non-leguminous plant cell cultures. *Nature*. 253: 350–51, 1975.

Charyulu, P.B.B.N., D.N. Nayak and V.R. Rao. $^{15}N_2$ incorporation by rhizosphere soil. Influence of variety, organic matter and combined nitrogen. *Plant & Soil*. 59: 399–405, 1981.

Day, J.M., M.C.P. Neves and J. Döbereiner. Nitrogenase activity of the roots of tropical grasses. *Soil Biol. Biochem*, 7: 107–12, 1975.

De Bont, J.A.M., Kuk-ki-Lee and D. Bouldin,. Environ. Role Nitrogen-Fixing Blue-green Algae-Symbiotic Bacteria, Int. Symp., Uppsala. 1976.

Döbereiner. J. Nitrogen fixing bacteria in the rhizosphere p. 36–120. *In:* A. Quispell(ed) The biology of nitrogen fixation. North Holland Pub. Co., Amsterdam-Oxford. 1974.

Döbereiner, J. (Plant genotype effects on nitrogn fixation in grasses). *In:* Genetic Diversity in Plants. V. Prospects of Breeding for Physiological Characters. Amir Muhammed; R. Aksel and R.V. von Borstel (eds), Plenum Press, New York, USA. pp. 325–34, 1977

Döbereiner, J. Nitrogen fixation associated with non-leguminous plants. *In:* Genetic Engineering for Nitrogen Fixation. A. Hollaender; R.H. Burris; P.R. Day; R.W.F. Hardy; D.R. Helinski; M.R. Lamborg; L. Ovens and R.C. Valentine (eds), Plenum Press, New York, USA. pp. 451–61, 1977b.

Döbereiner, J., J.M. Day and P.J. Dart. Nitrogenase activity in the rhizosphere of sugarcane and some other tropical grasses. *Plant & Soil*, 37: 191–16, 1972.

Dommergues, Y., J. Balandreau; G. Rinaudo and P. Weinhard, *Soil Biol. Biochem*. 5: 83–89, 1973.

Du, Q.Y. and C.Y. Fan. (A preliminary report on the establishment of a nitrogen fixation system in crown galls on tomato plants). *Acta Botanica Sinica*. 23: 453–58, 1981.

Dunican., K., A.B. Tierney and F.O. Yara. *In*: Proc. of the 1st Int. Symp. on Nitrogen Fixation" Washington State Univ. Press, Pullman. 1976.

Eskew, D.L., A.R.J. Eaglesham and A.A. App. Heterotrophic $^{15}N_2$ fixation and distribution of newly fixed nitrogen in a rice-flood soil system. *Pl. Physiol.* 68: 48–52, 1981.

Foster, R.C. and A.D. Rovira. Ultrastructure of wheat rhizosphere. *New Phytol.* 76: 343–52, 1976.

Fujü, T., Y. Sano, S. Iyama and Y. Hirota. (Nitrogen fixing ability in rice). *Jap. J. Genet.* 56: 438, 1977.

Fujii, T., Y. Sano, S. Iyama and Y. Hirota. Nitrogen fixation in the rhizosphere of rice. *In:* Annual Report No. 29, 1978 (II), National Inst. of Genetics, Japan. pp. 101–03, 1979.

Godse, D.B., P.J. Dart and K.P. Hebbar. Nitrogen fixing bacteria associated with sorghum. (Abstr). *In:* Sorghum in Eighties. Vol. 2. Oxford & IBH Pub. Co. New Delhi. p. 757, 1982.

Hill, W.A., P. Bacon-Hill, S.M. Crossman and C. Stevens. Characterization of N_2-fixing bacteria associated with sweet potato roots. *Can. J. Microbiol.* 29 (8): 860–62, 1983.

Hirota, Y., T. Fujii, Y. Sano and S. Iyama. (Nitrogen fixation in rhizosphere of rice plant). *In*: Japan, National Institute of Genetics. Annual Report No 28. p.126, 1977.

Hirota, Y., T. Fujii, Y. Samo and S. Iyama. Nitrogen fixation in the rhizosphere of rice. *Nature* 276: 416–17, 1978.

Imam, M.K. and F.H. Badawy. Response of three potato cultivars to inoculation with *Azotobacter*. *Potato Research.* 21: 1–8, 1978.

Inst. Trop. Agron. Res. and Food Crops, France. Annual Report. p.219, 1978.

ICRISAT. *Pennisetum.* Annual Report, 1977–78. p. 295, 1978.

Ito, O., C. Cabrara and I. Watanabe. Fixation of dinitrogen-15 associated with rice plants. *Appl. Environ. Microbiol.* 39: 554–58, 1980.

Iyama, S.,Y. Sano and T. Fujii. Diallel analysis of nitrogen fixation in the rhizosphere of rice. *Plant Sci. Letters.* 30: 129–35., 1983.

Jagnow, G. Nitrogen fixing bacteria associated with graminaceous roots with special reference to *Spirillum lipoferum*, Beijerinck. *Z. Pflanzen-ernahrung Bodenkunde* 142 (3): 399–410, 1979.

Kavimardan, S.K., M.L. Kumari and N.S. Subba Rao. Non-symbiotic nitrogen fixing bacteria in the rhizosphere of wheat, maize and sorghum. *Indian Acad. Sci. Proc B.* 87B (11): 299-302, 1978.

Kumari, M.L., S.K. Kavimardan and N.S. Subba Rao. Occurrence of nitrogen fixing *Spirillum* in roots of rice, sorghum, maize and other plants. *Indian J. Expt. Biol.* 14: 638-9 1976.

Lamberg, M.R. Biological nitrogen fixation: A fertilizer strategy potentially beneficial for the poor in developing countries. *In*: Linking Research to Crop Production. (eds) R.C. Staples and R.J. Kuhr (eds), Plenum Press, New York. pp. 115–36, 1979.

Larson, R.I. and J.L. Neal, Jr. Environ. Role Nitrogen Fixing Blue-Green Algae Symbiotic Bacteria. Int. Symp. Uppsala. 1976.

Lee, K.K., T. Castro and T. Yoshida. Nitrogen fixation throughout growth, and varietal differences in nitrogen fixation by the rhizosphere of rice planted in pots. *Pl. & Soil.* 48: 613–9, 1977.

Manga, V.K., B. Venkateswarlu and M.B.L. Saxena. Combining ability and heterosis for non-symbiotic nitrogen fixation in pearl millet. *Indian J. agric. Sci.* 57: 135–37, 1987.

Millet, E., Y. Avivi and M. Feldrnan. Yield response of various wheat genotypes to inoculation with *Azospirillum brasilense. Plant & Soil.* 80: 216–66, 1984.

Moore, A.W. Azolla: ecology and agronomic significance. *Bot. Rev.* 35: 17, 1969.

Nelson, A.D., L.E. Barber; J. Tjepkema, S.A. Russell, R. Powelson, H.J. Evans and R.J. Seidler. Niotrogen fixation associated with grasses in Oregon. *Can. J. Microbiol.* 22: 523–30, 1977.

Neyra, C.A. and J. Döbereiner. Nitrogen fixation in grasses *Adv. Agron.* 29: 1–38, 1976.

Ohta, K., Y. Sano; T. Fuju and S. Iyama. Variation in nitrogen fixing activity among wild and cultivated rice strains. *Jap. J. Breed.* 34: 29–35, 1984.

Okon, Y., S.L. Albrecht and R.H. Burris. *J. Bacteriol.* 127: 1248–54, 1977.

Pedersen, W.L., K. Chakrabarty, R.V. Klucas and K. Vidaver, Nitrogen fixation (acetylene reduction) associated with roots of winter wheat and sorghum in Nebraska. Appl. *Environ. Microbiol.* 35: 129–35, 1978.

Qui, Y., X. Mo, Y. Zhang, Z. Li and C. You. Some properties of the nitrogen fixing associative symbiosis of *Alcaligens faecalis* A-15 with rice plants. *In:* (eds) C. Veeger and W.E. Newton; Advances in Nitrogen Fixation Research. Nijhoft/Junk Publishers, The Hague. p. 64, 1984.

Qui, Y., S. Shon, Z. Mo, S.Ye, X. Cai, C. Mao, Y. Chen, S.He and R. Deng. Study of nitrogen fixing bacteria associated with rice root. II. The characteristics of nitrogen fixation by *Alcaligens faecalis* strain A-15 and *Enterobacter cloacae* E26. *Acta Microbiol Sin.* 21: 473–76, 1981.

Rai, R., V. Prasad and I.C. Shukla. Interaction between finger millet (*Eleusine coracana*) genotypes and drug resistant mutants of *Azospisillum brasilene* in calcarious soil. *J. agric. Sci.,* U.K. 102: 521–29, 1984.

Raju, P.N., H.J. Evans and R.J. Seidler. An asymbiotic nitrogen fixing bacterium from the root environment of corn. *Proc. Natl. Acad. Sci. USA.,* 69: 3474–78, 1972.

Rao, R.V.S. and P.J. Dart. Nitrogen fixation associated with sorghum and millet. *In:* Associated N_2-fixation. Vol. I. Proc. of the int workshop on associative N_2-fixation, July 2-6, 1979. Piracicaba, Brazil. P.B. Vose and A.P. Ruschel (eds), CRC Press, 1981. pp. 169–77.

Rennie, R.J. N_2-fixation in cereals. *Can. Agric.* 29 (3/4): 4–9, 1983.

Rennie, R.J. and R.I. Larson. Dinitrogen fixation associated with disonic chromosome substitution lines of spring wheat. *Can. J. Bot.* 57: 2771–75, 1979.

Rinaudo, G.J. (Nitrogen fixation in the rhizosphere of rice: importance of varietal types). Cahiers ORSTOM, Biologie. 12: 117–19, 1977.

Rinaudo, G., J. Balardreau and Y. Dommergues. Algal and bacterial nonsymbiotic nitrogen fixation in paddy soils. *Plant & Soil* Spec. Vol. 1971: 471–79, 1971.

Rovira, A.D. Interactions between plant roots and soil microorganisms. *Annu. Rev. Microbiol.* 19: 241–66, 1965.

Ruschel, A.P. and R. Ruschel. Varietal differences affecting nitrogenase activity in the rhizosphere of sugar cane. In: Proc. 16th Cong. ISSCT. 1941–1947, 1977.

Sano, Y., T. Fujii, S. Iyama, Y. Hirota and K. Komagata. Nitrogen fixation in the rhizosphere of cultivated and wild rice strains. *Crop Sci.* 21: 758–61, 1981.

Schank, S.C., R.L. Smith and J.R. Milan. Acetylene reduction activity of several sorghum and N_2 fixing bacteria associations. (Abstr.) *In:* Sorghum in Eighties, Vol.2, Oxford & IBH Pub.Co., New Delhi. p. 752, 1982.

Scowcroft, W.R. and A.H. Gibson. Nitrogen fixation by *Rhizobium* associated with tobacco and cowpea cell cultures. *Nature* 253: 351–52, 1975.

Sloger. C. and L.D. Owens. Int. Symp. Nitrogen Fixation, 2nd Salamanca. 1976.

Smith, R.L., J.H. Bouton, S.C. Schank, K.H. *Quisenberry,* M.E. Tyler, J.R., Milam, M.H. Gaskins and R.C. Littell. Nitrogen fixation in grasses inoculated with *Spirillum lipoferum. Science* 193: 1003–05, 1976.

Talley, S.N., B.S. Talley and D.W. Rains. Nitrogen fixation by *Azolla* in rice fields. *In:* Genetic Engineering for Nitrogen Fixation. A. Hollaender (ed), Plenum Press, N.Y. p. 259, 1977.

Taylor, R.W. Response of two grasses to inoculation with *Azospirillum* sp. in a Bahamian soil. *Trop. Agric.* 56: 361–65, 1979.

Tilak, K.V., B.R. and B.N. Murthy. Occurrence of *Azospirillum* in association with the roots and stems of different cultivars of barley (*Hordium vilgare*). *Curr. Sci.* 50: 496–98, 1981.

Tjepkema, J.D. and R.H. Burris. Nitrogenase activity associated with some Wisconsin prairie grasses. *Plant & Soil.* 45: 81–94, 1976.

Trinick, M.J. *In*: Proc. Ist Int. Symp. Nitrogen Fixation; W.E. Newton and C.J. Nyman (eds), Vol. 2. Washington State Univ. Press, Pullman. pp. 507–17, 1975.

Van Berkum, P. and B.B. Bohlool. Evaluation of nitrogen fixation by bacteria in association with roots of tropical grasses. *Microbiol Res.* 44: 491–517, 1980.

Verdade, F.C. Proc. Biol. Ecol. Nitrogen (IBP Conf.) Davis, Calif. 1967.

Von Bulow, J.F.W. and J. Döbereiner. Potential for nitrogen fixation in maize genotypes in Brazil. *Proc. Nat. Acad. Sci.* USA 72: 2389–93, 1975.

Von Bulow, J.F.W., J. Döbereiner and J.A. Podesta. Reuniao Bras. Milho Sorgo, 11th, Piracicaba, Sao Paulo, 1976.

Wani, S.P., S. Chandrapalaiah and P.J. Dart. Response of pearl millet cultivars to inoculation with nitrogen fixing bacteria. *Exp. Agric.* 21: 175–82, 1985.

Wani, S.P., P.J. Dart, S.C. Chandrapalaiah and M.N. Upadhyaya. Nitrogen fixation associated with sorghum. (Abstr.) *In*: Sorghum in Eighties, Vol. 2, Oxford & IBH Pub. Co, New Delhi. p. 756, 1982.

Watanabe, I. Environ. Role Nitrogen - Fixing Blue-Green Algae, Symbiotic Bacteria, Int. Symp. Upsala. 1976.

Watanabe, I. Biological nitrogen fixation associated with wetland rice. *In*: A.H. Gibson and W.E. Newton (eds). Current perspectives in nitrogen fixation. Elsevier Publishers, Amsterdam. pp. 313–16, 1981.

Watanabe, I. and Kuk-ki-Lee. Int. Symp. Biol. Nitrogen Fixation Farming Syst. Humid Trop., Int. Inst. Trop. Agric. (IITA), Ibadan, Nigeria. 1975.

Weaver, R.W., S.F. Wright, M.W. Varanka, O.E. Smith and E.C. Holt. Dinitrogen fixation (C_2H_2) by established forage grasses in Texas. *Agron. J.* 72; 965–68, 1980.

Yoshida, T. Res. Results Rep., Soil Microbiol., Int. Rice Res. Inst. (IRRI), Los Banos, Laguna, Philippines. 1971.

Yoshida, T. and R.R. Ancajas. Nitrogen fixing activity in upland and flooded rice fields. *Soil Sci. Soc. Am. Proc.* 37: 42–46, 1973.

Yoshida, T. and T. Yoneyama. Atmospheric dinitrogen fixation in the flooded rice rhizosphere as determined by the ^{15}N isotopes technique. *Soil Sci. Pl. Nutr.* 26: 551–59, 1980.

Yoo, I.D. T. Fujii, Y. Sano, K. Komagata, T. Yoneyama, S. Iyama and Y. Hirota. Dinitrogen fixation of rice-*Klebsella* associations. *Crop Sci.* 26: 297–301, 1986.

You, C.B. and Y.S. Qui. (Nitrogen fixation by *Alcaligenes faecalis* on rice seedlings). *Scientia Agricultura Sinica.* No. 6: 1–5, 1982.

CHAPTER 11

YIELD STABILITY

INTRODUCTION

Stable yield of a variety is the most important breeding objective, especially in risk environments of developing countries. In the developed countries where the necessary inputs are easily and cheaply available, stability is not the priority. In the stress and variable environments of developing countries, an ideal cultivar should be adapted to a wide range of growing conditions in the given production areas, with above average yield and below average variance across environments. Stable performances in crop yield and quality traits over a wide range of growing conditions are desirable from a standpoint of management and marketing.

Minimum genotype $\times$ environment ($G \times E$) interactions are desired for a cultivar to be commercially successful, and must perform well across the range of environments in which the cultivar may be grown. The presence of $G \times E$ interactions reduces the correlation between phenotype and genotype, and makes it difficult to judge the genetic potential of a genotype. Plant breeders grow performance tests at different locations in different years in the target area, and data obtained from these tests are used to determine the magnitude of $G \times E$ interactions. Stability of a cultivar refers to its constancy in performance across environments and is affected by the presence of $G \times E$ interactions. In the presence of significant $G \times E$ interactions, stability parameters are estimated to determine the superiority of individual genotypes across the range of environments.

In practical terms, plant breeders are/should be concerned with performance in suboptimal environments. Tolerance to problem soils and drought stress have been associated with greater yield stability. There is variation in the reaction of different sorghum hybrids to drought and high temperature stress (Sullivan, 1972; Stout et al., 1978). Yields of genotypes with greater stress tolerance are more stable in areas where that specific stress condition occurs regularly. Identifying specific stress tolerance mechanisms that enhance yield stability within a region could provide valuable guide-lines for crop improvement.

A variety which maintains its deviation from the mean yield value of the site at all sites tested is considered ecologically stable. Or in other words, a genotype is considered environmentally stable when a unit change in genotype performance corresponds to a unit change in environmental index (Mahill et al., 1984). Two basic concepts of phenotypic stability are distinguished for a stable genotype: (i) low variance under different environmental conditions, and (ii) little interaction with environments as measured by covalence; that is, the genotypic contribution to the genotype × environment interaction sum of squares (Becker, 1981). Thus the aim should be to produce appropriate genotypes for a given ecological area which have high yield stability rather than highest yield (Nedelea et al., 1984). In developing countries where the availability of agricultural inputs is scarce and unpredictable, stability is needed most. In order to further reduce the risk, intercropping or mixed cropping with stable varieties/species is suggested, and is in fact being practiced from a long time.

The adaptability or plasticity of a variety is seen by its response to climate, edaphic, meteorological and agronomic conditions (for example, fertilizer rate and crop density). The adaptable varieties react strongly and quickly to improved conditions, but quickly lower their response under poor adaptability, react less strongly to environmental changes and are more suited to poor conditions. Their yield reduction is less than that of adaptable varieties. Effects of such factors as altitude, temperature, precipitation and soil type on the stability of four potato varieties were studied (Rasochova, 1987). Temperature and precipitation had the greatest effect on starch content and tuber weight. Skinnes and Buras (1987) studied the effects of constant temperatures ranging between 9 ° and 24 ° C on seven wheat genotypes which improved during the stages from heading to 20 per cent grain moisture. Genotype × temperature interactions were strong for grain set, moderate for final grain weight and low for earliness of maturity. The stability in grain set and in length of grain filling period is associated with yield stability.

With sugarcane, Sharma and Bharaj (1982) noted that different genotypes were stable for different traits, and suggested that such genotypes could be crossed among themselves. But the genotypes stable for a large number of quantitative characters can be exploited directly. The variety COJ 67 having a high sucrose content in poor environments can be used as a donor parent (Sharma and Bharaj, 1983). Both plant cane and ratoon crop's stability of sugar yield/ha can be predicted from the stability of cane yield/ha. Brix stability may give some indication of the stabilities for percentage sucrose and sugar concentration (Kang et al., 1987).

It should be possible to breed for high yields combined with high stability (Huhn and Leon, 1985) though the general concept is of a negative correlation. Kilchevskii et al. (1988) also reported that yield tends to be independent of stability, and it should be possible to combine high yield and good stability in a single genotype. Thus in order to avoid the dangers of eliminating stable varieties early in the course of selection for yield, selection should be carried under both favourable and limiting environments. With maize, Crossa et al. (1988b) reported that varieties derived from Antigua Republic Dominicana were stable in unfavourable environments, whereas selections from Blanco Cristalino-1 and Blanco Dentado-2 were stable in both low and high production sites. Combination of the environmental factors in specific test locations, allowed selection of varieties that were stable in other regions of the world.

On the other hand, Cammack (1984) reports that in general, high yielding genotypes show a trend towards adaptation to favourable environments or instability for yield. If this is true, stable varieties for unfavourable environments have to be different. Therefore, the school of breeders arguing for testing lines only in favourable high-yield (high input) environments have to think twice. Virk et al. (1985) recommend that in addition to high yield, testing for stability of yield in variable environments should be made an integral part of the variety release procedure. In fact, Kilchevskii and Khotyleva (1985a) and Leon (1986) have developed methods for simultaneous selection for yield and yield stability.

STABILITY DETERMINATION

A genotype may be considered stable: (i) if its environment variance is small, (ii) if its response to environments is parallel to the mean response of all genotype, in a trial, or (iii) if the residual mean square from a regression model on the environmental index is small. Unfortunately, these three concepts represent different aspects of stability and do not always provide a complete picture of the response. In the alternate approach of cluster analysis, the assimilatory measures define complete similarity in three different ways: (i) equality of genotype's response across locations, (ii) equality of all within location differences, and (iii) equality of all within location ratios. The advantage of the non-parametric approach is that a cluster's response characteristics can be assessed qualitatively, without the need for a mathematical characterization (Lin et al., 1986).

Methods available for estimating the magnitude of G × E interaction can be partitioned into components using regression analysis. This method was first proposed by Yates and Cochran (1938) and later modified by Plaisted and Peterson (1959); Finlay and Wilkinson (1963); Eberhart and Russell (1966); and, Perkins and Jinks (1968). It involves the regression of each genotype on an environmental index that is determined by the mean performance of all genotypes grown in each environment. Stability parameters are estimated from this regression analysis. Finlay and Wilkinson (1963) used mean yield of a genotype and the slope of its regression line to determine the stability of the genotype over the environments. This method was modified by Eberhart and Russell (1966) who added an extra parameter that measures the deviation from linear regression. Shukla (1972) redefined this estimate as stability variance and devised a test for the significance of the G × E component of a cultivar. Recently Zhang and Geng (1986) proposed a method that allows stability parameters of cultivars grown in different environments to be reparameterized and to be comparable.

The suitability of using regression and genotype grouping methods to evaluate yield stability in segregating populations of cowpea was determined by Ntare and Aken Ova (1985). F_3 and F_5 lines and bulks were used in the study. Significant G × E interactions were present in both generations. The two methods identified the same lines and bulks as stable, but the genotype grouping method would be more useful when a large number of genotypes are evaluated. Apart from determining the stability of cultivars, these techniques assist in identification of genotypes that respond to environmental change and therefore are likely to give satisfactory returns to added inputs such as fertilizers or pesticides. Hill (1975) suggested that the main advantage of regression techniques in studying G × E interactions is their ability to reduce complex interactions into an orderly series of linear responses.

A method of genetic analysis is proposed by Kilchevskii and Khotyleva (1985a) based on testing n genotypes in m environments, which allows the general and specific adaptive ability (GAA and SAA) of the genotypes to be calculated, together with their stability, and also allows environments to be compared for their capacity to differentiate genotypes (DAE, differential ability of environment). It is shown that variances of SAA and DAE give more accurate information on the stability of genotypes and the differential ability of environments than the corresponding variances for G × E interaction. GAA is determined by the mean value of a character in different environments and SAA by the deviation from GAA in a given environment (Kilchevskii and Khotyleva, 1985b). The method allows selection to be performed simultaneously for yield and stability on the basis of a single index, the breeding value of the genotype.

Later, Zhang and Geng (1986) proposed another method involving the following steps: (i) regression of a standard variety on environmental means, (ii) regression of varieties under test on a standard variety, and (iii) transformation by a process of reparameterization, of each test variety regression on the standard variety into the regression of the variety on the environmental means. Although intended for the analysis of data from complex designs, it is considered to possess some advantages over conventional procedures of simpler designs. Then Westcott (1987) proposed another method for the cases where genotype yields are measured across environments but no concomitant environment data are available. The method depends on the use of a novel measure of similarity between genotypes. Most of the stability information is obtained by sequential plotting of the first two principal coordinates of the similarity matrix. Stable genotypes are revealed as those points consistently remote from the majority.

Gerasimenko (1985) presented a mathematical model for evaluating the response of genotypes to environment by means of the ratio of their root mean square deviations. This method of estimation, unlike the method of regression on means and regression on a standard, is independent of the coefficients of correlation between the genotypes tested. Mariani et al. (1983) determined the efficiency of linear and multiphase regression methods for evaluating stability of grain yield and protein content. The best fit for grain yield was obtained by linear and three line regression models; for protein content the most efficient was the two line model.

Leon (1986) compared the following methods of simultaneous estimation of yield and yield stability: (i) comparison with the yield of a desired genotype, or of the highest-yielding genotype, (ii) use of an index combining yield and yield stability, and (iii) methods based on ranking. Methods of the first type applied to wheat trial data gave strong correlations with yield and showed nearly the same rank order as grain yield itself. The second method and certain ranking methods gave similar high correlations with yield and stability measures. Crossa (1988) also compared results obtained with two methods (special method based on principal coordinate analysis, and a modified conventional regression analysis method) for assessing yield stability of maize. Stability parameters determined by regression analysis varied for some varieties under these two conditions. The special method was considered more useful.

Singh et al. (1988) further determined the stability of two genotypes in intercropping using the method of joint regression. The advantages of the joint regression approach to the intercropping system are seen as the fact that it is based on actual yields of the component crops without any transformation and that it takes into account the covariability of both the crops grown together.

VARIABILITY

Different genomes respond differently to changes in the environment. Some give an average (stable) yield under both favourable and unfavourable environments while others give very high yield under favourable environments but perform poorly (below average) under unfavourable (marginal) environments. Literature is full of such instances which will not be catalogued here, rather only a few recent examples will be cited.

We know very well by now that the dwarf stature is related to higher productivity. Sudin (1985) studied the stability for plant height in rice genotypes. High stability for height was shown by the early varieties Novoselskii and Kakai 203. Midseason varieties showed higher variation within a variety, but the variety Kubanets 575 is highly stable and adaptable while Kuban 9 is highly stable but less adaptable. Late varieties showed most variations in height, with Lomello showing high stability and adaptability and Solyaito showing high stability and low adaptability. The dwarf plants, in general, were less stable.

From the nine populations of maize grown at 80 locations in several developing countries, Crossa et al. (1988a) report that the populations Mezcla Tropical Blanca and Amarilio Dentado produced selections with good stability in both favourable and unfavourable environments, while La Posta produced varieties that were stable only in favourable environments. Similarly, significant differences were noted among the wheat cultivars in both adaptability and stability, some being high yielding, stable and widely adapted (Yu et al., 1988). Out of the several sorghum varieties and hybrids grown under high and low input management, varieties SPV 99, SPV 138 and SPV 22 and hybrid CSH 1 were the most stable, while hybrids CSH 5 which was best under favourable environments was not stable (Palanisamy and Prasad, 1984). Further, Chhina and Phul (1987) report that the genotypes CSV 1 and CSH 5 were the most stable for grain yield while CSH 9 which gave the highest yield had below average stability. Although not stable for all traits, CSV 1 and CSV 5 are regarded as useful breeding material for developing stable, high yielding genotypes. Among the sugarcane genotypes, Lopez Lopez and Vega Rivero (1985) report that C 323-68. C24-68 and C 222-68 were the most stable and gave the highest sugar yields per unit area; C222-68 was particularly noted for its high sucrose content.

The pigeon pea varieties ICPL 87 for seed yield and yield-related plant structure characters and ICPL 81 for physiological characters showed the best stability under different environments (Katiyar and Sarial, 1987). Six pigeon pea genotypes were considered stable across environments (Singh et al., 1988). Some were stable in single crop systems but unstable when intercropped. Among the *mung* bean (*Vigna radiata*) cultivars, EGMG 174-3 and CES 55 were the most stable for yield, M 350, CES1D 21 and CES1F 5 were high yielders under the most favourable conditions, and CES 14 and MG 50-10 A (Y) gave relatively high yields under unfavourable conditions (Miah and Carangal, 1986). M 350 and CES 55 were stable under monoculture and EGMG 174-3 and CES1T 2 gave high yields under both monoculture and intercropping systems.

Among the eight dry bean lines evaluated (Vaid et al., 1985), four were high yielding but had low stability while the other four lines had high stability but average yield. The cowpea variety C125 may be used as a high-yielding, stable genotype, while CG 28 may be used in favourable environments and SC488 in unfavourable ones (Ranganatha, 1984). The pea (*Pisum sativum*) variety 6588-1 was considered stable in any environment for all characters, while T 163 proved suitable for favourable environments only (Singh et al., 1984). Among the chickpea varieties, Jain et al. (1984) report that K4 was suitable for high-yielding environments, and Pant 110 was promising for moderately fertile sites, while Kaka, NEC 240 and Pink 2 were suitable for low-yielding environments. Out of the 12 *Brassica juncea* genotypes, RLC 1005 gave the highest yield but showed low stability, while RLM 198 was the next highest yielding but moderately stable (Kumar et al., 1986). RLM 621 had a low coefficient of regression but was suited for poor environments. The groundnut varieties 34-2-2 and JSP 1 responded well under unfavourable conditions, but not so well under favourable conditions (Patel et al., 1983). This material could have been lost if not tested under poor environments.

SELECTION CRITERIA

The early flowering varieties with low yield are generally more stable (Yadav and Kumar, 1983). Saeed (1983) also reports that early and midseason sorghum genotypes were more yield stable than the late maturing genotypes. Also in maize, stability of yield was highest in the early genotypes (Martin del Campo Valle and Molina Galan, 1982). Hence earliness should be considered as one of the selection criteria for stability of grain yield.

Genotypic response to environment may depend upon the duration of plant growth and variation in seasonal conditions at certain critical stages of plant development (Saeed and Francis, 1984). Differences in yield stability among genotypes are largely a function of relative maturity duration. Early and medium maturing genotypes appear to be more stable than late maturing genotypes (Saeed and Francis, 1983). It is, therefore, questionable whether the genotypes that appear yield stable if planted early, are also stable if planted late in a given environment.

While selecting for yield stability one has to bear in mind that lines with excellent expression of a yield component are generally unstable for that component. Stability of yield components thus appears to be more important than the absolute expression of yield components for high yield in specific environments (Cammack et al., 1983).

Yield stability is associated with high 1000-grain weight, but high values for ears/m^2 or number of grains/ear are found in varieties with low stability (Nedelea et al., 1984). In relation to the dicot, pigeon pea, Shoran (1985) concluded that stability for pods/plant and 100-seed weight and plasticity for days to flowering and to 50 per cent flowering are the main components of stability for seed yield; and, should be looked into while selecting a stable variety.

Since harvest index is a ratio, it might be influenced to a lesser extent than grain yield by diverse environmental conditions. Thus the stability of HI under diverse environmental conditions should be an important selection criterion.

Indeterminate soybean cultivars possess good stability characteristics, showing average seed yield response to environments of varying productivity, and minimum deviations from regression (Beaver, 1980). Also the indeterminate dry bean genotypes such as black seeded 'La Vega' and white seeded 2W-33-2 produced greater than average mean seed yields, had an average or greater than average response to environments of varying levels of productivity, and possessed minimum deviations from regression. Although under favourable high-yield environments, determinate genotypes at appropriate plant populations almost always outyield their indeterminate counterparts (*see* the chapter on Determinate Habit), they do not fare well under stress environments, for example drought, as they possess no regeneration capacity or the compensation mechanism. However, in the determinate cereals, height becomes the main determinant. For instance, wheat varieties of moderate height were more stable than tall or dwarf varieties (Chaubey and Sastry, 1981). The yield component stability was greater in dense stands than in spaced plants (Royo and Romagosa, 1988). In *Triticum aestivum*, components measured on fertile tillers (>10 grains/ spike) were more stable than those measured over all tillers; components determined on a whole plant basis were more stable than those measured on the main culm. In *T. turgidum* var. *durum*, yield components were more stable in rain-fed than in irrigated trials. Components measured on the whole plant basis were more stable than those measured on fertile or main tillers. In general, the most stable components were 1000-grain weight and HI in *T. aestivum* and percentage of fertile tillers and 1000-grain weight in *T. durum*, suggesting that these could be used as selection criteria (Royo and Romagosa, 1988).

Quality scores are generally less responsive to environmental changes than yield *per se;* high-yielding cultivars, in general, are lower in stability (Geng et al., 1987). But the recent improvements in cotton have resulted in progressive changes in the direction of higher quality and greater stability in quality. Geng et al. report that breeding in the last 18 years has simultaneously improved yield and quality in Californian cotton.

Progress in yield improvement would be facilitated if selection for wide adaptability could be conducted in early generations. Recurrent selection techniques have been used to improve sorghum random-mating populations so as to derive improved inbred lines from these populations (Rose, 1978). Improvement in source populations could contribute to yield stability of the subsequently derived inbred lines and their hybrids. Gupta et al. (1986) report that in mustard, heterogenous populations (F_2, F_3 and binary mixtures) were more stable than the homogenous populations (parents and F_1).

Vallejo Delgado and Marquez Sanchez (1984) compared the convergent-divergent scheme of mass selection with the rotatory scheme in maize. The rotatory scheme produced a greater advance in adaptability than the other. If several cycles are completed, stabilization should be attained sooner under the rotatory scheme, but beyond a certain point, the convergent-divergent procedure should begin to catch up and eventually overtake the other. Further, Brown (1988) examined a new approach to genotypic selection in a plant breeding programme where the genotypes under assessment are grown in a number of environments. It is assumed that these environments are a random subset of all possible environments where the genotypes are likely to be grown. It involves estimating the probability that each genotype will, if grown at any location, exceed predefined target values for one or more characters. The multinormal probabilities are estimated from the genotype means and environmental variance of each variate. Where more than a single variate is to be considered, the correlation coefficients between variaties are also used in the estimation. It was found that the coefficients, obtained by correlating the predicted proportion of locations at which genotypes would exceed the set target values with the observed proportion of locations in a different year, were consistently higher than similar coefficients between observed proportions in different seasons. The latter were high enough for it to be concluded that the approach would be of use in practice. Such a method may, therefore, be used to identify genotypes which have a high probability of being suitable over a range of locations.

STABILITY MECHANISM

Stability of yield defined as the ability of a genotype to avoid substantial fluctuations in yield over a range of environments is a breeding objective difficult to achieve. The causes of yield stability often are unclear, and physiological, morphological and phenological mechanisms that impart stability are diverse. Mechanism of yield stability fall into three general categories: genetic heterogeneity, yield component compensation, and tolerance to stresses and suboptimal environments.

Genetic Heterogeneity

Allard and Bradshaw (1964) observed that genetic diversity in the form of heterogeneity or in the mixtures of different genotypes leads to stability under a range of environmental conditions. Although heterozygous populations tend to be more stable than single cross hybrids because of population buffering, several studies in grain sorghum have indicated no large differences in yield stability among various genetic combinations including single crosses, three-way crosses, and population blends (Jowett, 1972; Majisu and Doggett, 1972; Patanothai and Atkins, 1974; Reich and Atkins, 1970). Eberhart and Russell (1969) reported that some single cross maize hybrids were higher yielding and as stable as any double crosses, and that stability seemed mainly a property of inbred parents.

Yield Component Compensation

The stability of yield components in sorghum has been researched by Heinrich et al. (1983). They found that the stable hybrids had 40 per cent more seeds/m^2 than the non-stable hybrids under poor environments. This only partially explained their 70 per cent yield advantage; the balance was accounted for by seed size. The most stable hybrids maintained heads/m^2, seeds/head, and seed weight in poor environments. Tolerance of stable genotypes to stress conditions and the maintenance of all yield components at relatively high levels appeared more important than the compensation among the components.

Differences in 100-seed weight between stable and unstable hybrids (responsive) in low yield environments contributed to yield stability, adding evidence to the importance of large seed weight as one objective in a breeding programme for low input agriculture (Heinrich et al., 1985). Earlier studies (Crosbie and Mock, 1981; Eastin et al., 1971) have also shown that greater seed weight may contribute to higher yields. Higher seed weight has been attributed to longer grain filling period (Crosbie and Mock, 1981) and faster grain filling rate (Sofield et al., 1977).

Stress Tolerance and Suboptimal Environments

Tolerance to climatic, edaphic and biological stresses enhance stability. Research with rice has indicated that where periodic drought is a problem, both drought resistance and capacity to recover from drought are necessary for yield stability (IRRI, 1977). Effects of a particular stress may depend on genotype and growth stage of the plant. For example, moisture stress has differing effects on sorghum genotypes and typically has less effects on yields during the vegetative stage than during the boot to bloom stage (FAO, 1979). Because yield is a product of yield components, and these generally are the product of sequential development processes, the timing of critical stresses may be specified by examining yield component functions. Susceptibility to stress could be indicated by reduction in the yield component developing during the period from stress, and tolerance by less reduction in that yield component or compensation by a non-stressed yield component.

Out of the 24 wheat varieties examined (Rasal et al., 1988), the varieties HD 2190, NI 5429 and NI 8635 produced high grain yield and low regression coefficient values showing their suitability for poor environments. The variety NI 8625 with average regression coefficient values and high mean grain yield is adaptable for a wide range of environmental conditions. Among 11 black gram (*Vigna mungo*) varieties, RU 77-10 and RU 76-5-4 gave high yields and linear regression coefficients close to unity, indicating high stability (Deshmukh et al., 1987). The variety T9 is specifically adapted to poor environments.

Out of several Indian, Mexican and Italian wheat varieties and hybrids evaluated (Mahal et al., 1988), the performance and stability for grain yield/plant was the highest for varieties NP 404, Baijaga Yellow and Giorgio VZ 331. The overall best varieties were NP 404 and Baijaga Yellow and the hybrids Anhinga 's' and Mexicali 75. The cowpea varieties IT82E60 and Ife Brown are stable for grain yield under the arid environment of Nigerian Sahel Savanna (Okafor, 1986).

Stability of wheat varieties at different plant populations (2.25, 4.50 and 6.75×10^6 seeds/ha) was studied (Dottacil and Toman, 1987). Highest yields were generally obtained at the normal plant population (4.5×10^6), with a 4.5 per cent reduction at the lowest plant density and a 2.1 per cent reduction at the highest; but some varieties responded differently. Locations and year accounted for 37 per cent of variation in grain yield, while variety and variety $\times$ environment interaction accounted for 24.9 per cent. But sowing density accounted for only 0.3 per cent, due to compensation and self-thinning or autoregulation mechanisms operating within narrow ranges.

Some cowpea varieties give stable grain yield at different levels of soil fertility while others are responsive (Sulochana and Peter, 1987). The variety IIHR 6-1B gave a stable yield of 62.35 g/ plant and is recommended for cultivation on soils with different fertility levels.

Sowing time has a great influence on stability of a genotype. Francis et al. (1984) observed that among sorghum hybrids, some were stable in early planting but not in late planting and *vice versa*. However, there were a few hybrids that were equally stable in early and late plantings. Thus the influence of planting date cannot be ignored in evaluating yield stability, and a useful guide-line would be to plant all nurseries at the same time as commercial plantings in each area.

Balakrishnan and Natrajaratnam (1986) reported that pigeonpea cultivars CORG 5, UPAS 120 (both early) and CORG 11 and SA 1 (both late) are stable for photosynthetic rate, leaf: stem ratio, LAI and individual leaf area, and can be used for improving existing varieties for photosynthetic efficiency and thus grain yield.

STABILITY OF QUALITY CHARACTERS

Although adaptability of plants and stability of yield are our primary objectives, we do not want the quality of the produce, especially in terms of protein content and amino-acid profile, oil content, sugar content, fibre quality or things like aroma in scented rice, to reduce. To cite some examples, barley varieties K1012 and LL150 (two-rowed) and Notch 2 (six-rowed) have consistently higher protein percentages across environments (Varma et al., 1985). Varieties K 1012 and DL 150 (two-rowed) and DL 219 (six-rowed) also have high 100-grain weights. In an experiment with 32 soybean varieties, protein content was fairly stable while there were significant differences in oil content due to genotype and G × E interaction (Singh and Chaudhary, 1985). In French bean, protein content depended on genotype, yield, weather conditions and agronomic regime (Polyanskaya and Rogulina, 1986). But it was possible to select forms with high protein content over the different years, for example, N69-38, N79-124 and N80-283. Out of a large number of sorghum varieties and hybrids grown at four locations, seven hybrids exhibited average stability for lysine content, and three other hybrids exhibited stability for sugar content (Nayeem and Bapat, 1986). Among the several crosses made in wheat genotypes (Smocek et al., 1988), Regina × Una and KM 520-1-80 × Castan combined high yields with high grain quality.

Out of the 75 *Gossypium hirsutum* hybrids and their 25 female and three male parents grown in four contrasting environments, Acala SJ 1 × H 655 was the most stable hybrid for halo length, ginning outturn, seed index and lint index (Seth et al., 1987). Some of the mutants of Basmati (scented) rice were very stable, gave a high yield and did not lose their aroma for which Basmati rice is highly priced (Desai et al., 1986).

BREEDING OBJECTIVES AND RETRIEVAL OF VARIETAL EVALUATION

Breeding Objectives

A major goal of breeders is the development of strains which consistently give maximum economic yield across environments. Productivity of a population is a function of its adaptability which is a compromise of stability and flexibility. Stability does not imply general consistency of phenotype in varying environments but rather implies stability for yield and quality of economically important traits. A cultivar can achieve stability in two ways: (i) it may contain a number of genotypes (heterogeneity) each adapted to a somewhat different range of environments or (ii) each individual plant may be well buffered or adapted to a range of environments. The terms population buffering and individual buffering describe these two criteria. Genetically homogenous populations, such as pure line cultivars or single crosses, obviously depend heavily on individual buffering to stabilize productivity, whereas both criteria are available in genetically heterogenous populations (Allard and Bradshaw, 1964). However, Mahill et al. (1984) noted that heterogeneity of a strain or cultivar is not an apparent prerequisite for environmental buffering effects. Cumulative expression of genetic traits may play a more important role in maintaining environmental stability than relative heterozygosity.

Testing a number of environments would be more feasible with relatively a few lines in advanced stages of a breeding programme. In early generations such as the F_3, there are often large number of lines to test and consequently a few locations are used. But because of large G × E interactions, such a small number of sites may lead to elimination of promising materials.

Stability for yield components such as seed number and seed weight should be considered when breeding for yield stability. Genetic diversity and productivity are complexly related while genetic diversity and stability are more simply related. Genetic diversity consistently endow populations with stability (Allard, 1961). Harvest index is a more stable parameter than the yield components.

Retrieval of Varietal Evaluation

A specialized database system has been described by Martiniello (1988) which facilitates decisions by the farmer in the choice of variety to improve crop yield and reduce the effects of environmental interaction. The software enables a database to be planned, analyzed and built from varietal agronomic trials with cereals. On the basis of statistical analysis, the package provides a list of varieties recommended for cultivation in a specific environment. A further utilization of the package is for breeding and statistical interference to quantify the adaptability of varieties in a wide range of environments according to their stability parameters. The system was found to be flexible and user-friendly, and could be used for a number of different applications.

REFERENCES

Allard, R.W. Relationship between genetic diversity and consistency of performance in different environments. *Crop Sci.* 1: 127–33, 1961.

Allard, R.W. and A.D. Bradshaw. Implications of genotype-environment interactions in applied plant breeding. *Crop Sci.* 4: 503–08, 1964.

Balakrishnan, K. and N. Natrajaratnam. Identification of stable physiological traits for certain cultivars of pigeon pea in DMA. *Madras agric. J.* 73: 36–41, 1986.

Beaver, J.S. Response of determinate and indeterminate soybean (*Glycine max* (L) Merr) to different cultural practices and to environments of varying levels of productivity. *Diss. Abstr. Int. B.* 41: 2010 B, 1980.

Becker, H.C. Correlations among some statistical measures of phenotypic stability. *Euphytica* 30: 835–40, 1981.

Brown, J. An alternative approach to multivariate selection in plant breeding where genotypes are evaluated at many locations. *Theo. appl. Genet.* 76: 76–80, 1988.

Cammack, F.P. Stability, compensation and heritability of yield and yield components in winter wheat. *Diss. Abstr. Int. B.* 44: 2033 B, 1984.

Cammack, F.P., E.L. Smith and R.W. McNew. Yield and yield component stability for 10 winter wheat genotypes. *Agron. Abstr.* Madison, Wis., ASA, USA. p. 57, 1983.

Chaubey, J.S. and E.V.D. Sastry. Stability analysis of yield and yield components in some Indian and Mexican varieties of wheat. *Indian J. agric. Sci.* 51: 611–14, 1981.

Chhina, B.S. and P.S. Phul. Stability anlaysis for yield influencing traits in grain sorghum. *Crop Improvement.* 14: 36–41, 1987.

Crosbie, T.M. and J.J. Mock. Changes in physiological traits associated with grain yield improvement in three maize breeding programmes. *Crop Sci.* 21: 255–59, 1981.

Crossa, J. A comparison of results obtained with two methods for assessing yield stability. *Theo. appl. Genet.* 75: 406–07, 1988.

Crossa, J., B. Westcott and C. Gonzalez. The yield stability of maize genotypes across international environments: full season tropical maize. *Exp. Agric.* 24: 253–63, 1988a.

Crossa, J., B. Westcott and C. Gonzalez. Analyzing yield stability of maize genotypes using a special model. *Theo. appl. Genet.* 75: 863–68, 1988b.

Desai, N.D., M.U. Kukadia, M.R. Patel and R.B. Sarvaiya. Grain yield and its stability of some aromatic 'basmati' mutants of rice. *Gujrat Agric. Univ. Res. J.* 11: 76–78, 1986.

Deshmukh, R.B., S.B. Patil, A.H. Sonone and N.V. Shinde. Stability for grain yield in black gram. *J. Maharastra Agric. Univ.* 12: 163–64, 1987.

Dotracil, L. and K. Toman. (Effect of sowing density on yield stability and structure in different types of wheat varieties). *Rostlinna Vyroba.* 33: 1029–38, 1987.

Eastin, J.D., C.Y. Sullivan, W.M.Ross, M.D.Clegg and J.W. Maranville. Comparative developmental stages in sorghum hybrids and parent lines. *In:* Research in Physiology of Yield Management of Sorghum, Report No. 5. Sorghum programme., Univ. of Nebraska, Lincoln, pp. 168–81, 1971.

Eberhart, S.A. and W.A. Russell. Stability parameters for comparing varieties. *Crop Sci.* 6: 36–40, 1966.

Eberhart, S.A. and W.A. Russell. Yield and stability of a 10-line diallel of single cross and double cross maize hybrids. *Crop Sci.* 9: 357–61, 1969.

FAO. Yield response to water. FAO irrigation and drainage paper. FAO, Rome. No.33: 134–35, 1979.

Finlay, K.W. and G.N. Wilkinson. The analysis of adaptation in a plant breeding programme. *Aust. J. agric. Res.* 14: 742-54, 1963.

Francis, C.A., M. Saeed, L.A. Nelson and R. Moomaw. Yield stability of sorghum hybrids and random mating populations in early and late planting dates. *Crop Sci.* 24: 1109–12, 1984.

Geng, S., Q. Zhang and D.M. Bassett. Stability in yield and fibre quality of California cotton. *Crop Sci.* 27: 1004–10, 1987.

Gerasimenko, V.F. (Genetic mechanisms of ecological adaptability in winter bread wheat: ecological model of a quantitative character and criteria for evaluating the reaction of genotypes to environment). *Tsitologiya i Genetika.* 19: 359-64, 1985.

Gupta, M.L., K.S.Labana and B.S. Bawa. Phenotypic stability of different populations in Indian mustard. *J. Oilseeds Res.* 3: 184–88, 1986.

Heinrich, G.M., C.A. Francis and J.D. Eastin. Stability of grain sorghum yield components across diverse environments. *Crop Sci.* 23: 209–12, 1983.

Heinrich, G.M., C.A. Francis, J.D. Eastin and M. Saeed. Mechanisms of yield stability in sorghum. *Crop Sci.* 25: 1109–13, 1985.

Hill, J. Genotype-environment interactions: A challenge for plant breeding. *J. agric. Sci., U.K.* 85: 477–93, 1975.

Huhn, M. and J. Leon. Genotype-environment interactions and phenotypic stability of *Brassica napus. Z.Pflanzenzuchtung.* 95: 135–46, 1985.

IRRI. Drought resistance: field performance of rices in rain-fed culture. *In:* Annual report for 1977, IRRI, Los Banos, Philippines. pp. 89–97, 1977.

Jain, K.C., B.P. Pandya and K. Pande. Stability of yield components of chickpea genotypes. *Indian J. Genet. Pl. Breed.* 44: 159–63, 1984.

Jowett, D. Yield stability parameters from sorghum in East Africa. *Crop Sci.* 12: 314-17, 1972.

Kang, M.S., B. Glaz and J.D. Miller. Interrelationships among stabilities of important agronomic traits in sugarcane. *Theo. appl. Genet.* 74: 310–16, 1987.

Katiyar, R.P. and A.K. Sarial. Stability analysis and its application to pigeonpea development programme in Himachal hills. *Farm Science J.* 2: 53–63, 1987.

Kilchevskii, A.V. and L.V. Khotyleva. (Method for evaluating the adaptive capacity and stability of genotypes and the differential ability of the environment. I. Basis of the method). *Genetika,* USSR. 21: 1481–90, 1985 a.

Kilchevskii, A.V. and L.V. Khotyleva. Determining the adaptive capacity of genotypes and the differential capacity of the environment. *Doklady Akad Nauk Belorusskoi SSR.* 29: 374–76, 1985b.

Kilchevskii, A.V., L.V. Khotyleva and M.A. Fedin. (Association between yield and ecological stability in vegetable crop varieties). *Tsitotogiya i Genetika.* 22: 47–52, 1988.

Kumar, R., V.P. Gupta and M.L. Gupta. Phenotypic stability of yield and its components in Indian mustard. *J. Oilseeds Res.* 3: 251–54, 1986.

Leon, J. Methods of simultaneous estimation of yield and yield stability. *In:* Biometrics in Plant Breeding. M.J. Kearsey and C.P. Werner (eds). Eucarpia. Birmingham, U.K. pp. 299–308, 1986.

Lin, C.S., M.R. Binns and L.P. Lefkovitch. Stability analysis: where do we stand? *Crop Sci.* 26: 894–900, 1986.

Lopatina, L.M. (Planning ecological tests and evaluating the adaptability of varieties and hybrids with the aid of regression models). *Vestnik Seleskokhozyaistvennoi Nauki*, Moscow, USSR. No. 5: 71–76, 1986.

Lopatina, L.M., D. Yu. Papazov and V.A. Dragavtsev. (Genetic mechanisms for the establishment and variability of parameters of homeostasis in varieties of crop plants). *Genetika* USSR. 22: 2295–2302, 1986.

Lopez Lopez, E. and A. Vega Rivero. (Genotype-environment interaction in sugarcane (*Saccharum* spp.). I. Stability and/or adaptability in the basic yield characters in regional trials). Revista INICA, Instituto National de Investigacioues de la Cana de Azucar, Cuba. No. 2: 47–64, 1985.

Mahal, G.S., K.S. Gill and G.S. Bhullar. Stability parameters and performance of interregional crosses in durum wheat (*Triticum durum* Desf.). *Theo. appl. Genet.* 76: 438–42, 1988.

Mahill, J.F., J.N. Jenkins, J.C.Mc Carty, Jr. and W.L. Parrott. Performance and stability of double haploid lines of upland cotton derived via semigamy. *Crop Sci.* 24: 271–77, 1984.

Majisu, B.N. and H. Doggett. The yield stability of sorghum varieties and hybrids in East African environments. *East Afr. Agric. For. J.* 38: 179–92, 1972.

Mariani, B.M., P.N. Manmana and R. Stefanini. Efficiency of linear and multiphase regression methods to evaluate genotype-environment interaction for grain yield and protein content in Italian *durum* wheat varieties. *Zeitschuf fur Pflanzenzuchtung.* 90: 56–67, 1983.

Martin del Campo Valle, S. and J.D. Molina Galan. (Combining ability in three groups of maize populations in north central Mexico). *Agrociencia.* No. 47: 103–16, 1982.

Martiniello, P. Development of a database computer management system for retrieval on varietal field evaluation and plant breeding information in agriculture. *Computers and Electronics in Agriculture.* 2: 183–92, 1988.

Miah, M.N.L. and V.R. Carangal. Stability of selected *mung* bean (*Vigna radiata* (L) Wilczek) cultivars evaluated under different growing conditions. *Philippine J. Crop Sci.* 11(Suppl. 1): Abst. 3B–8a, 1986.

Nayeem, K.A. and D.R. Bapat. Phenotypic stability for protein, lysine and sugars in grain sorghum. *Indian J. Genet. Pl. Breed.* 46: 439–48, 1986.

Nedelea, G., A. Moisuc, R. Paraschioiu and V. Sonea. (Interaction between stability and yield and yield components in winter wheat). Lucrari Stuntifice, Institutal Agronomic Timisoara, Agronomie. 19: 65–72, 1984.

Ntare, B.R. and M. Aken'Ova. Yield stability in segregating populations of cowpea. *Crop Sci.* 25: 208–11, 1985.

Okafor, L.I. Adapting improved cowpea cultivars to the Nigerian arid zone. *Trop. Grain Legume Bull.* No. 33: 20–23, 1986.

Palanisamy, S. and M.N. Prasad. Stability for grain yield in certain varieties of sorghum. *Sorghum Newsl.* 27: 38, 1984.

Patanothai, A. and R.E. Atkins. Yield stability of single crosses and three-way hybrids of grain sorghum. *Crop Sci.* 14: 287–90, 1974.

Patel, V.J., A.S. Kawar, H.J. Joshi and B.K. Chovatia. Stability parameters for pod yield in groundnut. *Indian J. agric. Sci.* 53: 1071–73, 1983.

Perkins, J.M. and J.L. Jinks. Environmental and genotype-environmental components of variability. III. Multiple lines and crosses. *Heredity* 23: 339–56, 1968.

Plaisted, R.L. and L.E. Peterson. A technique for evaluating the ability of selections to yield consistently in different locations or seasons. *Am. Potato J.* 36: 381285, 1959.

Polyanskaya, L.I. and L.V. Rogulina. (Seed protein content in French bean). *Selektsiya i Semenovodstvo*, Ukrainiam SSR. No. 61: 63–65, 1986.

Ranganatha, A.R.G. Adaptability analysis in cowpea (*Vigna unguiculata* (L) Walp). *Mysore J. agric. Sci.* 18: 245, 1984.

Rasal, P.N., R.Y. Thete, S.D. Ugale and A.D. Dumbre. Stability analysis for yield in wheat. *J. Maharashtra Agric. Univ.* 13: 4–6, 1988.

Rasochova, M. (Influence of external conditions on the traits and characteristics of potato varieties). *In*: Abstr. Conf. papers and posters of the 10th triennial conf. of the European assoc. for potato research, Aalborg, Denmark, 26–31 July, 1987). N.E. Foldo, S.E. Hansen, N.K. Nielsen and R. Rasmussen. (eds). pp. 133–34, 1987.

Reich, V.H. and R.E. Atkins. Yield stability of four population types of grain sorghum, *Sorghum bicolor* (L) Moench, in different environments. *Crop Sci.* 10: 511–17, 1970.

Ross, W.M. Population breeding in sorghum phase 2. Proc. Corn and Sorghum Res. Conf. Proc. 33: 153–66, 1978.

Royo, C. and I. Romagosa. Yield component stability in *Triticum aestivum* L. and *T. turgidum.* var. *durum. Cereal Res. Commu.* 16: 77–83, 1988.

Saeed, M. Genotype × environment interactions and yield stability in relation to maturity in grain sorghum (*Sorghum bicolor* (L) Moench). *Diss. Abstr. Int. B.* 43: 3442 B, 1983.

Saeed, M. and C.A. Francis. Yield stability in relation to maturity in grain sorghum. *Crop Sci.* 23: 683–87, 1983.

Saeed, M. and C.A. Francis. Association of weather variables with genotype × environment interactions in grain sorghum. *Crop Sci.* 24: 13–16, 1984.

Seth, S., B.P.S. Lather, I.P. Singh, B.S. Chhabra and S.S. Siwach. Stability parameters in upland cotton. *Indian J. agric. Sci.*.57: 429–33, 1987.

Sharma, H.L. and T.S. Bharaj. Stability analysis of yield components in sugarcane. *Crop Improvement.* 9: 89–92, 1982.

Sharma. H.L. and T.S. Bharaj. Adaptability studies for quality traits in sugarcane (*Saccharum* sp.). *Sugar cane Breeder's Newsl.* No. 45: 86–92, 1983.

Shoran, J. Role of plant characters in determining adaptation in pigeon pea. *Int. Pigeon pea Newsl.* No. 4: 15–18, 1985.

Shukla, C.K. Some statistical aspects of partitioning genotype-environment components of variability. *Heredity* 29: 237–45, 1972.

Singh, M., B. Gilliver and M.R. Rao. Stability of genotypes in intercropping. *Biometrics.* 44: 561–70, 1988.

Singh, O. and B.D. Chaudhary. Stability analysis for protein and oil content in soybean. *J. Oilseeds Res.* 2: 57–61, 1985.

Singh, S.P., B.S. Yadav and V.G. Narsinghani. Stability of yield components in pea. *Indian J. agric. Sci.* 54: 608–12, 1984.

Skinnes, H. and T. Buras. Developmental stability in wheat to differences in the temperature during seed set and seed development. *Acta Agriculture Scandinavica.* 37: 287–97, 1987.

Smocek, J., L. Tvaruzek and Z. Ohnoutka. (Stability of combining ability for yield and grain quality in wheat.) *Genetika a Slechtani.* 24: 133–44, 1988.

Sofield, I., L.T. Evans, M.G. Cook and I.F.Wardlaw. Factors influencing the rate and duration of grain filling in wheat. *Aust. J. Pl. Physiol.* 4: 785–97, 1977.

Stout, D.G, T. Kannangara and G.S. Simpson. Drought resistance of *Sorghum bicolor.* 2. Water stress effects on growth. *Can.J.Pl.Sci.* 58: 225–33, 1978.

Sudin, V.M. (The adaptability and stability of plant height among rice varieties). Selektsiya i *Semenovodstvo* USSR.No.2: 11–12, 1985.

Sullivan, C.Y. Mechanism of heat and drought resistance in grain sorghum. *In*: Sorghum in Seventies, N.G.P. Rao and R.L. House (eds), Oxford & IBH Pub. Co., New Delhi. pp. 247–60, 1972.

Sulochana, K.A and K.V. Peter. Phenotypic stability in cowpea. *Indian J. agric. Sci.* 57: 425–26, 1987.

Vaid, K., V.P. Gupta and R.M. Singh. Stability analysis in dry beans. *Crop Improvement.* 12: 28–31, 1985.

Vallejo Delgado, H.L. and F. Marquez Sanchez. (Comparison of two mass selection procedures in breeding for adaptability in the maize (*Zea mays L.*) variety ZAC 58). *Agrociencia*, Mexico. No. 58: 29–43, 1984.

Varma, N.S., K.B.L. Jain and S.C. Gulathi. Genotype × environment interactions for protein content and grain weight in two-row and six-row barleys. *Indian J. Genet. Pl. Breed.* 45: 81–88, 1985.

Virk, D.S., P.S. Virk, G. Singh, G. Harinarayan and R.S. Saini. Stability analysis for comparing pearl millet hybrids. *SABRAO Journal.* 17: 129–33, 1985

Westcott, B. A method of assessing the yield stability of crop genotypes. *J. agric. Sci.* U.K. 108: 267–74, 1987.

Yadav, I.S. and D. Kumar. Association between stability parameters of productive traits in black gram. *Madras agric. J.* 70: 331–33, 1983.

Yates, F. and W.G. Cochran. The analysis of groups of experiments. *J. agric. Sci.* 28: 556–80, 1938.

Yu, S.R., Z.S. Wu and Z.P. Yang. (Evolutionary changes in yield and yield components of wheat cultivars grown in the Huainan region, Jiangsu province during 1970-1986). *Scientia Agricultural Sinica.* 21: 15–21, 1988.

Zang, Q.F. and S. Geng. A method of estimating varietal stability for data of long-term trials. *Theo. appl. Genet.* 71: 810–14, 1986.

AUTHOR INDEX

SUBJECT INDEX